The Master IC Cookbook

The Master IC Cookbook

Third Edition

Delton T. Horn

McGraw-Hill

New York   San Francisco   Washington, D.C.   Auckland   Bogotá
Caracas   Lisbon   London   Madrid   Mexico City   Milan
Montreal   New Delhi   San Juan   Singapore
Sydney   Tokyo   Toronto

Library of Congress Cataloging-in-Publication Data

Horn, Delton T.
    The master IC cookbook / Delton T. Horn.—3rd ed.
      p.   cm.
    Rev. ed. of: Master IC cookbook / Clayton L. Hallmark. 2nd ed.
1991.
    Includes index.
    ISBN 0-07-030564-1.—ISBN 0-07-030565-X (pbk.)
    1. Integrated circuits—Handbooks, manuals, etc. I. Hallmark,
Clayton L. Master IC cookbook. II. Title.
TK7874.H42 1997   96-40098
621.3815—dc21   CIP

McGraw-Hill

A Division of The McGraw-Hill Companies

1 2 3 4 5 6 7 8 9 0   AGM/AGM   9 0 2 1 0 9 8 7

ISBN 0-07-030565-X (PBK)
ISBN 0-07-030564-1 (HC)

The sponsoring editor for this book was Scott Grillo, the editing supervisor was Andrew Yoder, and the production supervisor was Suzanne Rapcavage. It was set in ITC Century Light by Jana Fisher through the services of Barry E. Brown (Broker—Editing, Design and Production).

Printed and bound by Quebecor/Martinsburg.

McGraw-Hill books are available at special quantity discounts to use as premiums and sales promotions, or for use in corporate training programs. For more information, please write to the Director of Special Sales, McGraw-Hill, 11 West 19th Street, New York, NY 10011. Or contact your local bookstore.

Contents

Introduction

This newly revised edition of *The Master IC Cookbook* is a one-source reference guide for data on hundreds of different integrated circuits, both digital and linear. This book is a handy addition to any electronics bookshelf or workbench, whether you are a professional technician or a hobbyist.

At this writing, all of the ICs described in this volume have been confirmed as commercially available from major hobbyist suppliers. However, remember that electronics is always a rapidly changing field, and some chips may become obsolete and unavailable without warning.

This new edition has been rearranged to make information on specific types of devices easier to find. The data is divided into ten sections, each spotlighting a particular type of integrated circuit. Digital devices are covered in Sections 1 through 3, and linear/analog devices are featured in Sections 4 through 9. Section 10 deals with the special case devices which combine both analog and digital circuitry and functions.

In each section, ICs are listed in numerical order. Any prefix numbers are generally ignored. In most cases, the prefix letters simply indicate the manufacturer. For example, National Semiconductor uses LM, Rohm uses BA and BU, Motorola often uses MC, and Signectics uses NE. In most cases, prefix letters can be safely ignored in parts selection. For example, an LM567 is exactly the same device as an NE567. The only difference is who actually manufactured that particular unit.

Suffix letters are used for various purposes. Sometimes they indicate a case style, an operating temperature rating, or an improved version of an older device. In most cases, suffix letters can also be ignored, but be aware that there are some exceptions, which will be noted in the chip description, when appropriate. When in doubt about chip compatibility, always check the relevant specification sheets, whether in this book, or those supplied by the manufacturer/supplier.

There is no way a book like this can be 100% comprehensive, or even fully updated. There are thousands of ICs on the market today, and a number will be introduced between the time this is written and the finished books roll off the presses. The most representative, popular, and useful devices are the primary focus in this book.

1
SECTION

74xx digital

In this section, the 74xx line of digital/logic devices will be covered. Devices range from simple logic gates, to flip-flops, counters, registers, and more exotic and complex special-purpose devices.

This standardized numbering scheme was first devised for TTL (Transistor-Transistor Logic) ICs. As various sub-families of TTL were developed, the same numbering system was used. Except for standard TTL, an identifying letter (or group of letters) is added after the "74" and before the device type number (xx) to indicate the logic sub-family. Devices of a given type number are pin-for-pin function compatible, regardless of the sub-family. (Devices of different sub-families might not be completely compatible electrically, however.) For example, the pin-out diagram for a 74HC45 would be identical to that of a 74L45 or a 7445.

Most, but not all, device types are available in each logic family, although some might be hard to find. All basic devices are available in standard TTL.

The same numbering scheme has also been adopted for a line of CMOS (Complementary Metal-Oxide Semiconductor) devices. Although the pinouts are identical, TTL and CMOS devices are usually not directly compatible, but they can be interfaced with some external circuitry.

TTL and its sub-families

Most TTL devices are designed to be operated from a well-regulated +5-V power supply only. Higher or lower voltages will, at best, result in erratic operation, and will, more likely, damage or destroy the semiconductor chips.

The logic levels of all the TTL products are fully compatible with each other. However, the input loading and output drive characteristics of each of these sub-families is different, and must be taken into consideration when mixing the TTL families in a single system.

The main difference between the various TTL sub-families lies in the trade-off between the power dissipation (how much current the device consumes in operation) and the propagation delay (how fast the output responds to a change in the states at its input or inputs) ratings. Generally, improving one of these ratings will result in a degradation of the other. Improvements in technology can narrow the difference somewhat, usually for a higher price per IC. There will still be a significant trade-off. The best choice will depend on the specific intended application.

Often there is little point in mixing TTL sub-families in a single circuit. For example, if a few high-speed chips are mixed in with some standard TTL ICs, the circuit's operating speed (as a whole) will probably be limited by the slowest device in the system. Different functional circuits within a large system might, however, have different speed/power consumption requirements, making mixing of sub-families desirable.

Standard TTL: 74xx

No code letter is used in device numbers for standard TTL ICs because originally this was the only type of chip that used this numbering system. Device numbers are in the 74xx format (where "xx" is the two- or three-digit identifier for the specific device function.)

Although standard TTL is the oldest logic family still in use, it is far from obsolete. It does quite well in the majority of applications, which have low to moderate operational demands.

Standard TTL offers moderate performance on both the power dissipation and propagation delay ratings. It is not great, but not bad on either specification. A typical standard TTL gate dissipates about 10 mW, with an average propagation delay of about 10 nS.

Standard TTL remains generally the most widely available and least expensive of all the logic families and sub-families currently on the market. Its biggest advantage is low cost. Virtually all device functions in the 74xx line are available in standard TTL.

Low-power TTL: 74Lxx

Low-power TTL devices, which are numbered 74Lxx (where "xx" is the two- or three-digit identifier for the specific device function) are intended for applications where power consumption must be kept at a minimum. Such devices are often used in portable, and other battery-operated circuits. In large systems that require a great many gates and other logic devices, minimal power dissipation would also be highly desirable. A gate in this logic sub-family typically uses only about $\frac{1}{10}$ of the power dissipated by a comparable standard TTL gate.

Unfortunately, consuming less power results in significantly slower operation. A typical low-power TTL gate dissipates only about 1 mW, but the propagation delay is typically about 30 nS, if not more. Notice that in most general-purpose applications, this would not be a noticeable disadvantage, but in some special applications, it could be highly undesirable or impractical. In some cases, the circuit might not work as intended, if at all, if gates with such long propagation delays are used.

Schottky TTL: 74Sxx

The other side of the specification trade-off is to improve the propagation delay at the cost of an increase in power dissipation. To achieve faster switching speeds than standard TTL, Schottky diodes can be used in the fabrication of the gate's circuit to prevent the gate's transistors from completely saturating, which happens in the standard devices. Obviously, it requires less time to achieve partial saturation than complete saturation. Schottky TTL devices are identified as 74Sxx, where "xx" is the two- or three-digit identifier for the specific device function.

Of course, the higher switching speed results in higher power consumption. A typical Schottky TTL gate dissipates about 20 mW, but offers a propagation delay of a mere 5 nS. Schottky TTL ICs are a good choice in applications where very high operating frequencies are used and a hefty power supply can provide ample current.

Low-power Schottky TTL: 74LSxx

Improved technology has permitted the creation of a TTL logic sub-family, which approximates the advantages of both the 74Lxx and 74Sxx, with just slightly inferior specifications. Not surprisingly, this low-power Schottky TTL line is designated as 74LSxx, where "xx" is the two- or three-digit identifier for the specific device function.

The power dissipation of a typical low-power Schottky TTL gate is about 2 mW, with a propagation delay of about 8 nS. Notice that this power-dissipation rating isn't quite as low as that of 74Lxx devices, and the propagation delay rating is not quite as good as that of 74Sxx devices, but both specifications are an appreciable improvement over standard TTL. However, 74LSxx devices are significantly more expensive than the three previously covered sub-families. When standard TTL will do the job, it is still economically desirable to use it.

Advanced Schottky TTL: 74ASxx

A sort of "super-Schottky" sub-family has been developed for even faster switching speeds. Devices in this line are numbered as 74ASxx, where "xx" is the two- or three-digit identifier for the specific device function, and "AS" represents "Advanced Schottky." The typical power dissipation for an Advanced Schottky gate is about 4 mW, with a propagation delay rating of about 1.5 nS.

Advanced low-power Schottky TTL: 74ALSxx

The next TTL sub-family is a fairly predictable development. Combining low-power techniques with Advanced Schottky circuitry has resulted in the advanced low-power Schottky TTL line, designated as 74ALSxx, where "xx" is the two- or three-digit identifier for the specific device function.

Typical power dissipation for a 74ALSxx gate is only about 1 mW, but the propagation delay is still an impressively fast 5 nS. A 74ALSxx gate can be operated at frequencies as high as 35 MHz.

FAST ICs: 74Fxxx

National Semiconductor has developed a different fabrication approach to create the FAST line. FAST (a registered trade-mark of National Semiconductor) stands for Fairchild Advanced Schottky TTL. FAST technology has been adopted by other manufacturers, such as Motorola.

FAST devices are designated as 74Fxx, where "xx" is the two- or three-digit identifier for the specific device function. A typical FAST gate dissipates about 4 mW, with a propagation delay of about 3 nS.

CMOS: 74Cxx

So far we have been looking only at variations on TTL (Transistor-Transistor Logic). But other approaches to logic gating circuitry are possible. The other important basic logic family today is CMOS (Complementary Metal-Oxide Semiconductor). Although CMOS devices can perform the same logic functions as TTL devices, they are not electrically compatible. A CMOS gate can not directly drive a TTL gate, or vice versa. They can be interfaced, but extra external circuitry is required. These two logic families do not use the same voltage levels to represent the LOW and HIGH states, so confusion can result. In effect, TTL and CMOS gates don't "speak the same language."

When CMOS chips first entered the market, they were numbered in the CD4xxx series (see Section 2). To prevent the need to redesign circuits or redraw schematics for existing circuits that needed to be converted to CMOS, ICs using this technology were also made available using the 74Cxx numbering scheme, where "xx" is the two- or three-digit identifier for the specific device function.

One of the biggest advantages of CMOS over TTL is much greater flexibility in the supply voltage. Most TTL devices require a tightly regulated 5-V supply voltage. CMOS gates can be operated off anything from 3 V to 15 V. Generally speaking, supply voltages of 9 V to 12 V will be the best choice.

Power dissipation in a CMOS gate varies with the operating frequency. At dc or low frequencies, almost no power at all is dissipated. At high frequencies, however, a CMOS gate can dissipate 10 to 15 mW, and sometimes even more.

The propagation delay of standard CMOS gates tends to be rather slow. Delays of 90 nS are not uncommon for 74Cxx gates.

High-speed CMOS: 74HCxx

The chief downfall of 74Cxx CMOS gates is the relatively slow propagation delays. Improved technology has resulted in a somewhat faster high-speed

CMOS sub-family, which is numbered 74HCxx, where "xx" is the two- or three-digit identifier for the specific device function. The propagation delay for a typical 74HCxx gate is about 20 nS. The power dissipation rating is typically less than 1 mW.

A 74HCxx gate can accept a somewhat narrower range of supply voltages than a 74Cxx device. This sub-family requires a supply voltage between 2 and 6 V. A good choice would be 5 V so that a standard power supply designed for TTL circuits can be used. However, the logic levels are not directly compatible with TTL gates. Assuming a 5-V supply voltage, a HIGH signal must exceed 3.15 V. TTL gates use 2.4 V as the cut-off level for a HIGH logic level, so the signal might not be reliably recognized by a high-speed CMOS gate. On the other hand, the output from a high-speed CMOS device should work fine as the input to a TTL (or a related sub-family) gate.

High-speed CMOS/TTL compatible: 74HCTxx

A variation on the high-speed CMOS sub-family is designed to be logic compatible with TTL. Devices in this sub-family are numbered 74HCTxx, where "xx" is the two- or three-digit identifier for the specific device function. The propagation delay for a typical 74HCTxx gate is about 40 nS. The power dissipation rating is typically less than 1 mW.

Like the 74HCxx sub-family, a 74HCTxx gate can accept a somewhat narrower range of supply voltages than a standard 74Cxx device. This sub-family requires a supply voltage between 2 to 6 V. A good choice would be 5 V so that a standard power supply designed for TTL circuits can be used. In this case, direct interfacing with TTL gates is possible. Assuming a 5-V supply voltage, a HIGH signal must exceed 2 V. TTL gates use 2.4 V as the cut-off level for a HIGH logic level so that the signal will be reliably recognized by a 74HCTxx gate.

Advanced CMOS: 74ACxx and 74ACTxx

An even more improved CMOS sub-family is the Advanced CMOS line, numbered 74ACxx, where "xx" is the two- or three-digit indicator of the device function. A typical 74ACxx gate dissipates less than 0.5 mW, but has a propagation delay of a mere 3 nS. It can operate at frequencies higher than 100 MHz. The logic levels are not necessarily directly compatible with TTL.

A variant form is numbered 74ACTxx, where "xx" is the two- or three-digit identifier for the specific device function. A typical 74ACtxx gate dissipates less than 0.5 mW, but has a propagation delay of about 5 nS. It can operate at frequencies higher than 100 MHz. The difference here is that logic levels are designed to be directly compatible with TTL.

Mixing TTL Families

Most TTL families are intended for use together, but this cannot be done indiscriminately. Each family of TTL devices has unique input and output characteristics optimized to get the desired speed or power features. Fast devices like 74S and 74H are designed with relatively low input and output impedances. The speed of these devices is determined primarily by fast rise and fall times internally, as well as at the input and output nodes. These fast transitions cause noise of various types in the system. Power and ground line noise is generated by the large currents needed to charge and discharge the circuit and load capacitances during the switching transitions. Signal line noise is generated by the fast output transitions and the relatively low output impedances, which tend to increase reflections.

The noise generated by these 74S and 74H can only be tolerated in systems designed with very short signal leads, elaborate ground planes, and good well-decoupled power distribution networks. Mixing the slower TTL families like 74 and 74LS with the higher speed families is also possible but must be done with caution. The slower speed families are more susceptible to induced noise than the higher speed families due to their higher input and output impedances. The low power Schottky 74LS family is especially sensitive to induced noise and must be isolated as much as possible from the 74S and 74H devices. Separate or isolated power and ground systems are recommended, and the LS input signal lines should not run adjacent to lines driven by 74S or 74H devices.

Mixing 74 and 74LS is less restrictive, and the overall system design need not be so elaborate. Standard two sided PC boards can be used with good decoupled power and ground grid systems. The signal transitions are slower and therefore generate less noise. However, good high-speed design techniques are still required, especially when working with counters, registers, or other devices with memory.

7400 quad 2-input NAND gate

This device consists of four 2-input NAND gates. Each gate can be used as an inverter (by shorting its inputs together), or two NAND gates can be cross-coupled to form bistable flip-flop circuits.

The NAND gate is a variation on the basic AND gate, delivering an inverted (false) output when all inputs are true. *NAND* is a contraction of *NOT AND*. Notice that the output is HIGH when either or both of the inputs is LOW. Essentially, a NAND gate is the result of using an active inverting element in the gate circuitry. As with a conventional AND gate, the NAND gate can have any number of inputs. (In practical digital circuitry, a NAND gate is simpler than an AND gate. Actual AND gates are usually formed by re-inverting the output of a NAND gate circuit.)

	74S	74H	74LS	74	74L
Typ. Delay Time (ns)	3	6	9.5	10	33
Typ. Power Per Gate (mW)	19	22	2	10	1

7400 Truth Table

INPUTS		OUTPUT
A	**B**	
0	0	1
0	1	1
1	0	1
1	1	0

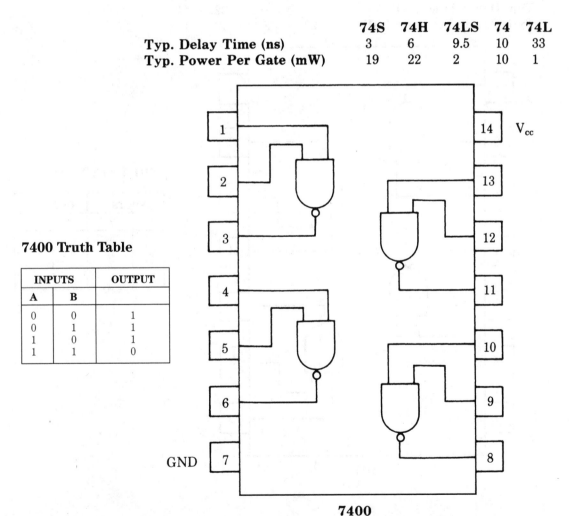

7400

7

7401 quad 2-input NAND gate

This device consists of four 2-input NAND gates. Each gate can be used as an inverter (by shorting its inputs together) or two NAND gates can be cross-coupled to form bistable flip-flop circuits.

The NAND gate is a variation on the basic AND gate, delivering an inverted (false) output when all inputs are true. *NAND* is a contraction of *NOT AND*. Notice that the output is HIGH when either or both of the inputs is LOW. Essentially a NAND gate is the result of using an active inverting element in the gate circuitry. As with a conventional AND gate, the NAND gate can have any number of inputs.

	74H	74LS	74	74L
Typ. Delay Time (ns)	8	16	22	41
Typ. Power Per Gate (mW)	22	2	10	1

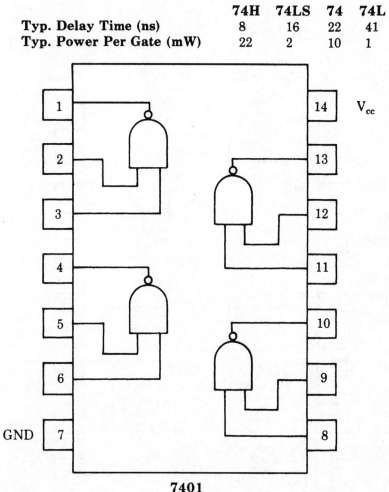

7401

7401 Truth Table

INPUTS		OUTPUT
A	**B**	
0	0	1
0	1	1
1	0	1
1	1	0

7402 quad 2-input NOR gate

This device consists of four 2-input NOR gates. Each gate can be used as an inverter (by shorting its inputs together).

The NOR gate is a variation on the basic OR gate, delivering a true output only when all inputs are false. *NOR* is a contraction of *NOT OR*. Notice that the output is LOW when either or both of the inputs is HIGH. Essentially, a NOR gate is the result of using an active inverting element in the gate circuitry.

As with a conventional OR gate, the NOR gate can have any number of inputs.

	74S	74LS	74	74L
Typ. Delay Time (ns)	3.5	10	10	33
Typ. Power Per Gate (mW)	29	2.75	14	1.5

7402 Truth Table

INPUTS		OUTPUT
A	**B**	
0	0	1
0	1	0
1	0	0
1	1	0

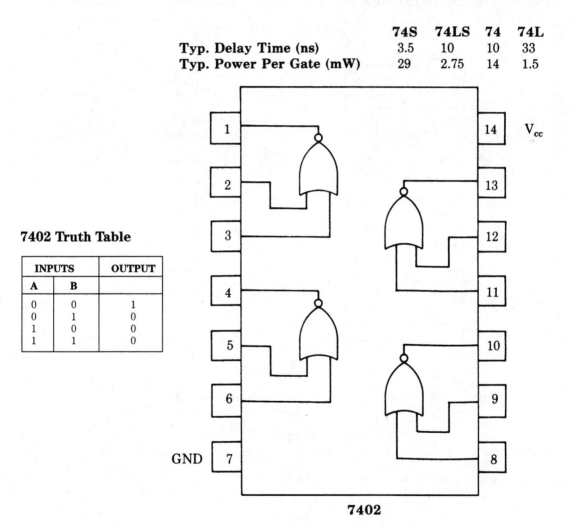

7402

7403 quad 2-input NAND gate with open collector outputs

This device consists of four 2-input NAND gates. Unlike the 7400 or the 7401, the 7403 features open collector outputs. The NAND gate is a variation on the basic AND gate, delivering an inverted (false) output when all inputs are true.

	74S	74LS	74	74L
Typ. Delay Time (ns)	5	16	22	41
Typ. Power Per Gate (mW)	17.2	2	10	1

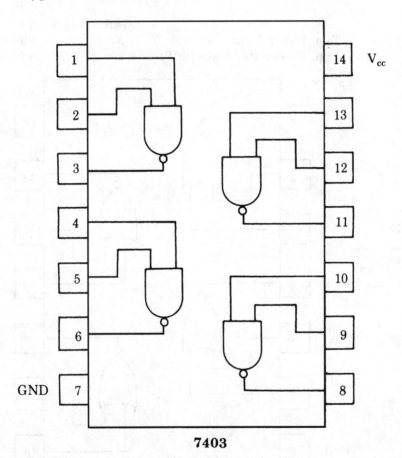

7403

7403 Truth Table

INPUTS		OUTPUT
A	**B**	
0	0	1
0	1	1
1	0	1
1	1	0

7404 hex inverter

The 7404 contains eight single-input/single-output inverter/buffer stages. Only the power-supply connections are in common. As logic elements are combined in series to perform complex logic functions, voltage levels tend to degrade. At some point, the circuitry might not be able to reliably distinguish between HIGH and LOW states. Thus, amplifiers (or buffers) are sometimes needed to restore voltages or currents to their proper levels. In the case of the 7404, an inversion function is added to the basic buffer. A LOW input results in a HIGH output and a HIGH input will produce a LOW output.

An inverter/buffer is often simply referred to as an *inverter*. The buffer is assumed.

	74S	74H	74LS	74	74L
Typ. Delay Time (ns)	3	6	9.5	10	33
Typ. Power Per Gate (mW)	19	22	2	10	1

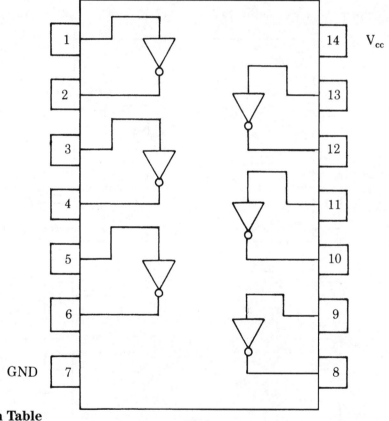

7404

7404 Truth Table

INPUT	OUTPUT
0	1
1	0

7405 hex inverter with open collector outputs

The 7405, like the 7404, contains eight single-input/single-output inverter/buffer stages. Only the power-supply connections are in common. The difference offered by the 7405 is that open collector-type outputs are used.

As logic elements are combined in series to perform complex logic functions, voltage levels tend to degrade. At some point, the circuitry might not be able to reliably distinguish between HIGH and LOW states. Thus, amplifiers (or buffers) are sometimes needed to restore voltages or currents to their proper levels. In the case of the 7405, an inversion function is added to the basic buffer. A LOW input results in a HIGH output and a HIGH input will produce a LOW output.

An inverter/buffer is often simply referred to as an *inverter*. The buffer is assumed.

	74S	74H	74LS	74
Typ. Delay Time (ns)	5	8	16	22
Typ. Power Per Gate (mW)	17.5	22	2	10

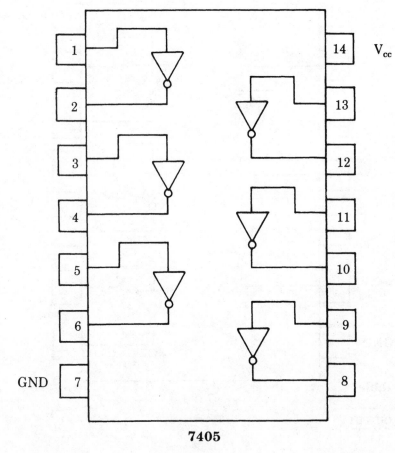

7405

7405 Truth Table

INPUT	OUTPUT
0	1
1	0

7406 hex inverter buffer/driver
with open collector outputs

The 7406, like the 7404, contains eight single-input/single-output inverter/buffer stages. Only the power supply connections are in common. The differences offered by the 7406 are that open collector-type outputs are used and heftier amplification is supplied, hence the "driver" designation.

As logic elements are combined in series to perform complex logic functions, voltage levels tend to degrade. At some point, the circuitry might not be able to reliably distinguish between HIGH and LOW states. Thus, amplifiers (or buffers) are sometimes needed to restore voltages or currents to their proper levels. In the case of the 7406, an inversion function is added to the basic buffer. A LOW input results in a HIGH output and a HIGH input will produce a LOW output.

An inverter/buffer is often simply referred to as an *inverter*. The buffer is assumed.

7406 Truth Table

INPUT	OUTPUT
0	1
1	0

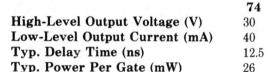

	74
High-Level Output Voltage (V)	30
Low-Level Output Current (mA)	40
Typ. Delay Time (ns)	12.5
Typ. Power Per Gate (mW)	26

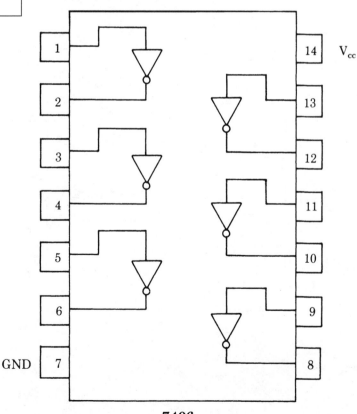

7406

7407 hex buffer/driver with open collector outputs

The 7407 contains eight single-input/single-output noninverting hefty buffers, each with an open collector output. As logic elements are combined in series to perform complex logic functions, voltage levels tend to degrade. At some point, the circuitry might not be able to reliably distinguish between HIGH and LOW states. Thus, amplifiers (or buffers) are sometimes needed to restore voltages or currents to their proper levels. In the case of the 7407, there is no logic inversion, so the output state always matches the input state.

	74
High-Level Output Voltage (V)	30
Low-Level Output Current (mA)	40
Typ. Delay Time (ns)	13
Typ. Power Per Gate (mW)	21

7407 Truth Table

INPUT	OUTPUT
0	0
1	1

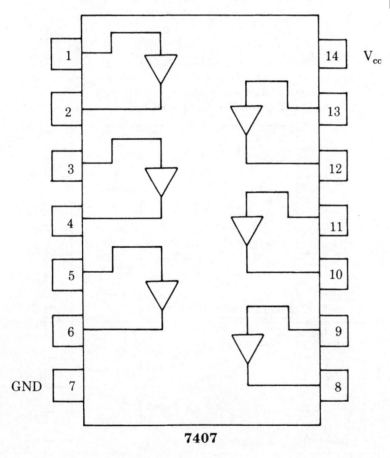

7407

7408 quad 2-input AND gate

The 7408 contains four functionally independent two-input/single-output AND gates. Only the power supply connections are common to the four gates.

An AND gate is the noninverted form of the somewhat more common NAND gate. Reinverting the output of a NAND gate will create an AND gate function.

The output of an AND gate is HIGH if, and only if, all inputs are HIGH.

7408 Truth Table

INPUTS		OUTPUT
A	**B**	
0	0	0
0	1	0
1	0	0
1	1	1

	74LS	74
Typ. Delay Time (ns)	12	15
Typ. Power Per Gate (mW)	4.25	19

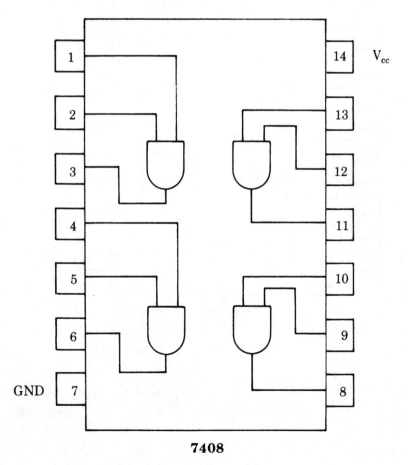

7408

7410 triple 3-input NAND gate

The 7410 contains three functionally independent three-input/single-output NAND gates. Only the power supply connections are common to the three gates.

A NAND gate is the inverted form of the AND gate. The output of a NAND gate is LOW if, and only if, all inputs are HIGH.

	74S	74H	74LS	74	74L
Typ. Delay Time (ns)	3	6	9.5	10	33
Typ. Power Per Gate (mW)	19	22	2	10	1

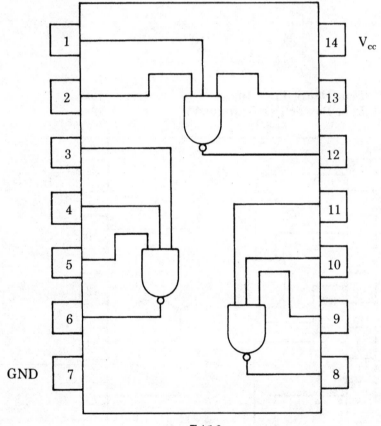

7410

7410 Truth Table

INPUTS			OUTPUT
A	**B**	**C**	
0	0	0	1
0	0	1	1
0	1	0	1
0	1	1	1
1	0	0	1
1	0	1	1
1	1	0	1
1	1	1	0

7411 triple 3-input AND gate

The 7411 contains three functionally independent three-input/single-output AND gates. Only the power supply connections are common to the three gates.

An AND gate is the noninverted form of the somewhat more common NAND gate. Reinverting the output of a NAND gate will create an AND gate function. The output of an AND gate is HIGH if, and only if, all inputs are HIGH.

	74S	74H	74LS
Typ. Delay Time (ns)	4.75	8.2	12
Typ. Power Per Gate (mW)	31	40	4.25

7411 Truth Table

INPUTS			OUTPUT
A	B	C	
0	0	0	0
0	0	1	0
0	1	0	0
0	1	1	0
1	0	0	0
1	0	1	0
1	1	0	0
1	1	1	1

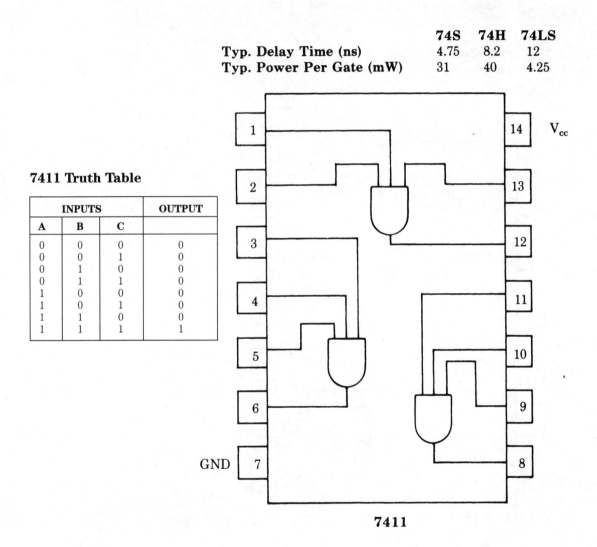

7411

7412 triple 3-input NAND gate

The 7412 contains three functionally independent three-input/single-output NAND gates. Only the power supply connections are common to the three gates. The difference between the 7412 and the 7410 is that this device uses open collector outputs.

A NAND gate is the inverted form of the AND gate. The output of a NAND gate is LOW if, and only if, all inputs are HIGH.

	74LS
Typ. Delay Time (ns)	16
Typ. Power Per Gate (mW)	2

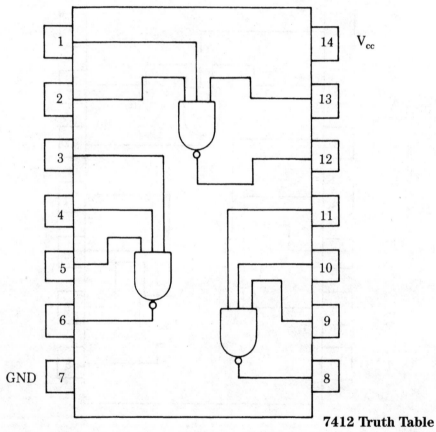

7412

7412 Truth Table

INPUTS			OUTPUT
A	**B**	**C**	
0	0	0	1
0	0	1	1
0	1	0	1
0	1	1	1
1	0	0	1
1	0	1	1
1	1	0	1
1	1	1	0

7413 dual 4-input NAND Schmitt trigger

The 7413 contains a pair of four-input NAND gates, which incorporate built-in Schmitt triggers, permitting them to respond reliably to slowly changing or erratic input signals. These devices have a greater noise margin than conventional NAND gates. The output signals are sharply defined and jitter-free.

A Schmitt trigger uses positive feedback, and, in effect, "speeds up" a slowly changing input transition. Different threshold voltages are used to trigger HIGH-to-LOW and LOW-to-HIGH transitions, creating a hysteresis band to avoid any ambiguity or jitter from input signals in the "grey" area between a clear-cut HIGH or an unquestionable LOW.

The hysteresis in the 7413 is determined by internal resistance ratios. It is typically about 800 mV, and is reasonably insensitive to temperature or supply voltage variations.

As with any NAND gate, the output is LOW if, and only if, all inputs are HIGH. If any one or more of the inputs is LOW, the output of the gate will be HIGH.

	74	74LS	74F
Typical Hysteresis (V)	0.8	0.8	0.8
Typical Delay Time (nS)	15	15	5

7413 Truth Table

INPUTS				OUTPUT
A	**B**	**C**	**D**	
0	x	x	x	1
x	0	x	x	1
x	x	0	x	1
x	x	x	0	1
1	1	1	1	0

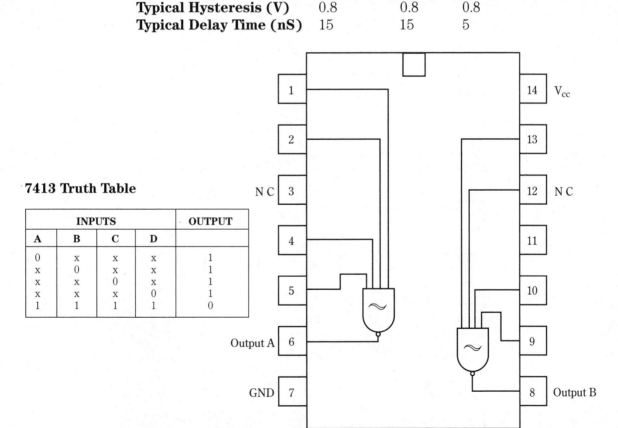

7413

19

7414 Hex Schmitt trigger

The 7414 contains six single-input/single-output logic inverters that accept standard and even somewhat sub-standard digital logic signals (of the appropriate logic family) and provides standard output levels. They are capable of cleaning up noisy or distorted logic signals, transforming slowly changing input signals into sharply defined, jitter-free output signals. In addition, they have a much greater noise margin than conventional inverters.

Each of the six inverter circuits in the 7414 contains a Schmitt trigger followed by a Darlington level shifter, and a phase splitter driving a TTL (or comparable) totem-pole output. The Schmitt trigger uses positive feedback to effectively speed up slow input transitions, and provide different input threshold voltages for positive-going and negative-going transitions. This hysteresis between the positive-going and negative-going input thresholds (typically about 800 mV) is determined internally by on-chip resistor ratios, and is essentially insensitive to temperature and supply voltage variations. Use of such hysteresis results in much greater immunity to noise and voltage fluctuations in a weak input signal.

	74	74LS
Typ. Hysteresis (V)	0.8	0.8
Typ. Delay Time (ns)	15	15

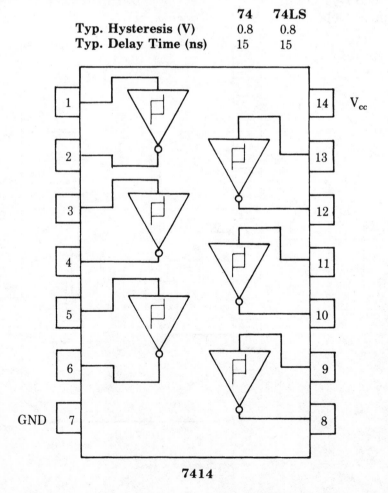

7414

7415 Triple 3-input and gate
with open collector outputs

The 7415 contains three functionally independent three-input/single-output AND gates. Only the power supply connections are common to the three gates. The difference between the 7415 and the 7411 is that this device uses open collector outputs.

The output of an AND gate is HIGH if, and only if, all inputs are HIGH.

	6	20
Typ. Delay Time (ns)	6	20
Typ. Power Per Gate (mW)	28	4.25

7415 Truth Table

INPUTS			OUTPUT
A	**B**	**C**	
0	0	0	0
0	0	1	0
0	1	0	0
0	1	1	0
1	0	0	0
1	0	1	0
1	1	0	0
1	1	1	1

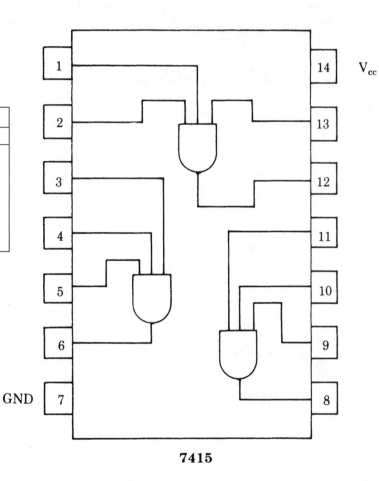

7415

21

7416 Hex inverter buffer/driver
with open collector outputs

The 7416 contains six single-input/single-output inverter/buffer driver stages. Only the power supply connections are in common. The chief difference over most other inverter packages offered by the 7416 is that open collector-type outputs are used.

As logic elements are combined in series to perform complex logic functions, voltage levels tend to degrade. At some point, the circuitry might not be able to reliably distinguish between HIGH and LOW states. Thus, amplifiers (or buffers) are sometimes needed to restore voltages or currents to their proper levels. In the case of the 7416, an inversion function is added to the basic buffer. A LOW input results in a HIGH output and a HIGH input will produce a LOW output.

An inverter/buffer is often simply referred to as an *inverter*. The buffer is assumed. The driver designation for this device indicates somewhat greater than usual amplification.

	74
High-Level Output Voltage (V)	15
Low-Level Output Current (mA)	40
Typ. Delay Time (ns)	12.5
Typ. Power Per Gate (mW)	26

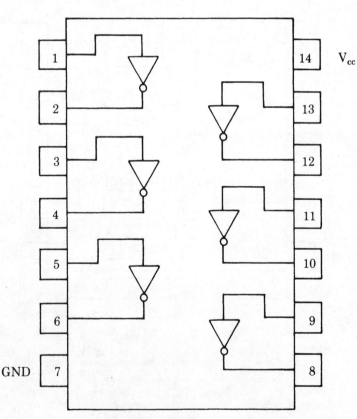

7416

7416 Truth Table

INPUT	OUTPUT
0	1
1	0

7417 Hex buffer/driver with open collector outputs

The 7417 contains six single-input/single-output noninverting hefty buffers —each with an open collector output. The driver designation for this device indicates somewhat greater than usual amplification.

As logic elements are combined in series to perform complex logic functions, voltage levels tend to degrade. At some point, the circuitry might not be able to reliably distinguish between HIGH and LOW states. Thus, amplifiers (or buffers) are sometimes needed to restore voltages or currents to their proper levels. In the case of the 7417, there is no logic inversion, so the output state always matches the input state.

7417 Truth Table

INPUT	OUTPUT
0	0
1	1

	74
High-Level Output Voltage (V)	15
Low-Level Output Current (mA)	40
Typ. Delay Time (ns)	13
Typ. Power Per Gate (mW)	21

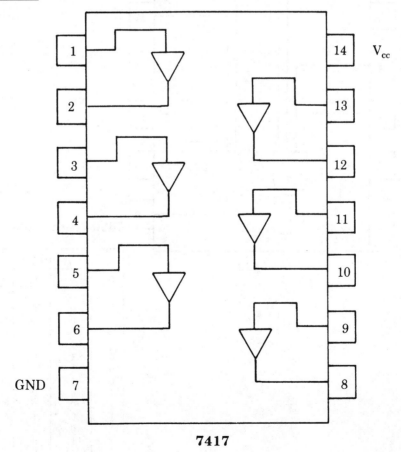

7417

7420 dual 4-input NAND gate

The 7420 contains a pair of functionally independent four-input/single-output NAND gates. Only the power supply connections are common to the three gates.

A NAND gate is the inverted form of the AND gate. The output of a NAND gate is LOW if, and only if, all inputs are HIGH.

	74S	74H	74LS	74
Typ. Delay Time (ns)	3	6	9.5	10
Typ. Power Per Gate (mW)	19	22	2	10

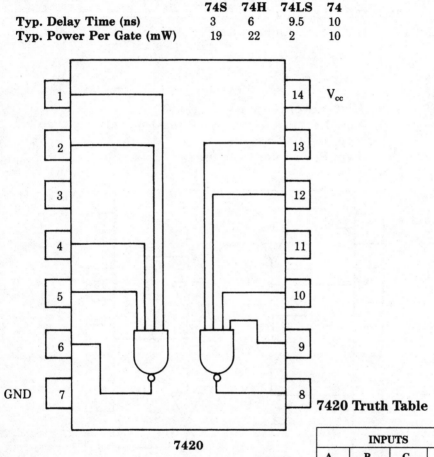

7420

7420 Truth Table

INPUTS				OUTPUT
A	**B**	**C**	**D**	
0	0	0	0	1
0	0	0	1	1
0	0	1	0	1
0	0	1	1	1
0	1	0	0	1
0	1	0	1	1
0	1	1	0	1
0	1	1	1	1
1	0	0	0	1
1	0	0	1	1
1	0	1	0	1
1	0	1	1	1
1	1	0	0	1
1	1	0	1	1
1	1	1	0	1
1	1	1	1	0

7421 dual 4-input AND gate

The 7421 contains two functionally independent four-input/single-output AND gates. Only the power supply connections are common to the three gates.

An AND gate is the noninverted form of the somewhat more common NAND gate. The output of an AND gate is HIGH if, and only if, all inputs are HIGH.

	74H	74LS
Typ. Delay Time (ns)	8.2	12
Typ. Power Per Gate (mW)	40	4.25

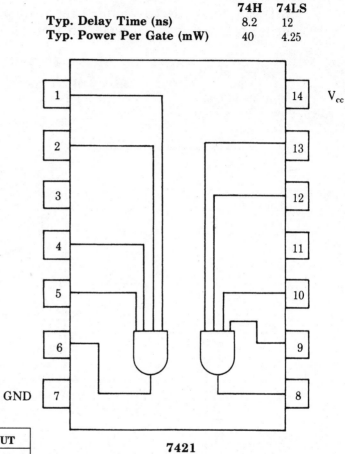

7421

7421 Truth Table

INPUTS				OUTPUT
A	B	C	D	
0	0	0	0	0
0	0	0	1	0
0	0	1	0	0
0	0	1	1	0
0	1	0	0	0
0	1	0	1	0
0	1	1	0	0
0	1	1	1	0
1	0	0	0	0
1	0	0	1	0
1	0	1	0	0
1	0	1	1	0
1	1	0	0	0
1	1	0	1	0
1	1	1	0	0
1	1	1	1	1

7422 dual 4-input NAND gate
with open collector outputs

The 7422, like the 7420, contains a pair of functionally independent four-input/single-output NAND gates. Only the power supply connections are common to the three gates. The main difference of the 7422 is that it features open collector outputs.

A NAND gate is the inverted form of the AND gate. The output of a NAND gate is low if, and only if, all inputs are high.

	74S	74H	74LS
Typ. Delay Time (ns)	5	8	16
Typ. Power Per Gate (mW)	17.5	22	2

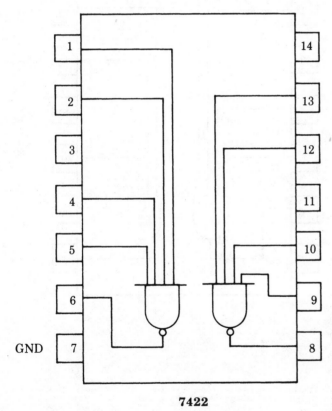

7422

7422 Truth Table

INPUTS				OUTPUT
A	**B**	**C**	**D**	
0	0	0	0	1
0	0	0	1	1
0	0	1	0	1
0	0	1	1	1
0	1	0	0	1
0	1	0	1	1
0	1	1	0	1
0	1	1	1	1
1	0	0	0	1
1	0	0	1	1
1	0	1	0	1
1	0	1	1	1
1	1	0	0	1
1	1	0	1	1
1	1	1	0	1
1	1	1	1	0

7425 dual 4-input NOR gate with strobe

The 7425 contains a pair of functionally independent four-input/single-output NOR gates. Only the power supply connections are common to the three gates. An unusual feature of the 7425 is that each NOR gate has an extra "strobe" input to turn it on or off.

A NOR gate is the inverted form of the OR gate. The output of a NOR gate is HIGH if, and only if, all inputs are LOW.

	74
Typ. Delay Time (ns)	10.5
Typ. Power Per Gate (mW)	23

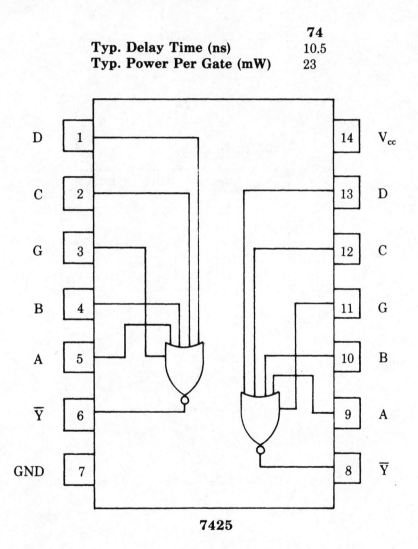

7425

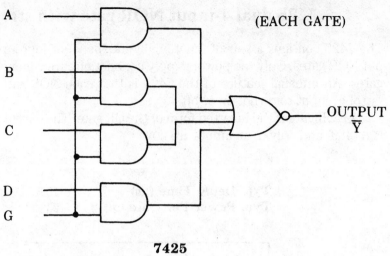

(EACH GATE)

7425

7425 Truth Table

INPUTS				OUTPUT
A	**B**	**C**	**D**	
0	0	0	0	1
0	0	0	1	0
0	0	1	0	0
0	0	1	1	0
0	1	0	0	0
0	1	0	1	0
0	1	1	0	0
0	1	1	1	0
1	0	0	0	0
1	0	0	1	0
1	0	1	0	0
1	0	1	1	0
1	1	0	0	0
1	1	0	1	0
1	1	1	0	0
1	1	1	1	0

7426 quad 2-input NAND gate with open collector outputs

The 7426 is another package of four two-input NAND gates, each with open collector outputs. A NAND gate is the inverted form of the AND gate. The output of a NAND gate is LOW if, and only if, all inputs are HIGH.

	74	74LS
High-Level Output Voltage (V)	15	15
Low-Level Output Current (mA)	16	8
Typ. Delay Time (ns)	13.5	16
Typ. Power Per Gate (mW)	10	2

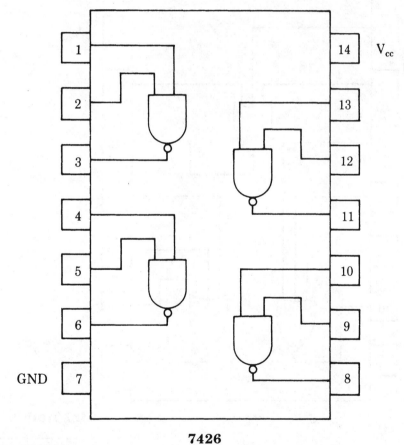

7426

7426 Truth Table

INPUTS		OUTPUT
A	B	
0	0	1
0	1	1
1	0	1
1	1	0

7427 triple 3-input NOR gate

The 7427 contains three functionally independent three-input/single-output NOR gates. Only the power supply connections are common to the three gates.

A NOR gate is the inverted form of the OR gate. The output of a NOR gate is HIGH if, and only if, all inputs are LOW.

	74	74LS
Typ. Delay Time (ns)	8.5	10
Typ. Power Per Gate (mW)	22	4.5

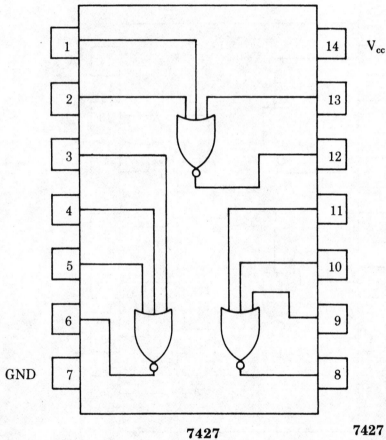

7427

7427 Truth Table

INPUTS			OUTPUT
A	**B**	**C**	
0	0	0	1
0	0	1	0
0	1	0	0
0	1	1	0
1	0	0	0
1	0	1	0
1	1	0	0
1	1	1	0

7428 quad 2-input NOR buffer

The 7428 contains four functionally independent two-input/single-output NOR gates. Only the power supply connections are common to the three gates. Each section of this chip is a buffer (logic amplifier), as well as a gate. This eliminates the need for external buffer stages in some applications.

A NOR gate is the inverted form of the OR gate. The output of a NOR gate is HIGH if, and only if, all inputs are LOW.

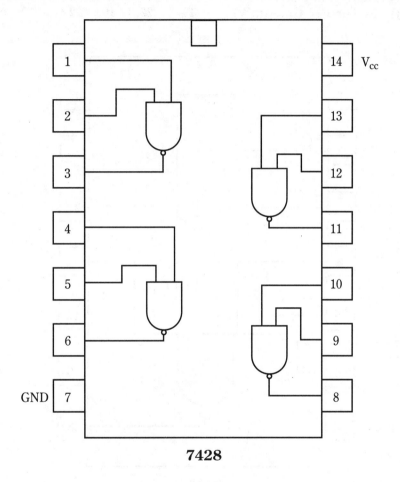

7428

7428 Truth Table

INPUTS		OUTPUT
A	B	
0	0	1
0	1	0
1	0	0
1	1	0

7430 8-input NAND gate

The 7430 contains a single eight-input/single-output NAND gate. A NAND gate is the inverted form of the AND gate. The output of a NAND gate is LOW if, and only if, all inputs are HIGH.

	74S	74H	74	74LS	74L
Typ. Delay Time (ns)	3	6	10	17	33
Typ. Power Per Gate (mW)	19	22	10	2.4	1

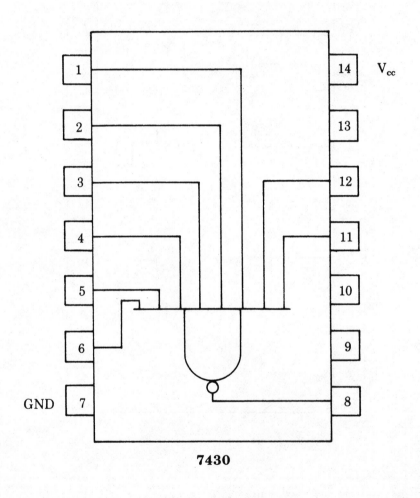

7430

7430 Truth Table

INPUTS								OUTPUT
A	B	C	D	E	F	G	H	
0	0	0	0	0	0	0	0	1
0	0	0	0	0	0	0	1	1
0	0	0	0	0	0	1	0	1
0	0	0	0	0	0	1	1	1
0	0	0	0	0	1	0	0	1
0	0	0	0	0	1	0	1	1
0	0	0	0	0	1	1	0	1
0	0	0	0	0	1	1	1	1
0	0	0	0	1	0	0	0	1
0	0	0	0	1	0	0	1	1
0	0	0	0	1	0	1	0	1
0	0	0	0	1	0	1	1	1
0	0	0	0	1	1	0	0	1
0	0	0	0	1	1	0	1	1
0	0	0	0	1	1	1	0	1
0	0	0	0	1	1	1	1	1
0	0	0	1	0	0	0	0	1
0	0	0	1	0	0	0	1	1
0	0	0	1	0	0	1	0	1
0	0	0	1	0	0	1	1	1
0	0	0	1	0	1	0	0	1
0	0	0	1	0	1	0	1	1
0	0	0	1	0	1	1	0	1
0	0	0	1	0	1	1	1	1
0	0	0	1	1	0	0	0	1
0	0	0	1	1	0	0	1	1
0	0	0	1	1	0	1	0	1
0	0	0	1	1	0	1	1	1
0	0	0	1	1	1	0	0	1
0	0	0	1	1	1	0	1	1
0	0	0	1	1	1	1	0	1
0	0	0	1	1	1	1	1	1
0	0	1	0	0	0	0	0	1
0	0	1	0	0	0	0	1	1
0	0	1	0	0	0	1	0	1
0	0	1	0	0	0	1	1	1
0	0	1	0	0	1	0	0	1
0	0	1	0	0	1	0	1	1
0	0	1	0	0	1	1	0	1
0	0	1	0	0	1	1	1	1
0	0	1	0	1	0	0	0	1
0	0	1	0	1	0	0	1	1
0	0	1	0	1	0	1	0	1
0	0	1	0	1	0	1	1	1
0	0	1	0	1	1	0	0	1
0	0	1	0	1	1	0	1	1
0	0	1	0	1	1	1	0	1
0	0	1	0	1	1	1	1	1
0	0	1	1	0	0	0	0	1
0	0	1	1	0	0	0	1	1
0	0	1	1	0	0	1	0	1
0	0	1	1	0	0	1	1	1
0	0	1	1	0	1	0	0	1
0	0	1	1	0	1	0	1	1
0	0	1	1	0	1	1	0	1
0	0	1	1	0	1	1	1	1
0	0	1	1	1	0	0	0	1
0	0	1	1	1	0	0	1	1

7430 Truth Table (Continued)

A	B	C	D	E	F	G	H	OUTPUT
0	0	1	1	1	0	1	0	1
0	0	1	1	1	0	1	1	1
0	0	1	1	1	1	0	0	1
0	0	1	1	1	1	0	1	1
0	0	1	1	1	1	1	0	1
0	0	1	1	1	1	1	1	1
0	1	0	0	0	0	0	0	1
0	1	0	0	0	0	0	1	1
0	1	0	0	0	0	1	0	1
0	1	0	0	0	0	1	1	1
0	1	0	0	0	1	0	0	1
0	1	0	0	0	1	0	1	1
0	1	0	0	0	1	1	0	1
0	1	0	0	0	1	1	1	1
0	1	0	0	1	0	0	0	1
0	1	0	0	1	0	0	1	1
0	1	0	0	1	0	1	0	1
0	1	0	0	1	0	1	1	1
0	1	0	0	1	1	0	0	1
0	1	0	0	1	1	0	1	1
0	1	0	0	1	1	1	0	1
0	1	0	0	1	1	1	1	1
0	1	0	1	0	0	0	0	1
0	1	0	1	0	0	0	1	1
0	1	0	1	0	0	1	0	1
0	1	0	1	0	0	1	1	1
0	1	0	1	0	1	0	0	1
0	1	0	1	0	1	0	1	1
0	1	0	1	0	1	1	0	1
0	1	0	1	0	1	1	1	1
0	1	0	1	1	0	0	0	1
0	1	0	1	1	0	0	1	1
0	1	0	1	1	0	1	0	1
0	1	0	1	1	0	1	1	1
0	1	0	1	1	1	0	0	1
0	1	0	1	1	1	0	1	1
0	1	0	1	1	1	1	0	1
0	1	0	1	1	1	1	1	1
0	1	1	0	0	0	0	0	1
0	1	1	0	0	0	0	1	1
0	1	1	0	0	0	1	0	1
0	1	1	0	0	0	1	1	1
0	1	1	0	0	1	0	0	1
0	1	1	0	0	1	0	1	1
0	1	1	0	0	1	1	0	1
0	1	1	0	0	1	1	1	1
0	1	1	0	1	0	0	0	1
0	1	1	0	1	0	0	1	1
0	1	1	0	1	0	1	0	1
0	1	1	0	1	0	1	1	1
0	1	1	0	1	1	0	0	1
0	1	1	0	1	1	0	1	1
0	1	1	0	1	1	1	0	1
0	1	1	0	1	1	1	1	1
0	1	1	1	0	0	0	0	1
0	1	1	1	0	0	0	1	1
0	1	1	1	0	0	1	0	1
0	1	1	1	0	0	1	1	1
0	1	1	1	0	1	0	0	1

7430 Truth Table (Continued)

INPUTS								OUTPUT
A	B	C	D	E	F	G	H	
0	1	1	1	0	1	0	1	1
0	1	1	1	0	1	1	0	1
0	1	1	1	0	1	1	1	1
0	1	1	1	1	0	0	0	1
0	1	1	1	1	0	0	1	1
0	1	1	1	1	0	1	0	1
0	1	1	1	1	0	1	1	1
0	1	1	1	1	1	0	0	1
0	1	1	1	1	1	0	1	1
0	1	1	1	1	1	1	0	1
0	1	1	1	1	1	1	1	1
1	0	0	0	0	0	0	0	1
1	0	0	0	0	0	0	1	1
1	0	0	0	0	0	1	0	1
1	0	0	0	0	0	1	1	1
1	0	0	0	0	1	0	0	1
1	0	0	0	0	1	0	1	1
1	0	0	0	0	1	1	0	1
1	0	0	0	0	1	1	1	1
1	0	0	0	1	0	0	0	1
1	0	0	0	1	0	0	1	1
1	0	0	0	1	0	1	0	1
1	0	0	0	1	0	1	1	1
1	0	0	0	1	1	0	0	1
1	0	0	0	1	1	0	1	1
1	0	0	0	1	1	1	0	1
1	0	0	0	1	1	1	1	1
1	0	0	1	0	0	0	0	1
1	0	0	1	0	0	0	1	1
1	0	0	1	0	0	1	0	1
1	0	0	1	0	0	1	1	1
1	0	0	1	0	1	0	0	1
1	0	0	1	0	1	0	1	1
1	0	0	1	0	1	1	0	1
1	0	0	1	0	1	1	1	1
1	0	0	1	1	0	0	0	1
1	0	0	1	1	0	0	1	1
1	0	0	1	1	0	1	0	1
1	0	0	1	1	0	1	1	1
1	0	0	1	1	1	0	0	1
1	0	0	1	1	1	0	1	1
1	0	0	1	1	1	1	0	1
1	0	0	1	1	1	1	1	1
1	0	1	0	0	0	0	0	1
1	0	1	0	0	0	0	1	1
1	0	1	0	0	0	1	0	1
1	0	1	0	0	0	1	1	1
1	0	1	0	0	1	0	0	1
1	0	1	0	0	1	0	1	1
1	0	1	0	0	1	1	0	1
1	0	1	0	0	1	1	1	1
1	0	1	0	1	0	0	0	1
1	0	1	0	1	0	0	1	1
1	0	1	0	1	0	1	0	1
1	0	1	0	1	0	1	1	1
1	0	1	0	1	1	0	0	1
1	0	1	0	1	1	0	1	1
1	0	1	0	1	1	1	0	1

7430 Truth Table (Continued)

INPUTS								OUTPUT
A	B	C	D	E	F	G	H	
1	0	1	0	1	1	1	1	1
1	0	1	1	0	0	0	0	1
1	0	1	1	0	0	0	1	1
1	0	1	1	0	0	1	0	1
1	0	1	1	0	0	1	1	1
1	0	1	1	0	1	0	0	1
1	0	1	1	0	1	0	1	1
1	0	1	1	0	1	1	0	1
1	0	1	1	0	1	1	1	1
1	0	1	1	1	0	0	0	1
1	0	1	1	1	0	0	1	1
1	0	1	1	1	0	1	0	1
1	0	1	1	1	0	1	1	1
1	0	1	1	1	1	0	0	1
1	0	1	1	1	1	0	1	1
1	0	1	1	1	1	1	0	1
1	0	1	1	1	1	1	1	1
1	1	0	0	0	0	0	0	1
1	1	0	0	0	0	0	1	1
1	1	0	0	0	0	1	0	1
1	1	0	0	0	0	1	1	1
1	1	0	0	0	1	0	0	1
1	1	0	0	0	1	0	1	1
1	1	0	0	0	1	1	0	1
1	1	0	0	0	1	1	1	1
1	1	0	0	1	0	0	0	1
1	1	0	0	1	0	0	1	1
1	1	0	0	1	0	1	0	1
1	1	0	0	1	0	1	1	1
1	1	0	0	1	1	0	0	1
1	1	0	0	1	1	0	1	1
1	1	0	0	1	1	1	0	1
1	1	0	0	1	1	1	1	1
1	1	0	1	0	0	0	0	1
1	1	0	1	0	0	0	1	1
1	1	0	1	0	0	1	0	1
1	1	0	1	0	0	1	1	1
1	1	0	1	0	1	0	0	1
1	1	0	1	0	1	0	1	1
1	1	0	1	0	1	1	0	1
1	1	0	1	0	1	1	1	1
1	1	0	1	1	0	0	0	1
1	1	0	1	1	0	0	1	1
1	1	0	1	1	0	1	0	1
1	1	0	1	1	0	1	1	1
1	1	0	1	1	1	0	0	1
1	1	0	1	1	1	0	1	1
1	1	0	1	1	1	1	0	1
1	1	0	1	1	1	1	1	1
1	1	1	0	0	0	0	0	1
1	1	1	0	0	0	0	1	1
1	1	1	0	0	0	1	0	1
1	1	1	0	0	0	1	1	1
1	1	1	0	0	1	0	0	1
1	1	1	0	0	1	0	1	1
1	1	1	0	0	1	1	0	1
1	1	1	0	0	1	1	1	1
1	1	1	0	1	0	0	0	1

7430 Truth Table (Continued)

INPUTS								OUTPUT
A	B	C	D	E	F	G	H	
1	1	1	0	1	0	0	1	1
1	1	1	0	1	0	1	0	1
1	1	1	0	1	0	1	1	1
1	1	1	0	1	1	0	0	1
1	1	1	0	1	1	0	1	1
1	1	1	0	1	1	1	0	1
1	1	1	0	1	1	1	1	1
1	1	1	1	0	0	0	0	1
1	1	1	1	0	0	0	1	1
1	1	1	1	0	0	1	0	1
1	1	1	1	0	0	1	1	1
1	1	1	1	0	1	0	0	1
1	1	1	1	0	1	0	1	1
1	1	1	1	0	1	1	0	1
1	1	1	1	0	1	1	1	1
1	1	1	1	1	0	0	0	1
1	1	1	1	1	0	0	1	1
1	1	1	1	1	0	1	0	1
1	1	1	1	1	0	1	1	1
1	1	1	1	1	1	0	0	1
1	1	1	1	1	1	0	1	1
1	1	1	1	1	1	1	0	1
1	1	1	1	1	1	1	1	0

7432 quad 2-input OR gate

The 7432 contains four functionally independent two-input/single-output OR gates. Only the power supply connections are common to the entire chip. The output of an OR gate is HIGH whenever one or more of the inputs are HIGH. The output is LOW if, and only if, all inputs are LOW.

	74
Typ. Delay Time (ns)	12
Typ. Power Per Gate (mW)	24

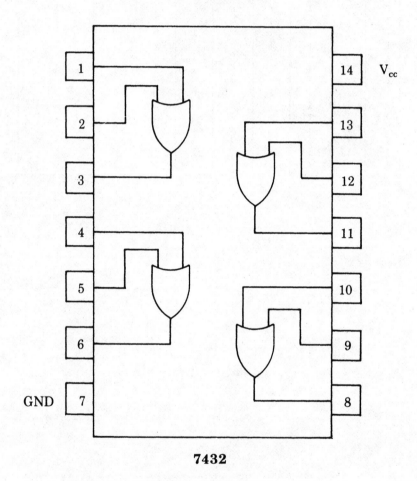

7432

7432 Truth Table

INPUTS		OUTPUT
A	**B**	
0	0	0
0	1	1
1	0	1
1	1	1

7437 quad 2-input NAND buffer

The 7437 contains four functionally independent two-input/single-output NAND gates. Only the power supply connections are common to the entire chip.

A NAND gate is the inverted output form of an AND gate. NAND is a contraction of NOT AND. The output is LOW if, and only if, all inputs are HIGH. As long as at least one input is LOW, the NAND gate's output will be HIGH.

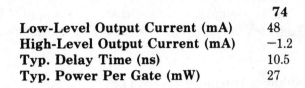

	74
Low-Level Output Current (mA)	48
High-Level Output Current (mA)	−1.2
Typ. Delay Time (ns)	10.5
Typ. Power Per Gate (mW)	27

7437 Truth Table

INPUTS		OUTPUT
A	B	
0	0	0
0	1	1
1	0	1
1	1	1

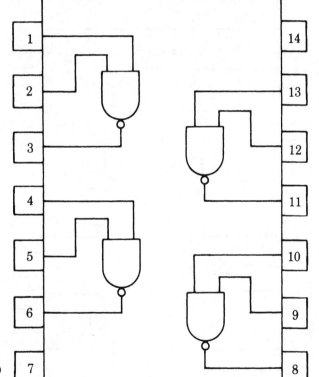

7437

7438 quad 2-input NAND buffer, open collector

The 7438 is functionally pretty similar to a standard two-input NAND gate, except that the output is in open collector form. External components are required to pull the output up to the supply voltage, rather than being handled internally within the chip itself. This added flexibility is desirable in some specialized applications. An open collector output cannot be used to drive another input directly. Some form of external pull-up is required for proper operation of the gate.

The buffer designates greater than usual amplification and signal boost. This eliminates the need for external buffer stages in some applications.

Maximum Output Current (LOW): 64 mA
Maximum Output Current (HIGH): 250 μA

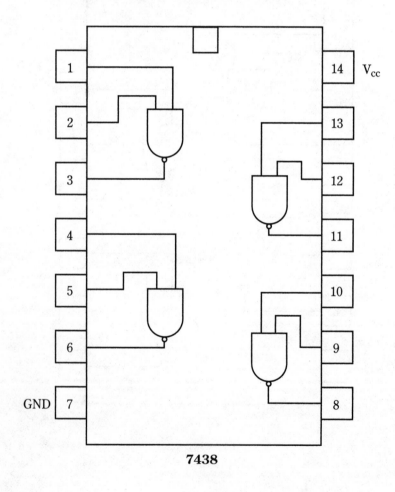

7438

7438 Truth Table

INPUTS		OUTPUT
A	B	
0	0	1
0	1	1
1	0	1
1	1	0

7440 Quad 2-input NAND buffer, open collector

The 7440 is functionally somewhat similar to a standard two-input NAND gate, except that the output is in open collector form. External components are required to pull the output up to the supply voltage, rather than being handled internally within the chip itself. This added flexibility is desirable in some specialized applications. An open-collector output cannot be used to drive another input directly. Some form of external pull-up is required for proper operation of the gate. The buffer designation indicates greater than usual amplification.

	74S	74H	74	74LS
Low-Level Output Current (mA)	60	60	48	24
High-Level Output Current (mA)	−3	−1.5	−1.2	−1.2
Typ. Delay Time (ns)	4	7.5	10.5	12
Typ. Power Per Gate (mW)	44	44	26	4.3

7440 Truth Table

INPUTS		OUTPUT
A	B	
0	0	1
0	1	1
1	0	1
1	1	0

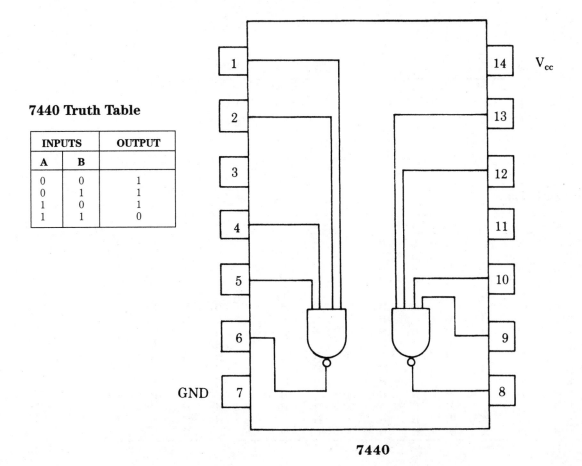

7440

7442 BCD-to-decimal decoder

The 7442 decoder accepts four active HIGH BCD inputs and provides 10 mutually exclusive active LOW outputs, as shown in the logic symbol. The active LOW outputs facilitate addressing other MSI units with active LOW enables. Of course, if active HIGH enables are required, external inverters can be added to the circuit.

The logic design of the 7442 ensures that all outputs are HIGH when binary codes greater than nine (invalid for BCD coding) are applied to the inputs. The most significant input bit (A_3) produces a useful inhibit function when the 7442 is used as a 1-of-8 decoder. The A_3 input can also be used as the data input in an eight-output demultiplexer application.

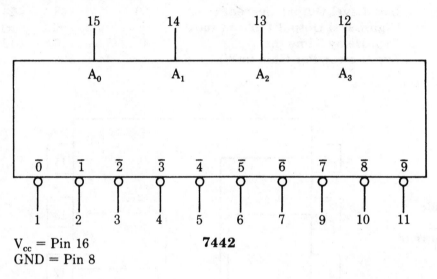

V_{cc} = Pin 16
GND = Pin 8

7442

Truth Table

A_3	A_2	A_1	A_0	$\bar{0}$	$\bar{1}$	$\bar{2}$	$\bar{3}$	$\bar{4}$	$\bar{5}$	$\bar{6}$	$\bar{7}$	$\bar{8}$	$\bar{9}$
L	L	L	L	L	H	H	H	H	H	H	H	H	H
L	L	L	H	H	L	H	H	H	H	H	H	H	H
L	L	H	L	H	H	L	H	H	H	H	H	H	H
L	L	H	H	H	H	H	L	H	H	H	H	H	H
L	H	L	L	H	H	H	H	L	H	H	H	H	H
L	H	L	H	H	H	H	H	H	L	H	H	H	H
L	H	H	L	H	H	H	H	H	H	L	H	H	H
L	H	H	H	H	H	H	H	H	H	H	L	H	H
H	L	L	L	H	H	H	H	H	H	H	H	L	H
H	L	L	H	H	H	H	H	H	H	H	H	H	L
H	L	H	L	H	H	H	H	H	H	H	H	H	H
H	L	H	H	H	H	H	H	H	H	H	H	H	H
H	H	L	L	H	H	H	H	H	H	H	H	H	H
H	H	L	H	H	H	H	H	H	H	H	H	H	H
H	H	H	L	H	H	H	H	H	H	H	H	H	H
H	H	H	H	H	H	H	H	H	H	H	H	H	H

7442

42

7445 BCD-to-decimal decoder/driver with open collector outputs

The 7445 decoder accepts BCD on the A_0 to A_3 address lines and generates 10 mutually exclusive active LOW outputs. When an input code greater than nine (invalid for BCD) is applied, all of the outputs are off (HIGH). This device can therefore be used as a 1-of-8 decoder with A_3 used as an active LOW enable.

The 7445 is similar to the 7442, except that it includes buffer/drivers for the outputs, and features open collector outputs. The 7445 can sink up to 20 mA while maintaining the standardized, guaranteed output LOW voltage (*Vol*) of 0.4 V, but it can sink up to 80 mA with a guaranteed *Vol* of less than 0.9 V, permitting greater versatility in some applications. The 7445 features an output breakdown voltage of 30 V; it is ideally suited for use as a lamp or solenoid driver.

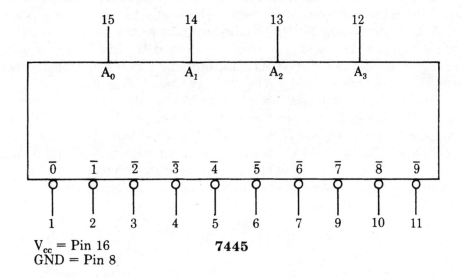

V_{cc} = Pin 16
GND = Pin 8

7445

Truth Table

A_3	A_2	A_1	A_0	$\overline{0}$	$\overline{1}$	$\overline{2}$	$\overline{3}$	$\overline{4}$	$\overline{5}$	$\overline{6}$	$\overline{7}$	$\overline{8}$	$\overline{9}$
L	L	L	L	L	H	H	H	H	H	H	H	H	H
L	L	L	H	H	L	H	H	H	H	H	H	H	H
L	L	H	L	H	H	L	H	H	H	H	H	H	H
L	L	H	H	H	H	H	L	H	H	H	H	H	H
L	H	L	L	H	H	H	H	L	H	H	H	H	H
L	H	L	H	H	H	H	H	H	L	H	H	H	H
L	H	H	L	H	H	H	H	H	H	L	H	H	H
L	H	H	H	H	H	H	H	H	H	H	L	H	H
H	L	L	L	H	H	H	H	H	H	H	H	L	H
H	L	L	H	H	H	H	H	H	H	H	H	H	L
H	L	H	L	H	H	H	H	H	H	H	H	H	H
H	L	H	H	H	H	H	H	H	H	H	H	H	H
H	H	L	L	H	H	H	H	H	H	H	H	H	H
H	H	L	H	H	H	H	H	H	H	H	H	H	H
H	H	H	L	H	H	H	H	H	H	H	H	H	H
H	H	H	H	H	H	H	H	H	H	H	H	H	H

7445

7446A, 7447A BCD-to-7 segment decoder/driver

The 7446A and 7447A seven-segment decoders accept a four-bit BCD code input and provide the appropriate outputs for selection of segments in a seven-segment matrix display (LED or LCD), which is used to represent decimal numerals 0 through 9. The seven (inverted) outputs $(\bar{a}, \bar{b}, \bar{c}, \bar{d}, \bar{e}, \bar{f}, \bar{g})$ of the decoder select the corresponding segments in the matrix, as shown.

The 7446A and 7447A have provisions for automatic blanking of the leading- and/or trailing-edge zeros in a multi-digit decimal number, resulting in an easily readable decimal display conforming to normal writing practice. In an eight-digit mixed integer fraction decimal representation (that is, including a decimal point), using the automatic blanking capability, "0070.0500" would be displayed as "70.05." Leading-edge zero suppression is obtained by connecting the ripple blanking output (BI/RBO) of a decoder to the ripple blanking input (RBI) of the next lower-stage device. The most significant decoder stage should have the RBI input grounded. Because suppression of the least significant integer zero in a number is not usually desired, the RBI input of this decoder stage should be left open. A similar procedure for the fractional part of a display will provide automatic suppression of trailing-edge zeros.

The decoder has an active low-input lamp test, which overrides all other input combinations, and enables a check to be made on possible display malfunctions. The BI/RBOP terminal of the decoder can be OR-tied with a modulating signal via an isolating buffer to achieve pulse-duration intensity modulation. A suitable signal for this purpose can be generated by forming a variable frequency multivibrator with a cross-coupled pair of open collector gates.

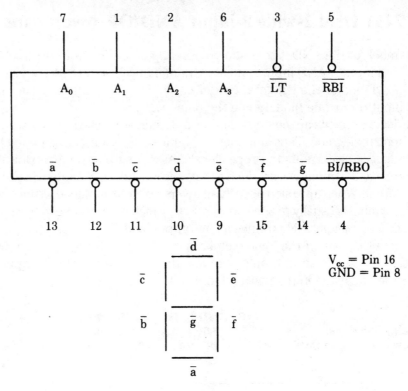

7446

V_{cc} = Pin 16
GND = Pin 8

FUNCTION TABLE

DECIMAL OR FUNCTION	INPUTS							OUTPUTS						
	$\overline{LT}$	$\overline{RBI}$	A_3	A_2	A_1	A_0	$\overline{BI/RBO}^{(b)}$	$\overline{a}$	$\overline{b}$	$\overline{c}$	$\overline{d}$	$\overline{e}$	$\overline{f}$	$\overline{g}$
0	H	H	L	L	L	L	H	L	L	L	L	L	L	H
1	H	X	L	L	L	H	H	H	L	L	H	H	H	H
2	H	X	L	L	H	L	H	L	L	H	L	L	H	L
3	H	X	L	L	H	H	H	L	L	L	L	H	H	L
4	H	X	L	H	L	L	H	H	L	L	H	H	L	L
5	H	X	L	H	L	H	H	L	H	L	L	H	L	L
6	H	X	L	H	H	L	H	H	H	L	L	L	L	L
7	H	X	L	H	H	H	H	L	L	L	H	H	H	H
8	H	X	H	L	L	L	H	L	L	L	L	L	L	L
9	H	X	H	L	L	H	H	L	L	L	H	H	L	L
10	H	X	H	L	H	L	H	H	H	H	L	L	H	L
11	H	X	H	L	H	H	H	H	H	L	L	H	H	L
12	H	X	H	H	L	L	H	H	L	H	H	H	L	L
13	H	X	H	H	L	H	H	L	H	H	L	H	L	L
14	H	X	H	H	H	L	H	H	H	H	L	L	L	L
15	H	X	H	H	H	H	H	H	H	H	H	H	H	H
$\overline{BI}^{(b)}$	X	X	X	X	X	X	L	H	H	H	H	H	H	H
$\overline{RBI}^{(b)}$	H	L	L	L	L	L	L	H	H	H	H	H	H	H
$\overline{LT}$	L	X	X	X	X	X	H	L	L	L	L	L	L	L

H = High voltage level
L = Low voltage level
X = Don't care

7447

7451 Dual 2-wide 2-input AND/OR/invert gate

The symbol for the AND gate is shown here. By definition, for output C to be HIGH, both input A AND input B must be HIGH; hence the term *AND gate*. Notice the symbol for the OR gate. By definition, for C to be HIGH, either input A OR input B must be HIGH; hence the name *OR gate*.

As logic elements are combined in series to perform logic functions, voltage levels tend to degrade. Thus amplifiers, or buffers, are sometimes needed to restore voltages or currents to proper levels. Assuming that no inversion is encountered, a HIGH input will produce a HIGH output and a LOW input will produce a LOW output. When inversion occurs, an inversion circle is placed at the output, and the symbol is usually termed an *inverter*, which reverses the logic state. In this case, a HIGH input will produce a LOW output, and vice versa.

Notice that one of the logic elements in the 7451 has four inputs, while the other has six. Both work in essentially the same way. The four input version is the one listed in the truth table.

	74S	74H	74	74LS	74L
Typ. Delay Time (ns)	3.5	6.5	10.5	12.5	43
Typ. Power Per Gate (mW)	28	29	14	2.75	1.5

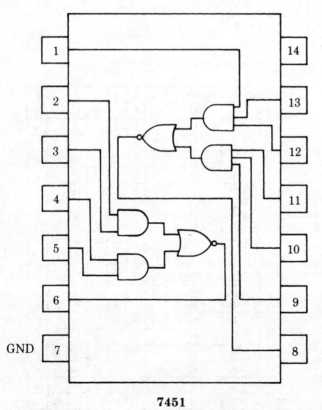

7451

7451 Truth Table

INPUTS				OUTPUT
A	**B**	**D**	**D**	
0	0	0	0	1
0	0	0	1	1
0	0	1	0	1
0	0	1	1	0
0	1	0	0	1
0	1	0	1	1
0	1	1	0	1
0	1	1	1	0
1	0	0	0	1
1	0	0	1	1
1	0	1	0	1
1	0	1	1	0
1	1	0	0	0
1	1	0	1	0
1	1	1	0	0
1	1	1	1	0

7453 Expandable 4-Wide 2-Input AND/OR/invert gate

The symbol for the AND gate is shown here. By definition, for output C to be true, inputs A and B must be true; hence, the term *AND gate*.

Note symbol for the OR gate. By definition, for C to be true either input A or input B must be true, hence the name *OR gate*.

As logic elements are combined in series to perform logic functions, voltage levels tend to degrade. Thus, amplifiers are sometimes needed to restore voltages or currents to proper levels. Assuming no inversion is encountered, a true input will produce a true output, and a false input will produce a false output. When inversion occurs, an inversion circle is placed at the output, and the symbol is usually termed an *inverter*.

	74H	74
Typ. Delay Time (ns)	6.6	10.5
Typ. Power Per Gate (mW)	41	23

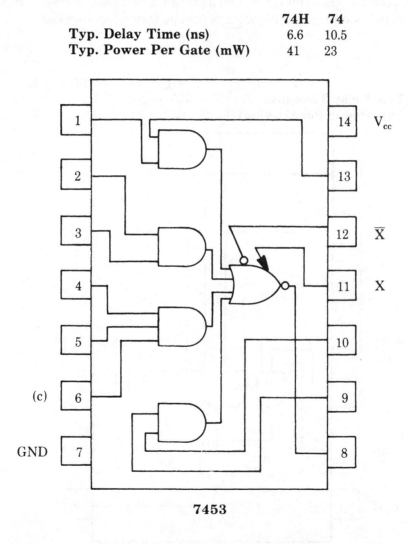

7453

7454 4-Wide 2- & 3-Input AND/OR/invert gate

The symbol for the AND gate is shown here. By definition, for the output (C) to be true, both inputs (A and B) must be true; hence, the term *AND gate*.

Note also the symbol for the OR gate. By definition, for the output (C) to be true, either input A or input B must be true; hence, the name *OR gate*.

The output of the OR gate is inverted (indicated by the small circle), so C becomes C̄ (Not C); hence the term invert gate, or inverter.

As logic elements are combined in series to perform logic functions, voltage levels tend to degrade. Thus, amplifiers (or buffers) are sometimes needed to restore voltages or currents to their proper levels. Assuming no inversion is encountered, a true input will produce a true output, and a false input will produce a false output. When inversion occurs, an inversion circle is placed at the output, and the symbol is usually termed an *inverter*.

	74H	74	74LS	74L
Typ. Delay Time (ns)	6.5	10.5	12.5	43
Typ. Power Per Gate (mW)	41	23	4.5	1.5

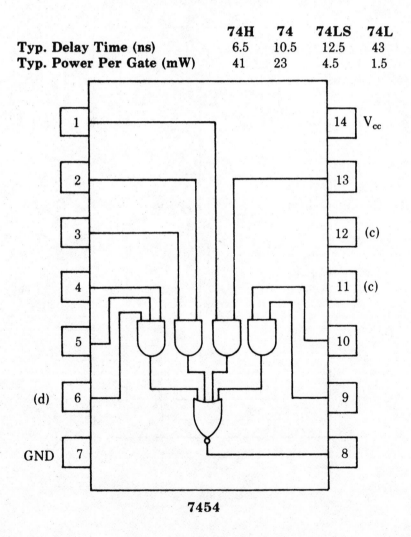

7454

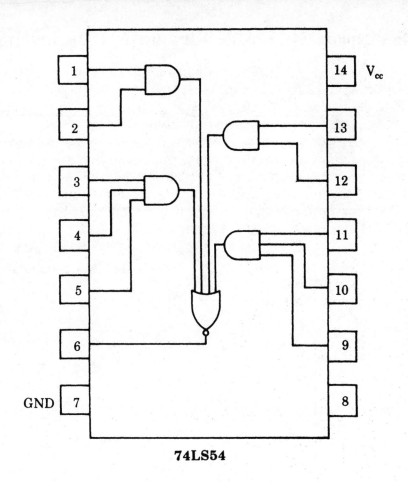

74LS54

7455 Expandable 2-Wide 4-Input AND/OR/invert gate

The symbol for the AND gate is shown here. By definition, for output C to be true, inputs A and B must be true; hence, the term *AND gate*.

Note the symbol for the OR gate. By definition, for C to be true either input A or input B must be true, hence the name *OR gate*.

As logic elements are combined in series to perform logic functions, voltage levels tend to degrade. Thus, amplifiers are sometimes needed to restore voltages or currents to proper levels. Assuming no inversion is encountered, a true input will produce a true output, false input will produce a false output. When inversion occurs, an inversion circle is placed at the output, and the symbol is usually termed an *inverter*.

	74H	74LS (Not Expandable)
Typ. Delay Time (ns)	6.8	12.5
Typ. Power Per Gate (mW)	30	2.75

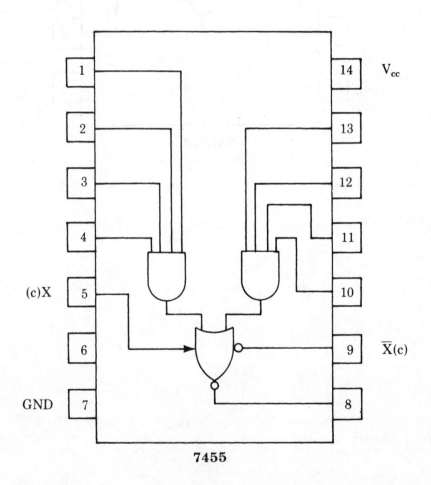

7455

7664 4-2-32 Input AND/OR/invert gate

The symbol for the AND gate is shown here. By definition, for output C to be HIGH, both input A AND input B must be HIGH; hence the term *AND gate*. Notice the symbol for the OR gate. By definition, for C to be HIGH, either input A OR input B must be HIGH; hence the name *OR gate*.

As logic elements are combined in series to perform logic functions, voltage levels tend to degrade. Thus amplifiers, or buffers, are sometimes needed to restore voltages or currents to proper levels. Assuming no inversion is encountered, a HIGH input will produce a HIGH output and a LOW input will produce a LOW output. When inversion occurs, an inversion circle is placed at the output, and the symbol is usually termed an *inverter*, which reverses the logic state. In this case, a HIGH input will produce a LOW output, and vice versa.

The inputs are ANDed together in groups of two, three, or four, then the outputs of these internal gates are NORed together. The logic function is an expanded form of what was encountered in the truth table for the 7451.

	74S	74H	74	74LS	74L
Typ. Delay Time (ns)	3.5	6.6	10.5	12.5	43
Typ. Power Per Gate (mW)	29	41	23	4.5	1.5

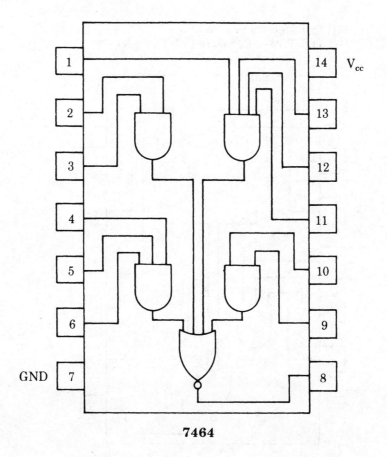

7464

7465 4-2-32 Input AND/OR
invert gate with open collector output

The symbol for the AND gate is shown here. By definition, for output C to be HIGH, both input A AND input B must be HIGH; hence the term *AND gate*. Notice the symbol for the OR gate. By definition, for C to be HIGH, either input A OR input B must be HIGH; hence the name *OR gate*.

As logic elements are combined in series to perform logic functions, voltage levels tend to degrade. Thus amplifiers, or buffers, are sometimes needed to restore voltages or currents to proper levels. Assuming no inversion is encountered, a HIGH input will produce a HIGH output and a LOW input will produce a LOW output. When inversion occurs, an inversion circle is placed at the output, and the symbol is usually termed an *inverter*, which reverses the logic state. In this case, a HIGH input will produce a LOW output, and vice versa.

The inputs are ANDed together in groups of two, three, or four, then the outputs of these internal gates are NORed together. The logic function is an expanded form of what was encountered in the truth table for the 7451. The 7465 is functionally identical to the 7464, except this device has an open collector output at pin 8.

	74S
Typ. Delay Time (ns)	5.5
Typ. Power Per Gate (mW)	36

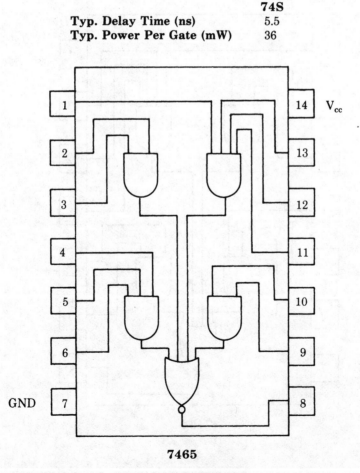

7465

52

7473 Dual JK flip-flop

The '73 is a dual flip-flop with individual JK, clock and direct reset inputs. The 7473 and 74H73 are positive pulse-triggered flip-flops. JK information is loaded into the master while the clock is HIGH and transferred to the slave on the HIGH-to-LOW clock transition. For these devices, the J and K inputs should be stable while the clock is HIGH for conventional operation. The 74LS73 is a negative edge-triggered flip-flop. The J and K inputs must be stable one setup time prior to the HIGH-to-LOW Clock transition for predictable operation.

The Reset ($\overline{R_D}$) is an asynchronous active LOW input. When LOW, it overrides the clock and data inputs, forcing the Q output LOW and the $\overline{Q}$ output HIGH.

	74H	74	74L	74LS
Typ. Max. Clock Frequency (MHz)	30	20	6	45
Typ. Power Per Flip-Flop (mW)	80	50	3.8	10
Setup Time (ns)	0	0	0	20
Hold Time (ns)	0	0	0	0

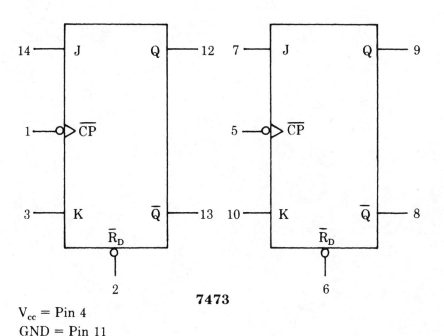

7473

V_{cc} = Pin 4
GND = Pin 11

Truth Table

OPERATING MODE	INPUTS				OUTPUTS	
	$\overline{R}_D$	$\overline{CP}$ (d)	J	K	Q	$\overline{Q}$
Asynchronous Reset (Clear)	L	X	X	X	L	H
Toggle	H	⊓	h	h	$\overline{q}$	q
Load "0" (Reset)	H	⊓	l	h	L	H
Load "1" (Set)	H	⊓	h	l	H	L
Hold "no change"	H	⊓	l	l	q	$\overline{q}$

H High voltage level steady state
L Low voltage level steady state
h High voltage level one setup time prior to the high-to-low clock transition [c]
l Low voltage level one setup time prior to the high-to-low clock transition [c]
X Don't care
q Lower case letters indicate the state of the referenced output prior to the high-to-low clock transition
⊓ Positive clock pulse

54

7474 Dual D-type flip flop

The '74 is a dual positive edge-triggered D-type flip-flop featuring individual data, clock, set and reset inputs, and complementary Q and $\overline{Q}$ outputs. Set ($\overline{S}_D$) and reset ($\overline{R}_D$) are asynchronous active LOW inputs and operate independently of the clock input. Information on the data (D) input is transferred to the Q output on the LOW-to-HIGH transition of the clock pulse. The D inputs must be stable one setup time prior to the LOW-to-HIGH clock transition for predictable operation. Although the clock input is level sensitive, the positive transition of the clock pulse between the 0.8 V and 2.0 V levels should be equal to or less than the clock to output delay time for reliable operation.

	74S	74H	74LS	74	74L
Typ. Max. Clock Frequency (MHz)	110	43	33	25	6
Typ. Power Per Flip-Flop (mW)	75	75	10	43	4
Setup Time (ns)	3	15	25	20	50
Hold Time (ns)	2	5	5	5	15

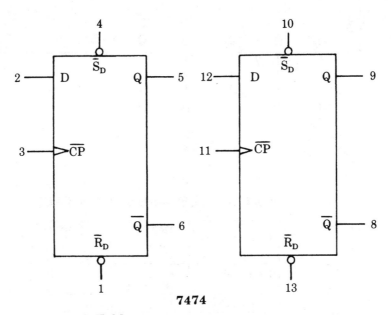

7474

Truth Table

OPERATING MODE	INPUTS				OUTPUTS	
	$\overline{S}_D$	$\overline{R}_D$	CP	D	Q	$\overline{Q}$
Asynchronous Set	L	H	X	X	H	L
Asynchronous Reset (Clear)	H	L	X	X	L	H
Undetermined (c)	L	L	X	X	H	H
Load "1" (Set)	H	H	↑	h	H	L
Load "0" (Reset)	H	H	↑	l	L	H

H = High voltage level steady state
h = High voltage level one setup time prior to the low-to-high clock transition
L = Low voltage level steady state
l = Low voltage level one setup time prior to the low-to-high clock transition
X = Don't care

55

7475 Dual 2-bit transparent latch

The '75 has two independent 2-bit transparent latches. Each 2-bit latch is controlled by an active high enable input (E). When E is HIGH, the data enters the latch and appears at the Q output. The Q outputs follow the data inputs as long as E is HIGH. The data (on the D) input one setup time before the HIGH-to-LOW transition of the enable will be stored in the latch. The latched outputs remain stable as long as the enable is LOW.

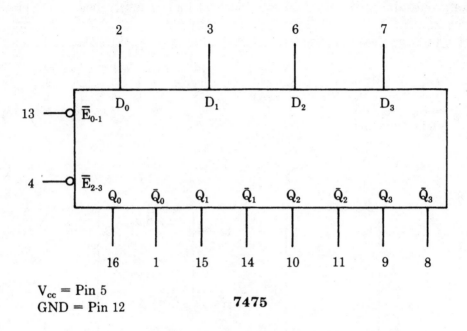

V_{cc} = Pin 5
GND = Pin 12

7475

MODE SELECT — FUNCTION TABLE

OPERATING MODE	INPUTS		OUTPUTS	
	$\overline{E}$	D	Q	$\overline{Q}$
Data Enabled	H	L	L	H
	H	H	H	L
Data Latched	L	X	q	$\overline{q}$

H = High voltage level
L = Low voltage level
X = Don't care
q = Lower case letters indicate the state of referenced output one setup time prior to the high-to-low enable transition.

7476 Dual JK flip-flop

The '76 is a dual JK flip-flop with individual J, K, clock, set, and reset inputs. The 7476 and 74H76 are positive pulse-triggered flip-flops. JK information is loaded into the master while the clock is HIGH and transferred to the slave on the HIGH-to-LOW clock transition. The J and K inputs must be stable while the clock is HIGH for conventional operation.

The 74LS76 is a negative edge-triggered flip-flop. The J and K inputs must be stable only one setup time prior to the HIGH-to-LOW clock transition. The set ($\overline{S}_D$) and reset ($\overline{R}_D$) are asynchronous active LOW inputs. When LOW, they override the clock and data inputs forcing the outputs to the steady state levels, as shown in the truth table.

	74H	74	74LS
Typ. Max. Clock Frequency (MHz)	30	20	45
Typ. Power Per Flip-Flop (mW)	80	50	10
Setup Time (ns)	0	0	20
Hold Time (ns)	0	0	0

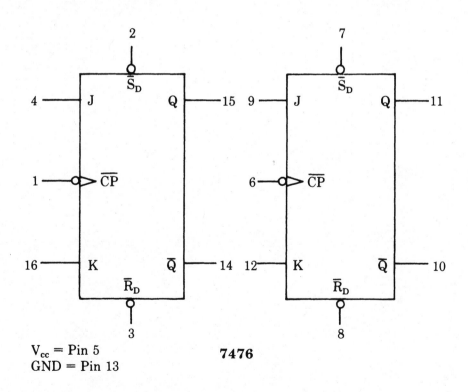

V_{cc} = Pin 5
GND = Pin 13

7476

Truth Table

OPERATING MODE	INPUTS					OUTPUTS	
	$\overline{S}_D$	$\overline{R}_D$	$\overline{CP}$ (d)	J	K	Q	$\overline{Q}$
Asynchronous Set	L	H	X	X	X	H	L
Asynchronous Reset (Clear)	H	L	X	X	X	·L	H
Undetermined (c)	L	L	X	X	X	H	H
Toggle	H	H	⊓	h	h	$\overline{q}$	q
Load "0" (Reset)	H	H	⊓	l	h	L	H
Load "1" (Set)	H	H	⊓	h	l	H	L
Hold "no change"	H	H	⊓	l	l	q	$\overline{q}$

H = High voltage level steady state
L = Low voltage level steady state
h = High voltage level one setup time prior to the high-to-low clock transition [c]
l = Low voltage level one setup time prior to the high-to-low clock transition [c]
X = Don't care
q = Lower case letters indicate the state of the referenced output prior to the high-to-low clock transition
⊓ = Positive clock pulse

7478 Dual JK edge-triggered flip-flop

The '78 is a dual JK negative edge-triggered flip-flop featuring individual J, K, set, common-clock, and common-reset inputs. The set ($\overline{S_D}$) and reset ($\overline{R_D}$) inputs, when LOW, set or reset the outputs as shown in the truth table, regardless of the levels at the other inputs. A HIGH level on the clock ($\overline{CP}$) input enables the J and K inputs and data to be accepted. The logic levels at the J and K inputs can be allowed to change while the $\overline{CP}$ is HIGH and the flip-flop will perform according to the truth table as long as minimum setup and hold times are observed. Output state changes are initiated by the HIGH-to-LOW transition of $\overline{CP}$.

	74H	74L	74LS
Typ. Max. Clock Frequency (MHz)	30	6	45
Typ. Power Per Flip-Flop (mW)	80	3.8	10
Setup Time (ns)	0	0	20
Hold Time (ns)	0	0	0

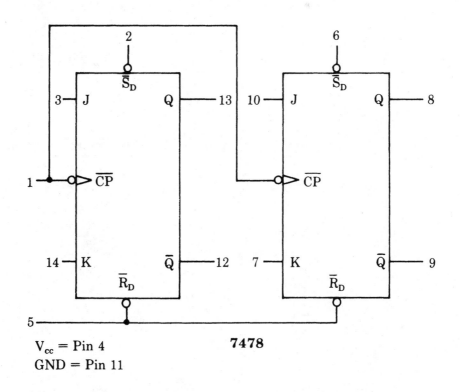

V_{cc} = Pin 4
GND = Pin 11

7478

59

Truth Table

OPERATING MODE	INPUTS					OUTPUTS	
	$\overline{S}_D$	$\overline{R}_D$	$\overline{CP}$	J	K	Q	$\overline{Q}$
Asynchronous Set	L	H	X	X	X	H	L
Asynchronous Reset (Clear)	H	L	X	X	X	L	H
Undetermined [c]	L	L	X	X	X	H	H
Toggle	H	H	.	h	h	$\overline{q}$	q
Load "0" (Reset)	H	H	.	l	h	L	H
Load "1" (Set)	H	H	.	h	l	H	L
Hold "no change"	H	H	.	l	l	q	$\overline{q}$

H = High voltage level steady state
h = High voltage level one setup time prior to the high-to-low clock transition
L = Low voltage level steady state
l = Low voltage level one setup time prior to the high-to-low clock transition
q = Lower case letters indicate the state of the referenced output prior to the high-to-low clock transition
X = Don't care

7483 4-Bit full adder

The '83 is a high-speed 4-bit binary full adder with internal carry lookahead. It accepts two 4-bit binary words (A_1 through A_4, B_1 through B_4) and a carry input (C_{IN}). The sum of the two 4-bit words is combined with the carry input and presented at the four sum outputs (Σ_1 through Σ_4) and the carry output (C_{OUT}). It operates with either HIGH or LOW operands (positive or negative logic).

Because of the symmetry of the binary add function, the '83 can be used with either all active HIGH operands (positive logic) or with all active LOW operands (negative logic). With active HIGH inputs, C_{IN} cannot be left open; instead, it must be held low when no carry in is intended. Interchanging the inputs of equal weight does not affect the operation, so C_{IN}, A_1 and B_1 can arbitrarily be assigned to pins 10, 11, 13, etc.

	74LS	74
Typ. Carry Time (ns)	10	10
Typ. Add Time (ns)	15	16
Typ. Power Per Bit (mW)	24	76

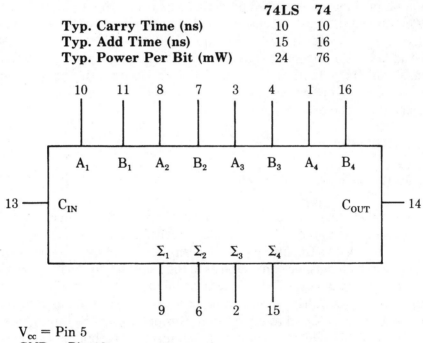

V_{cc} = Pin 5
GND = Pin 12

PINS	C_{IN}	A_1	A_2	A_3	A_4	B_1	B_2	B_3	B_4	Σ_1	Σ_2	Σ_3	Σ_4	C_{OUT}
Logic Levels	L	L	H	L	H	H	L	L	H	H	H	L	L	H
Active high	0	0	1	0	1	1	0	0	1	1	1	0	0	1
Active Low	1	1	0	1	0	0	1	1	0	0	0	1	1	0

$(10+9=19)$
$(\text{carry} + 5+6=12)$

7483

61

7485 4-Bit magnitude comparator

The 7485 is a four-bit magnitude comparator that can be expanded to almost any length. It compares two 4-bit binary, BCD, or other monotonic codes, and presents the three possible magnitude results ($A > B, A < B$, or $A = B$) at the outputs. The four-bit inputs are weighted A_0 through A_3 and B_0 through B_3, where A_3 and B_3 are the most significant bits.

The operation of the 7485 is described in the truth table, which shows all possible logic conditions. The upper part of the table describes the normal operation under all conditions that will occur in a single device or in a series expansion scheme. In the upper part of the table, the three outputs are mutually exclusive. In the lower part of the table, the outputs reflect the feed-forward conditions that exist in the parallel expansion scheme.

The expansion inputs $I(A > B)$, $I(A = B)$, and $I(A < B)$ are the least significant bit positions. When used for series expansion, the $A > B, A = B$, and $A < B$ outputs of the least significant word are connected to the corresponding $I(A > B)$, $I(A = B)$, and $I(A < B)$ inputs of the next higher stage. Stages can be added in this manner to any length, but a cumulative propagation delay penalty of about 15 nS (for standard TTL) is added with each additional stage. For proper operation, the expansion inputs of the least significant word should be tied as:

$I(A > B)$	HIGH
$I(A = B)$	LOW
$I(A < B)$	HIGH

The parallel expansion scheme demonstrates the most efficient general use of these comparators. In the parallel expansion scheme, the expansion inputs can be used as the fifth input bit position, except on the least significant device, which must be connected as in the serial scheme. The expansion inputs are used by labelling $I(A > B)$ as an A input, $I(A < B)$ as a B input, and setting $I(A = B)$ LOW. The 7485 can be used as a five-bit comparator only when the outputs are used to drive the A_0 through A_3 and B_0 through B_3 inputs of another 7485 chip. This parallel connection technique can be expanded to any number of bits, as shown in the Truth Table.

	74	74L
Typ. Compare Time (ns)	21	70
Typ. Total Power (mW)	275	20

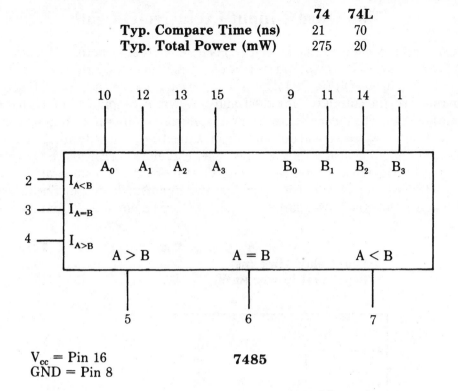

V_{cc} = Pin 16
GND = Pin 8

7485

Truth Table

COMPARING INPUTS				CASCADING INPUTS			OUTPUTS		
A_3,B_3	A_2,B_2	A_1,B_1	A_0,B_0	$I_{A>B}$	$I_{A<B}$	$I_{A=B}$	A>B	A<B	A=B
$A_3>B_3$	X	X	X	X	X	X	H	L	L
$A_3<B_3$	X	X	X	X	X	X	L	H	L
$A_3=B_3$	$A_2>B_2$	X	X	X	X	X	H	L	L
$A_3=B_3$	$A_2<B_2$	X	X	X	X	X	L	H	L
$A_3=B_3$	$A_2=B_2$	$A_1>B_1$	X	X	X	X	H	L	L
$A_3=B_3$	$A_2=B_2$	$A_1<B_1$	X	X	X	X	L	H	L
$A_3=B_3$	$A_2=B_2$	$A_1=B_1$	$A_0>B_0$	X	X	X	H	L	L
$A_3=B_3$	$A_2=B_2$	$A_1=B_1$	$A_0<B_0$	X	X	X	L	H	L
$A_3=B_3$	$A_2=B_2$	$A_1=B_1$	$A_0=B_0$	H	L	L	H	L	L
$A_3=B_3$	$A_2=B_2$	$A_1=B_1$	$A_0=B_0$	L	H	L	L	H	L
$A_3=B_3$	$A_2=B_2$	$A_1=B_1$	$A_0=B_0$	L	L	H	L	L	H
$A_3=B_3$	$A_2=B_2$	$A_1=B_1$	$A_0=B_0$	X	X	H	L	L	H
$A_3=B_3$	$A_2=B_2$	$A_1=B_1$	$A_0=B_0$	H	H	L	L	L	L
$A_3=B_3$	$A_2=B_2$	$A_1=B_1$	$A_0=B_0$	L	L	L	H	H	L

H = High voltage level
L = Low voltage level
X = Don't care

7485

7486 Quad 2-input Exclusive-OR gate

In an ordinary OR gate, the output (*C*) is HIGH if either or both of the inputs (*A* and *B*) are HIGH. The output is LOW only when all inputs are LOW.

An Exclusive-OR (X-OR) gate is a useful variation on the basic OR gate. In this case, the output (*C*) is HIGH if, and only if, either input (*A* or *B*) is HIGH, but not both. The output is LOW if both inputs are HIGH or if both inputs are LOW. In other words, a HIGH output indicates that the two inputs are different, but a LOW input indicates they are the same. Therefore, an Exclusive-OR gate can be considered a difference detector or a one-bit digital comparator.

The 7486 contains four independent two-input/single-output Exclusive-OR gates. Only the power-supply connections are common to all four gates.

	74S	74LS	74	74L
Typ. Delay Time (ns)	7	10	14	29
Typ. Total Power (mW)	250	30	150	15

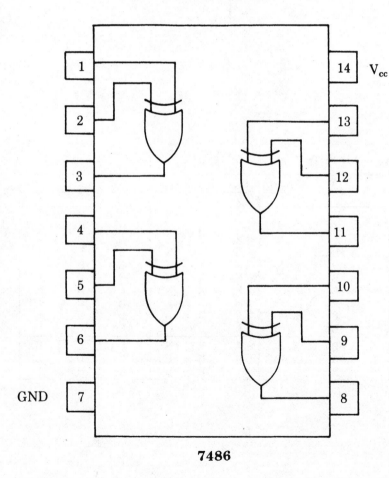

7486

Truth Table

INPUTS		OUTPUT
A	**B**	**Y**
L	L	L
L	H	H
H	L	H
H	H	L

L = Low voltage level
H = High voltage level

7489 64-Bit random-access memory with open collector outputs

The 7489 is a high-speed array of 64 memory cells organized as 16 words of 4 bits apiece. A one-of-16 address decoder selects a single word, which is specified by the four address lines (A_0 through A_3). A Read operation is initiated after the address lines are stable when the write-enable (*WE*) input is HIGH and the chip select-memory enable (*CS*) input is LOW. Data is read at the outputs inverted from the data that was written into the memory.

A Write operation requires that both the *WE* and *CS* inputs are LOW. The address inputs must be stable during the Write mode for predictable operation. When the Write mode is selected, the outputs are the complement of the data inputs. The selected memory cells are transparent to changes in the data during the Write mode. Therefore, data must be stable one setup time before the LOW-to-HIGH transition of *CE* or *WE*.

	74
Read Time (ns)	33
Write Time (ns)	48
Current Per Package (mA)	75

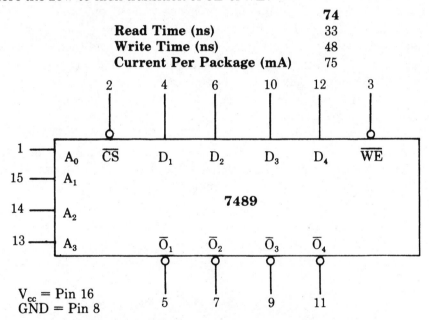

V_{cc} = Pin 16
GND = Pin 8

MODE SELECT — FUNCTION TABLE

OPERATING MODE	INPUTS		OUTPUTS	
	$\overline{CS}$	$\overline{WE}$	D_n	$\overline{O}_n$
Write	L	L	L	H
	L	L	H	L
Read	L	H	X	$\overline{Data}$
Inhibit Writing	H	L	L	H
	H	L	H	L
Store-Disable Outputs	H	H	X	H

7490 Decade counter

The 7490 is a four-bit ripple-type decade counter. The device consists of four master/slave flip-flops that are internally connected to provide a divide-by-two section, and a divide-by-five section. Each section has a separate clock input to initiate stage changes of the counter on the HIGH-to-LOW clock transition. State changes of the Q outputs do not occur simultaneously because of internal ripple delays. Therefore, decoded output signals are subject to decoding spikes and they should not be used for clocks or strobes. The Q_0 output is designed and specified to drive the rate fan-out, plus the $\overline{CP_1}$ input of the device.

A gated AND asynchronous master reset ($\overline{MR_1}$ and $\overline{MR_2}$) is provided, which overrides both clock and resets (clears) all the flip-flops. Also provided is a gated AND asynchronous master set ($\overline{MS_1}$ and $\overline{MS_2}$), which overrides the clocks and the $\overline{MR}$ inputs, setting the outputs to a value of nine (1001).

Because the output from the divide-by-two section is not internally connected to the succeeding stages, the device can be operated in various counting modes. In a BCD (8421) counter, the $\overline{CP_1}$ input must be externally connected to the Q_0 output. The $\overline{CP_0}$ input receives the incoming count, producing a BCD count sequence. In a symmetrical binary divide-by-10 counter, the Q_3 output must be connected externally to the $\overline{CP_0}$ input. The input count is then applied to the $\overline{CP_1}$, and a divide-by-10 square wave is obtained at output Q_0.

To operate as a divide-by-2, or a divide-by-five counter, no external interconnections are required. The first flip-flop is used as a binary element for the divide-by-two function ($\overline{CP_0}$ as the input, and Q_0 as the output). The $\overline{CP_1}$ input is used to obtain divide-by-five operation at the Q_3 output.

	74LS	**74**	**74L**
Count Frequency (MHz)	32	32	6
Parallel Load	set to 9	set to 9	set to 9
Clear	High	High	High
Typ. Total Power (mW)	40	160	20

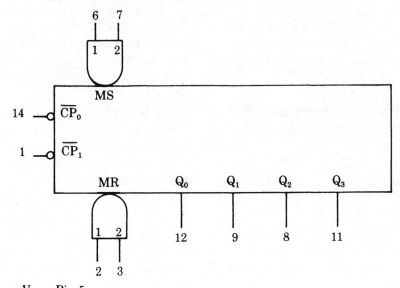

V_{cc} = Pin 5
GND = Pin 10

Truth Table

COUNT	OUTPUT			
	Q_0	Q_1	Q_2	Q_3
0	L	L	L	L
1	H	L	L	L
2	L	H	L	L
3	H	H	L	L
4	L	L	H	L
5	H	L	H	L
6	L	H	H	L
7	H	H	H	L
8	L	L	L	H
9	H	L	L	H

NOTE: Output Q_0 connected to Input $\overline{CP_1}$

Truth Table

RESET/SET INPUTS				OUTPUTS			
MR_1	MR_2	MS_1	MS_2	Q_0	Q_1	Q_2	Q_3
H	H	L	X	L	L	L	L
H	H	X	L	L	L	L	L
X	X	H	H	H	L	L	H
L	X	L	X		Count		
X	L	X	L		Count		
L	X	X	L		Count		
X	L	L	X		Count		

H = High voltage level
L = Low voltage level
X = Don't care

7490

7492 Divide-by-12 counter

The '92 is a 4-bit ripple-type divide-by-12 counter. The device consists of four master-slave flip-flops internally connected to provide a divide-by-2 section and a divide-by-6 section. Each section has a separate clock input to initiate state changes of the counter on the HIGH-to-LOW clock transition. State changes of the Q outputs do not occur simultaneously because of internal ripple delays. Therefore, decoded output signals are subject to decoding spikes and should not be used for clocks or strobes. The Q_0 output is designed and specified to drive the rated fallout plus the $\overline{CP}_1$ input of the device. A gated AND asynchronous master reset ($MR_1 \bullet MR_2$) is provided which overrides both clocks and resets (clears) all the flip-flops.

Because the output from the divide-by-2 section is not internally connected to the succeeding stages, the device can be operated in various counting modes. In a modulo-12, divide-by-12 counter, the $\overline{CP}_1$ input must be externally connected to Q_0 output. The $\overline{CP}_0$ input receives the incoming count, and Q_3 produces a symmetrical divide-by-12 square-wave output. In a divide-by-6 counter, no external connections are required. The first flip-flop is used as a binary element for the divide-by-2 function. The $\overline{CP}_1$ input is used to obtain divide-by-3 operation at the Q_1 and Q_2 outputs and divide-by-6 operation at the Q_3 output.

	74LS	74
Count Frequency (MHz)	32	32
Parallel Lead	None	None
Clear	High	High
Typ. Total Power (mW)	39	160

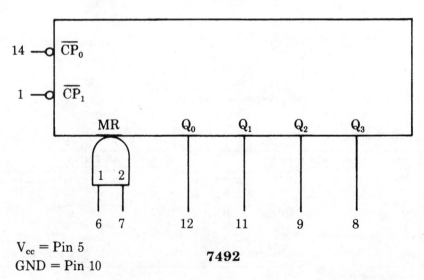

V_{cc} = Pin 5
GND = Pin 10

7492

MODE SELECTION

RESET INPUTS		OUTPUTS			
MR$_1$	MR$_2$	Q$_0$	Q$_1$	Q$_2$	Q$_3$
H	H	L	L	L	L
L	H	Count			
H	L	Count			
L	L	Count			

H = High voltage level
L = Low voltage level
X = Don't care

Truth Table

COUNT	OUTPUT			
	Q$_0$	Q$_1$	Q$_2$	Q$_3$
0	L	L	L	L
1	H	L	L	L
2	L	H	L	L
3	H	H	L	L
4	L	L	H	L
5	H	L	H	L
6	L	L	L	H
7	H	L	L	H
8	L	H	L	H
9	H	H	L	H
10	L	L	H	H
11	H	L	H	H

NOTE: Output Q$_0$ connected to Input $\overline{CP}_1$

7493 4-Bit binary ripple counter

The '93 is a 4-bit ripple-type binary counter. The device consists of four masterslave flip-flops internally connected to provide a divide-by-2 section and a divide-by-8 section. Each section has a separate clock input to initiate state changes of the counter on the HIGH-to-LOW clock transition. State changes of the Q outputs do not occur simultaneously because of internal ripple delays. Therefore, decoded output signals are subject to decoding spikes and should not be used for clocks or strobes. The Q_0 output is designed and specified to drive the rated fanout plus the $\overline{CP}_1$ input of the device. A gated AND asynchronous master reset ($MR_1 \bullet MR_2$) is provided, which overrides both clocks and resets (clears) all the flip-flops.

Because the output from the divide-by-2 section is not internally connected to the succeeding stages, the device can be operated in various counting modes. In a 4-bit ripple counter, the output Q_0 must be connected externally to input $\overline{CP}_1$. The input count pulses are applied to input $\overline{CP}_0$. Simultaneous divisions of 2, 4, 8, and 16 are performed at the Q_0, Q_1, Q_2, and Q_3 outputs, as shown in the truth table. As a 3-bit ripple counter, the input count pulses are applied to input $\overline{CP}_1$. Simultaneous frequency divisions of 2, 4, and 8 are available at the Q_1, Q_2, and Q_3 outputs. Independent use of the first flip-flop is available if the reset function coincides with the reset of the 3-bit ripple-through counter.

	74LS	74	74L
Count Frequency (MHz)	32	32	6
Parallel Load	None	None	None
Clear	High	High	High
Typ. Total Power (mW)	39	160	20

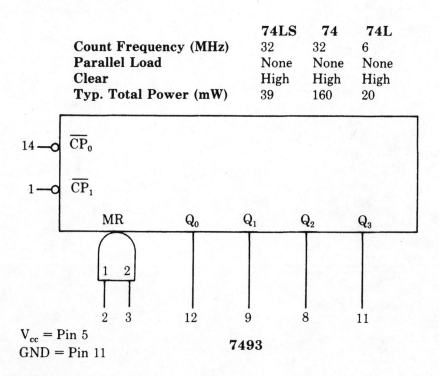

V_{cc} = Pin 5
GND = Pin 11

7493

MODE SELECTION

RESET INPUTS		OUTPUTS			
MR_1	MR_2	Q_0	Q_1	Q_2	Q_3
H	H	L	L	L	L
L	H	Count			
H	L	Count			
L	L	Count			

H = High voltage level
L = Low voltage level
X = Don't care

Truth Table

COUNT	OUTPUT			
	Q_0	Q_1	Q_2	Q_3
0	L	L	L	L
1	H	L	L	L
2	L	H	L	L
3	H	H	L	L
4	L	L	H	L
5	H	L	H	L
6	L	H	H	L
7	H	H	H	L
8	L	L	L	H
9	H	L	L	H
10	L	H	L	H
11	H	H	L	H
12	L	L	H	H
13	H	L	H	H
14	L	H	H	H
15	H	H	H	H

NOTE: Output Q_0 connected to Input $\overline{CP_1}$.

7495 4-Bit shift register

The '95 is a 4-bit shift register with serial and parallel synchronous operation modes. It has a serial data (D_S) and four parallel data (D_0 through D_3) inputs and four parallel outputs (Q_0 through Q_3). The serial or parallel mode of operation is controlled by a mode select input (S) and two clock inputs ($\overline{CP_1}$ and $\overline{CP_2}$). The serial (shift right) or parallel data transfers occur synchronously with the HIGH-to-LOW transition of the selected clock input.

When the mode select input (S) is high, $\overline{CP_2}$ is enabled. A HIGH-to-LOW transition on enabled $\overline{CP_2}$ loads parallel data from the D_0 through D_3 inputs into the register. When S is LOW, $\overline{CP_1}$ is enabled. A HIGH-to-LOW transition on enabled $\overline{CP_1}$ shifts the data from serial input D_S to Q_0 and transfers the data in Q_0 to Q_1, Q_1 to Q_2, and Q_2 to Q_3, respectively (shift right). Shift left is accomplished by externally connecting Q_3 to D_2, Q_2 to D_1, Q_1 to D_0, and operating the '95 in the parallel mode (S = HIGH).

In normal operations, the mode select should change states only when both clock inputs are LOW. However, changing S from HIGH-to-LOW while $\overline{CP_2}$ is low, or changing S from LOW-to-HIGH while $\overline{CP_1}$ is LOW will not cause any changes on the register outputs.

	74	74LS	74L
Shift Frequency (MHz)	25	25	6
Serial Data Input	D	D	D
Asynchronous Clear	None	None	None
Shift-Right Mode	Yes	Yes	Yes
Shift-Left Mode	No	No	No
Lead Mode	Yes	Yes	Yes
Hold Mode	No	No	No
Typ. Total Power (mW)	195	65	24

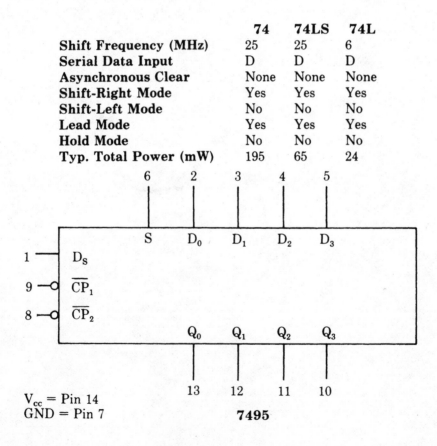

V_{cc} = Pin 14
GND = Pin 7

7495

MODE SELECT — FUNCTION TABLE

OPERATING MODE	INPUTS					OUTPUTS			
	S	$\overline{CP_1}$	$\overline{CP_2}$	D_s	D_n	Q_0	Q_1	Q_2	Q_3
Parallel Load	H	X	↓	X	l	L	L	L	L
	H	X	↓	X	h	H	H	H	H
Shift right	L	↓	X	l	X	L	q_0	q_1	q_2
	L	↓	X	h	X	H	q_0	q_1	q_2
Mode change	↑	L	X	X	X	no change			
	↑	H	X	X	X	undetermined			
	↓	X	L	X	X	no change			
	↓	X	H	X	X	undetermined			

H = High voltage level steady state.
L = Low voltage level steady state.
h = High voltage level one setup time prior to the high-to-low clock transition.
l = Low voltage level one setup time prior to the high-to-low clock transition.
X = Don't care.
q = Lower case letters indicate the state of the referenced output prior to the high-to-low clock transition.
↓ = High-to-low transition of clock or mode Select.
↑ = Low-to-high transition of mode Select.

74107 Dual JK flip-flop

The '107 is a dual flip-flop with individual JK, clock, and direct reset inputs. The 74107 is a positive pulse-triggered flip-flop. JK information is loaded into the master while the clock is HIGH and transferred to the slave on the HIGH-to-LOW clock transition. For these devices, the J and K inputs should be stable while the clock is HIGH for conventional operation.

The 74LS107 is a negative edge-triggered flip-flop. The J and K inputs must be stable one setup time prior to the HIGH-to-LOW clock transition for predictable operation. The reset ($\overline{R_D}$ is an asynchronous active LOW input. When LOW, it overrides the clock and data inputs, forcing the Q output LOW and the $\overline{Q}$ output HIGH.

	74LS
Typ. Max. Clock Frequency (MHz)	45
Typ. Power Per Flip-Flop (mW)	10
Setup Time (ns)	20
Hold Time (ns)	0

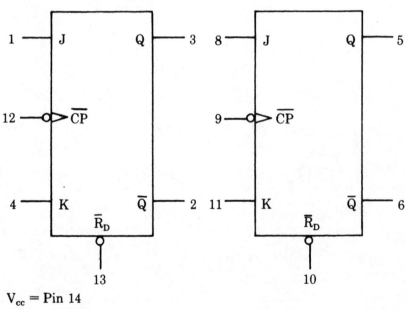

V_{cc} = Pin 14
GND = Pin 7

74107

MODE SELECT—TRUTH TABLE

OPERATING MODE	INPUTS				OUTPUTS	
	$\overline{R}_D$	$\overline{CP}$ (d)	J	K	Q	$\overline{Q}$
Asynchronous Reset (Clear)	L	X	X	X	L	H
Toggle	H	⊓	h	h	$\overline{q}$	q
Load "0" (Reset)	H	⊓	l	h	L	H
Load "1" (Set)	H	⊓	h	l	H	L
Hold "no change"	H	⊓	l	l	q	$\overline{q}$

H = High voltage level steady state.
L = Low voltage level steady state.
h = High voltage level one setup time prior to the high-to-low clock transition[c].
l = Low voltage level one setup time prior to the high-to-low clock transition[c].
X = Don't care.
q = Lower case letters indicate the state of the referenced output prior to high-to-low clock transition.
⊓ = Positive clock pulse.

74109 Dual JK̄ positive edge-triggered flip-flop

The '109 is a dual positive edge-triggered JK̄-type flip-flop that features individual J, K, clock, set, and reset inputs and also complementary Q and Q̄ outputs. Set ($\overline{S}_D$) and reset ($\overline{R}_D$) are asynchronous active LOW inputs and operate independently of the clock input.

The J and K̄ are edge-triggered inputs that control the state changes of the flip-flops as described in the mode-select truth table. The J and K̄ inputs must be stable just one setup time prior to the LOW-to-HIGH transition of the clock for predictable operation. The JK̄ design allows operation as a D flip-flop by tying the J and K̄ inputs together.

Although the clock input is level-sensitive, the positive transition of the clock pulse between the 0.8 V and 2.0 V levels should be equal to or less than the clock-to-output delay time for reliable operation.

	74LS	74
Typ. Max. Clock Frequency (MHz)	33	33
Typ. Power Per Flip-Flop (mW)	10	45
Setup Time (ns)	20	10
Hold Time (ns)	5	6

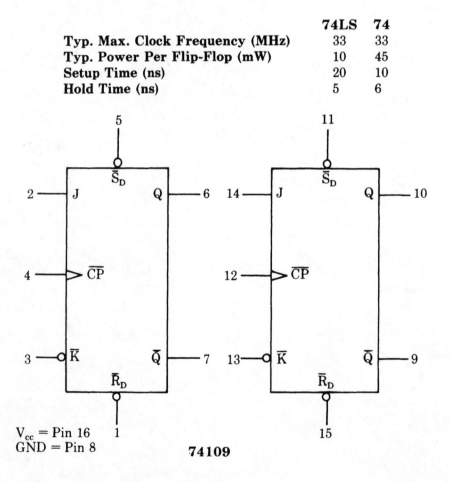

V_{cc} = Pin 16
GND = Pin 8

74109

MODE SELECT—TRUTH TABLE

OPERATING MODE	INPUTS					OUTPUTS	
	$\overline{S}_D$	$\overline{R}_D$	$\overline{CP}$	J	$\overline{K}$	Q	$\overline{Q}$
Asynchronous Set	L	H	X	X	X	H	L
Asynchronous Reset (Clear)	H	L	X	X	X	L	H
Undetermined[c]	L	L	X	X	X	H	H
Toggle	H	H	↑	h	l	$\overline{q}$	q
Load "0" (Reset)	H	H	↑	l	h	L	H
Load "1" (Set)	H	H	↑	h	l	H	L
Hold "no change"	H	H	↑	l	h	q	$\overline{q}$

H = High voltage level steady state

L = Low voltage level steady state

h = High voltage level one setup time prior to the low-to-high clock transition

l = Low voltage level one setup time prior to the low-to-high clock transition

X = Don't care

q = Lower case letters indicate the state of the referenced output prior to the low-to-high clock transition

↑ = Low-to-high clock transition

74112 Dual JK edge-triggered flip-flop

The '112 is a dual JK negative edge-triggered flip-flop featuring individual J, K, clock, set, and reset inputs. The set (S_D) and reset ($\overline{R_D}$) inputs, when LOW, set or reset the outputs as shown in the truth table, regardless of the levels at the other inputs.

A HIGH level on the clock ($\overline{CP}$) input enables the J and K inputs, and data will be accepted. The logic levels at the J and K inputs can be allowed to change while the $\overline{CP}$ is high and the flip-flop will perform according to the truth table as long as minimum setup and hold times are observed. Output state changes are initiated by the HIGH-to-LOW transition of $\overline{CP}$.

	74S	74LS
Typ. Max. Clock Frequency (MHz)	125	45
Typ. Power Per Flip-Flop (mW)	75	10
Setup Time (ns)	6	20
Hold Time (ns)	0	0

V_{cc} = Pin 16
GND = Pin 8

74112

MODE SELECT — TRUTH TABLE

OPERATING MODE	INPUTS					OUTPUTS	
	$\overline{S}_D$	$\overline{R}_D$	$\overline{CP}$	J	K	Q	$\overline{Q}$
Asynchronous Set	L	H	X	X	X	H	L
Asynchronous Reset (Clear)	H	L	X	X	X	L	H
Undetermined[c]	L	L	X	X	X	H	H
Toggle	H	H	↓	h	h	$\overline{q}$	q
Load "0" (Reset)	H	H	↓	l	h	L	H
Load "1" (Set)	H	H	↓	h	l	H	L
Hold "no change"	H	H	↓	l	l	q	$\overline{q}$

H = High voltage level steady state.
h = High voltage level one setup time prior to the high-to-low clock transition.
L = Low voltage level steady state.
l = Low voltage level one setup time prior to the high-to-low clock transition.
q = Lower case letters indicate the state of the referenced output one setup time prior to the high-to-low clock transition.
X = Don't care.

74113 Dual JK edge-triggered flip-flop

The '113 is a dual JK negative edge-triggered flip-flop featuring individual J, K, clock, set, and reset inputs. The asynchronous set ($\overline{S_D}$) input, when LOW, forces the outputs to the steady-state levels as shown in the truth table, regardless of the levels at the other inputs.

A HIGH level on the clock ($\overline{CP}$) input enables the J and K inputs, and data will be accepted. The logic levels at the J and K inputs can be allowed to change while the $\overline{CP}$ is HIGH and the flip-flop will perform according to the truth table as long as minimum setup and hold times are observed. Output state changes are initiated by the HIGH-to-LOW transition of $\overline{CP}$.

	74S	74LS
Typ. Max. Clock Frequency (MHz)	125	45
Typ. Power Per Flip-Flop (mW)	75	10
Setup Time (ns)	6	20
Hold Time (ns)	0	0

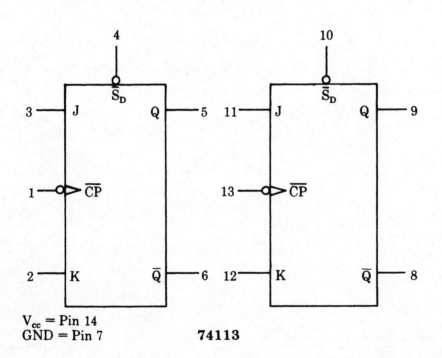

V_{cc} = Pin 14
GND = Pin 7

74113

MODE SELECT — TRUTH TABLE

OPERATING MODE	INPUTS				OUTPUTS	
	$\overline{S}_D$	$\overline{CP}$	J	K	Q	$\overline{Q}$
Asynchronous Set	L	X	X	X	H	L
Toggle	H	↓	h	h	$\overline{q}$	q
Load "0" (Reset)	H	↓	l	h	L	H
Load "1" (Set)	H	↓	h	l	H	L
Hold "no change"	H	↓	l	l	q	$\overline{q}$

H = High voltage level steady state.

h = High voltage level one setup time prior to the high-to-low clock transition.

L = Low voltage level steady state.

l = Low voltage level one setup time prior to the high-to-low clock transition.

q = Lower case letters indicate the state of the referenced output one setup time prior to the high-to-low clock transition.

X = Don't care.

74114 Dual JK edge-triggered flip flop

The '114 is a dual JK negative edge-triggered flip-flop featuring individual J, K, and set inputs and common clock and reset inputs. The set ($\overline{S}_D$) and reset ($\overline{R}_D$) inputs, when LOW, set or reset the outputs, as shown in the truth table, regardless of the levels at the other inputs.

A HIGH level on the clock ($\overline{CP}$) input enables the J and K inputs, and data will be accepted. The logic levels at the J and K inputs can be allowed to change while the $\overline{CP}$ is HIGH, and the flip-flop will perform according to the truth table as long as minimum setup and hold times are observed. Output state changes are initiated by the HIGH-to-LOW transition of $\overline{CP}$.

	74S	74LS
Typ. Max. Clock Frequency (MHz)	125	45
Typ. Power Per Flip-Flop (mW)	75	10
Setup Time (ns)	6	20
Hold Time (ns)	0	0

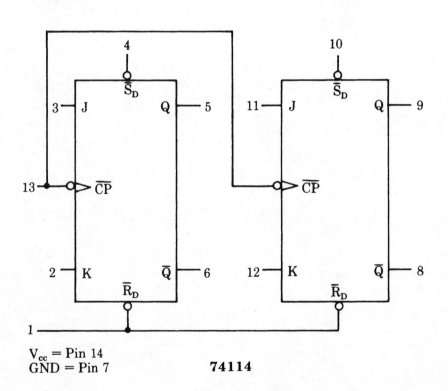

V_{cc} = Pin 14
GND = Pin 7

74114

MODE SELECT—TRUTH TABLE

OPERATING MODE	INPUTS					OUTPUTS	
	$\overline{S}_D$	$\overline{R}_D$	$\overline{CP}$	J	K	Q	$\overline{Q}$
Asynchronous Set	L	H	X	X	X	H	L
Asynchronous Reset (Clear)	H	L	X	X	X	L	H
Undetermined[c]	L	L	X	X	X	H	H
Toggle	H	H	↓	h	h	$\overline{q}$	q
Load "0" (Reset)	H	H	↓	l	h	L	H
Load "1" (Set)	H	H	↓	h	l	H	L
Hold "no change"	H	H	↓	l	l	q	$\overline{q}$

H = High voltage level steady state.
h = High voltage level one setup time prior to the high-to-low clock transition.
L = Low voltage level steady state.
l = Low voltage level one setup time prior to the high-to-low clock transition.
q = Lower case letters indicate the state of the referenced output one setup time prior to the high-to-low clock transition.
X = Don't care.

74121 Monostable multivibrator

These multivibrators feature dual active LOW-going edge inputs and a single active HIGH-going edge input, which can be used as an active HIGH enable input. Complementary output pulses are provided.

Pulse triggering occurs at a particular voltage level and is not directly related to the transition time of the input pulse. Schmitt-trigger input circuitry (TTL hysteresis) for the B input allows jitter-free triggering from inputs with transition rates as slow as 1 volt/second, providing the circuit with an excellent noise immunity of typically 1.2 volts. A high immunity to V_{CC} noise of typically 1.5 volts is also provided by internal latching circuitry. Once fired, the outputs are independent of further transitions of the inputs and are a function only of the timing components. Input pulses can be of any duration relative to the output pulse. Output pulse length can be varied from 20 nanoseconds to 28 seconds by choosing appropriate timing components. With no external timing components (i.e., R_{int} connected to V_{CC}, C_{ext} and R_{ext}/C_{ext} open), an output pulse of typically 30 or 35 nanoseconds is achieved that can be used as a dc-triggered reset signal. Output rise and fall times are TTL compatible and independent of pulse length.

Pulse width stability is achieved through internal compensation and is virtually independent of V_{CC} and temperature. In most applications, pulse stability will only be limited by the accuracy of external timing components.

Jitter-free operation is maintained over the full temperature and V_{CC} ranges for more than six decades of timing capacitance (10 pF to 10 μF) and more than one decade of timing resistance (2 kΩ to 40 kΩ). In circuits where pulse cutoff is not critical, tuning capacitance of up to 1000 μF and timing resistance of as low as 1.4 kΩ can be used.

	74
Positive Inputs	1
Negative Inputs	2
Output Pulse Range (ns to s)	40ns-28s
Typ. Total Power (mW)	90

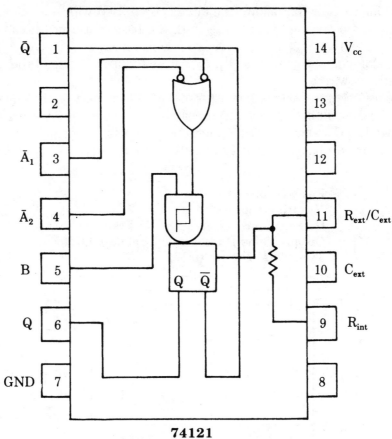

74121

FUNCTION TABLE

INPUTS			OUTPUTS	
$\overline{A}_1$	$\overline{A}_2$	B	Q	$\overline{Q}$
L	X	H	L	H
X	L	H	L	H
X	X	L	L	H
H	H	X	L	H
H	↓	H	⊓	⊔
↓	H	H	⊓	⊔
↓	↓	H	⊓	⊔
L	X	↑	⊓	⊔
X	L	↑	⊓	⊔

H = High voltage level
L = Low voltage level
X = Don't care
↑ = Low-to-high transition
↓ = High-to-low transition

74122 Retriggerable monostable multivibrator

The '122 is a retriggerable monostable multivibrator featuring output pulse width control by three methods. The basic pulse time is programmed by selection of external resistance and capacitance values. The '122 has an internal timing resistor that allows the circuit to be used with only an external capacitor, if so desired. Once triggered, the basic pulse width can be extended by retriggering the gated active LOW-going edge inputs ($\overline{A}_1$, $\overline{A}_2$) or the active HIGH-going edge inputs ($\overline{B}_1$, $\overline{B}_2$), or be reduced by use of the overriding active LOW reset.

To use the internal timing resistor of the '122, connect R_{int} to V_{CC}. For improved pulse width accuracy and repeatability, connect an external resistor between R_{ext}/C_{ext} and V_{CC} with R_{int} left open. To obtain variable pulse widths, connect an external variable resistance between R_{int} or R_{ext}/C_{ext} and V_{CC}.

	74LS
Positive Inputs	2
Negative Inputs	2
Direct Clear	Yes
Output Pulse Range (ns)	45ns-00 (inf.)
Typ. Total Power (mW)	30

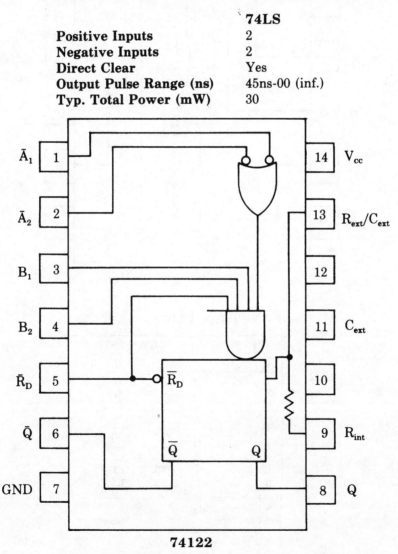

74122

FUNCTION TABLE

INPUTS					OUTPUTS	
$\overline{R}_D$	$\overline{A}_1$	$\overline{A}_2$	B_1	B_2	Q	$\overline{Q}$
L	X	X	X	X	L	H
X	H	H	X	X	L	H
X	X	X	L	X	L	H
X	X	X	X	L	L	H
H	L	X	↑	H	⊓	⊔
H	L	X	H	↑	⊓	⊔
H	X	L	↑	H	⊓	⊔
H	X	L	H	↑	⊓	⊔
H	H	↓	H	H	⊓	⊔
H	↓	↓	H	H	⊓	⊔
H	↓	H	H	H	⊓	⊔
↑	L	X	H	H	⊓	⊔
↑	X	L	H	H	⊓	⊔

H = High voltage level
L = Low voltage level
X = Don't care
↑ = Low-to-high input transition
↓ = High-to-low input transition
⊓ = Active high pulse
⊔ = Active low pulse

74123 Dual retriggerable monostable multivibrator

This retriggerable monostable multivibrator features dc triggering from gated active LOW inputs ($\bar{A}$) and active HIGH inputs (B), and also provides overriding direct-reset inputs. Complementary outputs are provided. The retrigger capability simplifies the generation of output pulses of extremely long duration. By triggering the input before the output pulse is terminated, the output pulse can be extended. The overriding reset capability permits any output pulse to be terminated at a predetermined time that is independent of the timing components, R and C.

	74	74L	74LS
Positive Inputs	1	1	1
Negative Inputs	1	1	1
Direct Clear	Yes	Yes	Yes
Output Pulse Range (ns)	45ns-00 (inf.)	90ns-00 (inf.)	45ns-00 (inf.)
Typ. Total Power (mW)	230	25	60

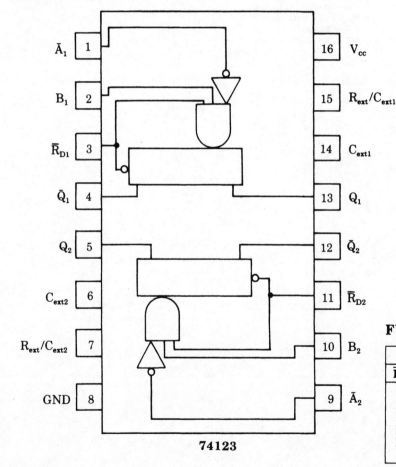

74123

FUNCTION TABLE

INPUTS			OUTPUTS	
$\bar{R}_D$	$\bar{A}$	B	Q	$\bar{Q}$
L	X	X	L	H
X	H	X	L	H
X	X	L	L	H
H	L	↑	⊓	⊔
H	↓	H	⊓	⊔
↑	L	H	⊓	⊔

H = High voltage level
L = Low voltage level
X = Don't care
↑ = Low-to-high transition
↓ = High-to-low transition

88

74125 Quad 3-state buffer

The output of each section of the '125 is the same state as the digital input (A) if the control input (C) is LOW. If, however, the control input (C) is made HIGH, the buffer's output will go into the HIGH-impedance third state. The state of the data input (A) will be irrelevant in this case.

	74
Typ. Delay Time (ns)	10
Typ. Power Per Gate (mW)	40

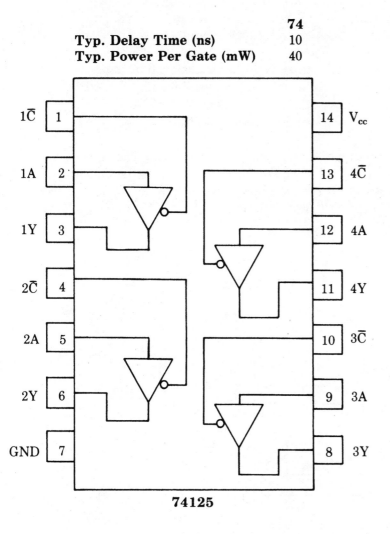

74125

Truth Table

INPUTS		OUTPUT
$\overline{C}$	A	Y
L	L	L
L	H	H
H	X	(Z)

L = Low voltage level
H = High voltage level
X = Don't care
(Z) = High impedance (off)

74126 Quad tri-state buffer

As logic elements are combined in series to perform complex logic functions, voltage levels tend to degrade. Thus, amplifiers, or buffers, are sometimes needed to restore voltages of currents to their proper levels for reliable operation. The 74126 contains four independent buffer stages. Only the power supply connections are common to all four buffers.

Each buffer is of the tri-state type. Ordinary logic devices have just two possible output states: LOW or HIGH. A tri-state device can be forced into a third, high-impedance state (via a special input), which effectively deletes the device from the circuit. The high-impedance state is achieved by applying a LOW state to the extra tri-state input (C). As long as this input is held LOW, it doesn't matter what happens at the main data input (A). The output (Y) will exhibit a high-impedance state, for all intents and purposes, the buffer is electrically turned off.

If a HIGH signal is placed on the tri-state input (C), the device will act like an ordinary buffer. The output (Y) will have the same logic state as the data input (A). That is, if the data input is LOW, the output will be LOW, and if the data input is HIGH, then the output will also be HIGH.

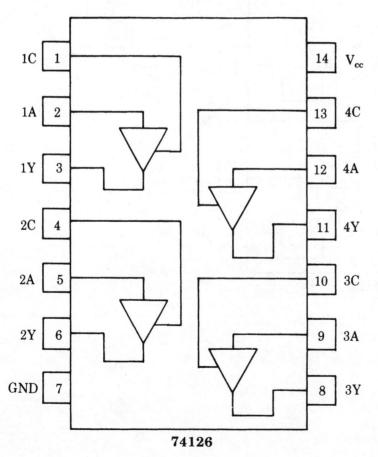

74126

Truth Table

INPUTS		OUTPUT
C	A	Y
H	L	L
H	H	H
L	X	(Z)

L = Low voltage level
H = High voltage level
X = Don't care
(Z) = High impedance (off)

74132 Quad 2-input NAND Schmitt trigger

The 74132 contains four two-input NAND gates that accept standard TTL input signals and provide standard TTL output levels. As logic devices, they function like any other NAND gate—the output is LOW if, and only if, all inputs are HIGH. As long as one or both of the inputs is LOW, the gates' output will be HIGH. Unlike most standard gates, however, those in the 74132 are capable of transforming slowly changing input signals into sharply defined, jitter-free output signals. In addition, they have greater noise margins than conventional NAND gates.

Each gate circuit within this chip contains a two-input Schmitt trigger, followed by a Darlington level shifter, and a phase splitter driving a TTL (or equivalent) totem-pole output. The Schmitt trigger uses positive feedback to effectively speed up slow-input transitions and provide different input threshold voltages for positive-going and negative-going transitions. This hysteresis between the positive-going and negative-going input thresholds (typically about 800 mV) is determined internally by on-chip resistor ratios, and is essentially insensitive to temperature and supply voltage fluctuations. As long as one input remains at a more positive voltage than $Vt+(max)$, the gate will respond to the transitions of the other input.

	74	74LS
Typ. Hysteresis (V)	0.8	0.8
Typ. Delay Time (ns)	15	15

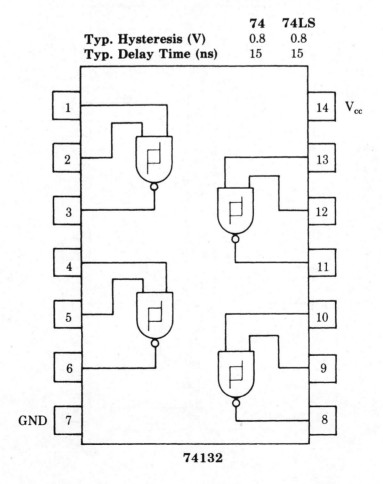

74132

74133 13-Input NAND gate

In principle, almost any standard logic gate can be expanded to offer as many inputs as desired, while still operating in the same basic way. The 74133 is a NAND gate with 13 inputs. The output (pin 9) will remain HIGH as long as at least one (or more) of the 13 inputs is LOW. The output goes LOW if, and only if, all 13 inputs are HIGH.

Thirteen inputs are offered because that is the maximum number of pins available on a standard 16-pin DIP housing, after the output and power supply (V_{cc} and GND) connections are accounted for. This arrangement is more versatile than it might appear to be at first glance. If fewer than 13 data inputs are required in a given application, just tie the unused inputs to a constant HIGH voltage, rendering them effectively transparent to the logic functioning of the gate. The normal NAND function will be preserved for the remaining (actively used) inputs.

	74S
Typ. Delay Time (ns)	3
Typ. Power Per Gate (mW)	19

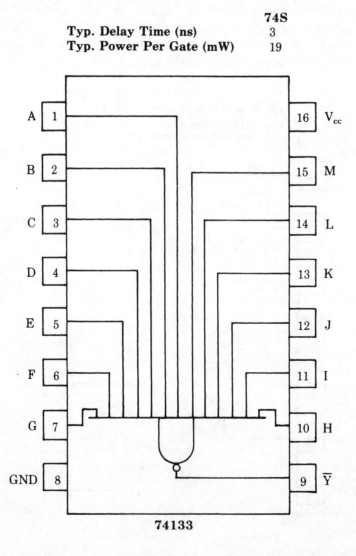

74133

74134 12-Input NAND gate with tri-state output

In principle, almost any standard logic gate can be expanded to offer as many inputs as desired, while still operating in the same basic way. The 74134 is a NAND gate with 12 inputs. Ignoring the tri-state function for the moment, the output (pin 9) will remain HIGH as long as at least one (or more) of the 12 inputs is LOW. The output goes LOW if, and only if, all 13 inputs are HIGH.

If fewer than 12 data inputs are required in a given application, just tie the unused inputs to a constant HIGH voltage, rendering them effectively transparent to the logic functioning of the gate. The normal NAND function will be preserved for the remaining (actively used) inputs.

The 74134 is very similar to the 74133. The only difference is that in this case, there are only 12 (instead of 13) data inputs. The remaining input ($\overline{OE}$, pin 15) is reassigned here to act as a control input for the tri-state function. If this pin is made HIGH, the output will be forced into the special high-impedance (off) state. The data at the other 12 inputs will be irrelevant in this case. If $\overline{OE}$ is held LOW, the 74134 will function normally as a NAND gate, which happens to have 12 inputs.

Truth Table

INPUTS		OUTPUT
D_0 - - - - D_{11}	$\overline{OE}$	$\overline{Y}$
H - - - - - H	L	L
one input = L	L	H
X - - - - - X	H	(Z)

H = High voltage level
L = Low voltage level
X = Don't care
(Z) = High impedance "off" state

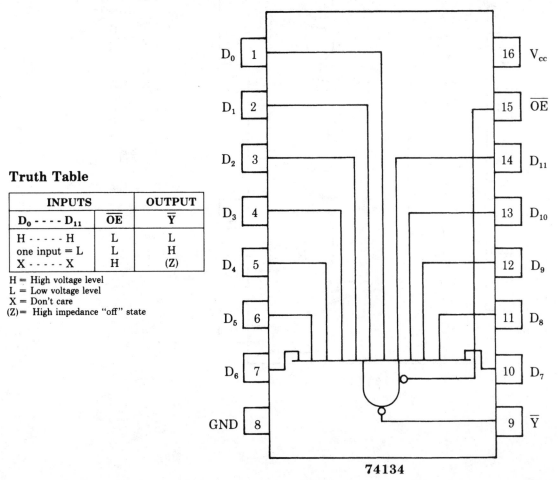

74134

74135 Quad Exclusive OR/NOR gate

In the 74135, Exclusive-OR gates are internally combined in pairs to produce an Exclusive-OR/NOR function, as defined in the Truth Table. Notice that one C input is shared for each of two pairs of outputs. The two joined sections therefore are not completely independent, as is usually the case in most multi-gate ICs.

	74S
Max. Supply Current (mA)	99
Typ. Propagation Delay (ns)	12

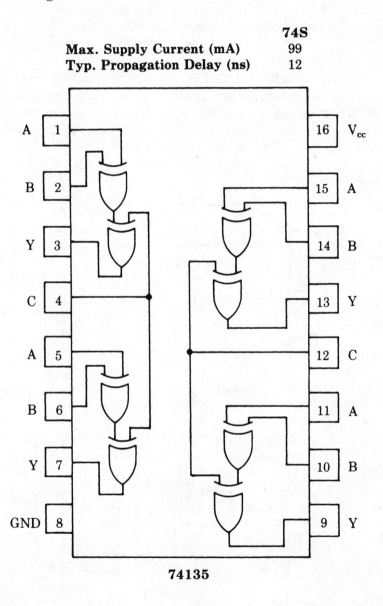

74135

Truth Table

INPUTS			OUTPUT
A	**B**	**C**	**Y**
L	L	L	L
L	H	L	H
H	L	L	H
H	H	L	L
L	L	H	H
L	H	H	L
H	L	H	L
H	H	H	H

H = High voltage level
L = Low voltage level

74136 Quad 2-input Exclusive-OR gate with open collector outputs

The basic functioning of the 74136 is the same as for the 7486, described earlier in this section. The difference is that the gates in this chip feature open collector outputs.

	74LS
Max. Supply Current (mA)	10
Max. Propagation Delay (ns)	30

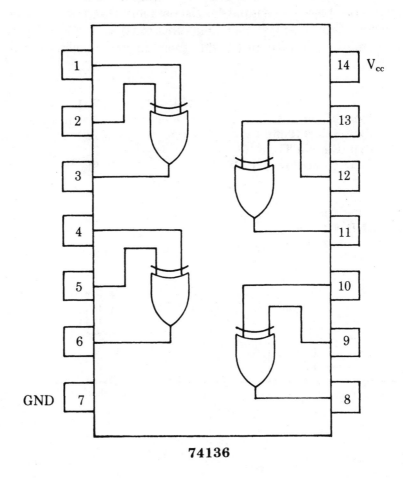

74136

Truth Table

INPUTS		OUTPUT
A	**B**	**Y**
L	L	L
L	H	H
H	L	H
H	H	L

L = Low voltage level
H = High voltage level

74138 1-of-8 decoder/demultiplexer

The 74138 decoder accepts three binary weighted inputs (A_0, A_1, and A_2) and, when enabled, provides eight mutually exclusive active LOW outputs ($\overline{0}$ through $\overline{7}$). The device features three enable inputs: two active LOW ($\overline{E_1}$ and $\overline{E_2}$), and one active HIGH ($\overline{E_3}$). Every output will be HIGH unless and $\overline{E_1}$ and $\overline{E_2}$ are LOW, and E3 is HIGH. This multiple enable function allows easy parallel expansion of the device to a 1-of-32 (5 lines to 32 lines) decoder with just four 74138s and one inverter.

The device can be also used as an eight-output demultiplexer by using one of the active LOW enable inputs as the data input, and the remaining enable inputs as strobes. Any enable inputs that are not used in the particular application must be permanently tied to their appropriate active-HIGH or active-LOW state.

	74S	74LS
Type of Output	Totem Pole	Totem Pole
Typ. Select Time (ns)	8	22
Typ. Enable Time (ns)	7	21
Typ. Total Power (mW)	225	31

V_{cc} = Pin 16
GND = Pin 8

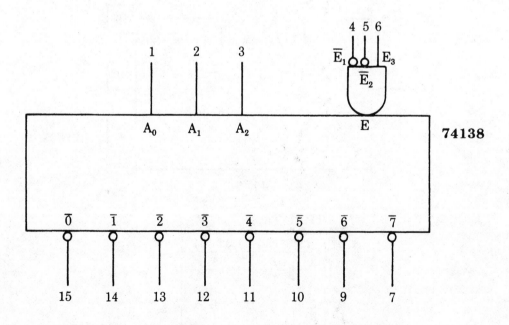

Truth Table

INPUTS						OUTPUTS							
$\overline{E_1}$	$\overline{E_2}$	E_3	A_0	A_1	A_2	$\overline{0}$	$\overline{1}$	$\overline{2}$	$\overline{3}$	$\overline{4}$	$\overline{5}$	$\overline{6}$	$\overline{7}$
H	X	X	X	X	X	H	H	H	H	H	H	H	H
X	H	X	X	X	X	H	H	H	H	H	H	H	H
X	X	L	X	X	X	H	H	H	H	H	H	H	H
L	L	H	L	L	L	L	H	H	H	H	H	H	H
L	L	H	H	L	L	H	L	H	H	H	H	H	H
L	L	H	L	H	L	H	H	L	H	H	H	H	H
L	L	H	H	H	L	H	H	H	L	H	H	H	H
L	L	H	L	L	H	H	H	H	H	L	H	H	H
L	L	H	H	L	H	H	H	H	H	H	L	H	H
L	L	H	L	H	H	H	H	H	H	H	H	L	H
L	L	H	H	H	H	H	H	H	H	H	H	H	L

NOTES
H = High voltage level
L = Low voltage level
X = Don't care

74139 Dual 1-of-4 decoder/demultiplexer

The 74139 is a high-speed dual 1-of-4 decoder/demultiplexer. This device has two independent decoders, each accepting two binary weighted inputs ($\overline{A_0}$ and $\overline{A_1}$), and providing four mutually exclusive active LOW outputs ($\overline{0}$ through $\overline{3}$). Each decoder has an active LOW enable ($\overline{E}$). When $\overline{E}$ is HIGH, every output is forced HIGH. This enable can be used as the data input for a 1-of-4 demultiplexer application.

Type of Output	74S	74LS
	Totem Pole	Totem Pole
Typ. Select Time (ns)	7.5	22
Typ. Enable Time (ns)	6	19
Typ. Total Power (mW)	300	34

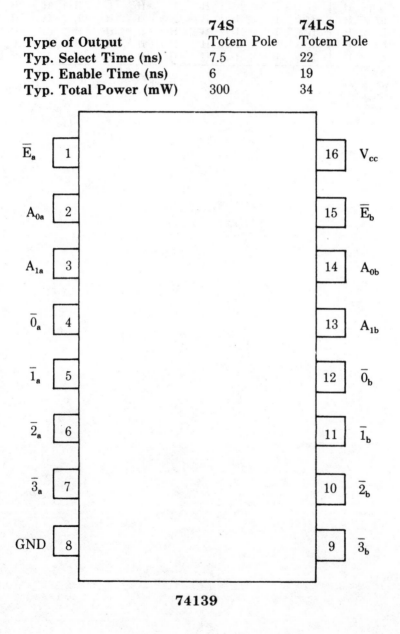

74139

74145 BCD-to-decimal decoder/ driver with open collector outputs

The 74145 is a one-of-10 decoder with open collector outputs. This decoder accepts BCD inputs on the A_0 to A_3 address lines, and generates 10 mutually exclusive active LOW outputs. When an input code greater than nine (1001) is applied, all outputs are HIGH. This device can therefore be used as a 1-of-8 decoder, with A_3 used as an active LOW enable.

The 74145 features an output breakdown voltage of 15 V, making this device ideal for use as a lamp or solenoid driver.

		74
	Output Sink Current (mA)	80
	Off-State Output Voltage (V)	15
V_{cc} = Pin 16	Typ. Total Power (mW)	215
GND = Pin 8	Blanking	Invalid Codes

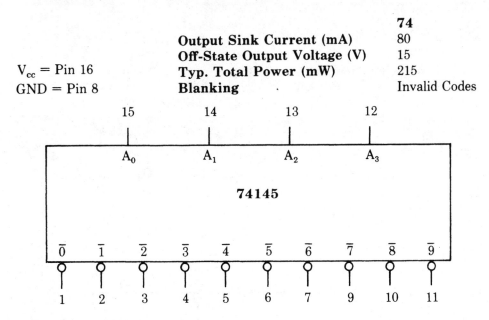

Truth Table

A_3	A_2	A_1	A_0	$\bar{0}$	$\bar{1}$	$\bar{2}$	$\bar{3}$	$\bar{4}$	$\bar{5}$	$\bar{6}$	$\bar{7}$	$\bar{8}$	$\bar{9}$
L	L	L	L	L	H	H	H	H	H	H	H	H	H
L	L	L	H	H	L	H	H	H	H	H	H	H	H
L	L	H	L	H	H	L	H	H	H	H	H	H	H
L	L	H	H	H	H	H	L	H	H	H	H	H	H
L	H	L	L	H	H	H	H	L	H	H	H	H	H
L	H	L	H	H	H	H	H	H	L	H	H	H	H
L	H	H	L	H	H	H	H	H	H	L	H	H	H
L	H	H	H	H	H	H	H	H	H	H	L	H	H
H	L	L	L	H	H	H	H	H	H	H	H	L	H
H	L	L	H	H	H	H	H	H	H	H	H	H	L
H	L	H	L	H	H	H	H	H	H	H	H	H	H
H	L	H	H	H	H	H	H	H	H	H	H	H	H
H	H	L	L	H	H	H	H	H	H	H	H	H	H
H	H	L	H	H	H	H	H	H	H	H	H	H	H
H	H	H	L	H	H	H	H	H	H	H	H	H	H
H	H	H	H	H	H	H	H	H	H	H	H	H	H

H = High voltage levels
L = Low voltage levels

74147 10- to 4-line priority encoder

The 74147 nine-input priority encoder accepts data from nine active LOW inputs ($\bar{I}_1$ through $\bar{I}_9$) and provides a binary representation on the four active LOW outputs (A_0 through A_3). A priority is assigned to each input, so that when two or more inputs are simultaneously active, the input with the highest priority is represented on the output. Higher value inputs have higher priority, with input line I_9 having the highest priority.

The 74147 provides the 10- to 4-line priority encoding function by use of the implied decimal zero. The zero is encoded when all nine data inputs are HIGH, forcing all four outputs HIGH (none active).

	74
Typ. Delay Time (ns)	10
Typ. Total Power (mW)	225

V_{cc} = Pin 16
GND = Pin 8

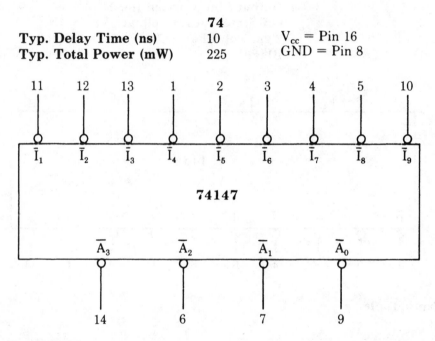

Truth Table

INPUTS									OUTPUTS			
$\bar{I}_1$	$\bar{I}_2$	$\bar{I}_3$	$\bar{I}_4$	$\bar{I}_5$	$\bar{I}_6$	$\bar{I}_7$	$\bar{I}_8$	$\bar{I}_9$	$\bar{A}_3$	$\bar{A}_2$	$\bar{A}_1$	$\bar{A}_0$
H	H	H	H	H	H	H	H	H	H	H	H	H
X	X	X	X	X	X	X	X	L	L	H	H	L
X	X	X	X	X	X	X	L	H	L	H	H	H
X	X	X	X	X	X	L	H	H	H	L	L	L
X	X	X	X	X	L	H	H	H	H	L	L	H
X	X	X	X	L	H	H	H	H	H	L	H	L
X	X	X	L	H	H	H	H	H	H	L	H	H
X	X	L	H	H	H	H	H	H	H	H	L	L
X	L	H	H	H	H	H	H	H	H	H	L	H
L	H	H	H	H	H	H	H	H	H	H	H	L

H = High voltage level
L = Low voltage level
X = Don't care

74148 Eight-input priority encoder

The 74148 eight-input priority encoder accepts data from eight active LOW inputs ($\bar{I}_0$ through $\bar{I}_7$) and provides a binary representation on the three active LOW outputs ($\bar{A}_0$ to $\bar{A}_2$). priority is assigned to each input so that when two or more inputs are simultaneously active, the input with the highest priority is represented on the output. Higher value inputs have higher priority, with input line $\bar{I}_7$ having the highest priority.

A HIGH on the input enable ($\overline{EI}$) will force all outputs to the inactive (HIGH) state, and allow new data to settle without producing erroneous information at the outputs.

A group signal output ($\overline{GS}$) and an enable output ($\overline{EO}$) are also provided, along with the three data outputs to facilitate system expansion. The $\overline{GS}$ output is active level LOW when any input is LOW. This indicates when any input is active. The EO output is active level LOW when all inputs are HIGH. Using the output enable along with the input enable allows priority coding of N input signals. Both $\overline{EO}$ and $\overline{GS}$ are active HIGH when the input enable is HIGH (device disabled).

		74
V_{CC} = Pin 16	**Typ. Delay Time (ns)**	12
GND = Pin 8	**Typ. Total Power (mW)**	130

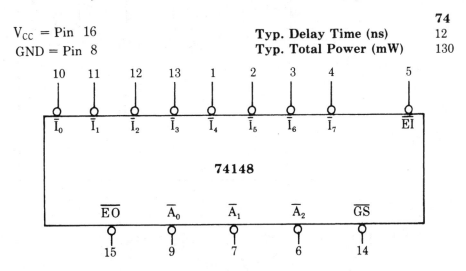

Truth Table

INPUTS									OUTPUTS				
$\overline{EI}$	$\bar{I}_0$	$\bar{I}_1$	$\bar{I}_2$	$\bar{I}_3$	$\bar{I}_4$	$\bar{I}_5$	$\bar{I}_6$	$\bar{I}_7$	$\overline{GS}$	$\bar{A}_0$	$\bar{A}_1$	$\bar{A}_2$	$\overline{Eo}$
H	X	X	X	X	X	X	X	X	H	H	H	H	H
L	H	H	H	H	H	H	H	H	H	H	H	H	L
L	X	X	X	X	X	X	X	L	L	L	L	L	H
L	X	X	X	X	X	X	L	H	L	H	L	L	H
L	X	X	X	X	X	L	H	H	L	L	H	L	H
L	X	X	X	X	L	H	H	H	L	H	H	L	H
L	X	X	X	L	H	H	H	H	L	L	L	H	H
L	X	X	L	H	H	H	H	H	L	H	L	H	H
L	X	L	H	H	H	H	H	H	L	L	H	H	H
L	L	H	H	H	H	H	H	H	L	H	H	H	H

H = High voltage level
L = Low voltage level
X = Don't care

101

74150 16-input multiplexer

The '150 is a logical implementation of a single pole, 16-position switch with the switch position controlled by the state of four select inputs: S_0, S_1, S_2 and S_3. The multiplexer output ($\overline{Y}$) inverts the selected data. The enable input ($\overline{D}$) is active LOW. When $\overline{E}$ is HIGH, the $\overline{Y}$ output is HIGH, regardless of all other inputs. In one package, the '150 provides the ability to select from 16 sources of data or control information.

Type of Output	**74** Standard
Typ. Delay, Data to Inverting Output (ns)	Standard
Typ. Delay Time, From Enable (ns)	18
Typ. Total Power (mW)	200

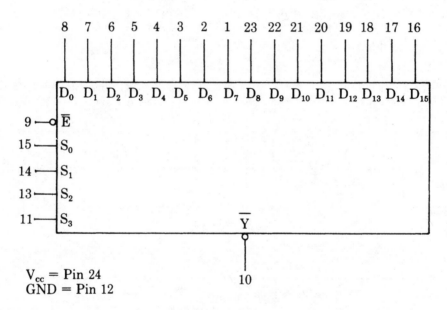

V_{cc} = Pin 24
GND = Pin 12

74150

Truth Table

INPUTS																					OUTPUT
S_3	S_2	S_1	S_0	$\overline{E}$	D_0	D_1	D_2	D_3	D_4	D_5	D_6	D_7	D_8	D_9	D_{10}	D_{11}	D_{12}	D_{13}	D_{14}	D_{15}	$\overline{Y}$
X	X	X	X	H	X	X	X	X	X	X	X	X	X	X	X	X	X	X	X	X	H
L	L	L	L	L	L	X	X	X	X	X	X	X	X	X	X	X	X	X	X	X	H
L	L	L	L	L	H	X	X	X	X	X	X	X	X	X	X	X	X	X	X	X	L
L	L	L	H	L	X	L	X	X	X	X	X	X	X	X	X	X	X	X	X	X	H
L	L	L	H	L	X	H	X	X	X	X	X	X	X	X	X	X	X	X	X	X	L
L	L	H	L	L	X	X	L	X	X	X	X	X	X	X	X	X	X	X	X	X	H
L	L	H	L	L	X	X	H	X	X	X	X	X	X	X	X	X	X	X	X	X	L
L	L	H	H	L	X	X	X	L	X	X	X	X	X	X	X	X	X	X	X	X	H
L	L	H	H	L	X	X	X	H	X	X	X	X	X	X	X	X	X	X	X	X	L
L	H	L	L	L	X	X	X	X	L	X	X	X	X	X	X	X	X	X	X	X	H
L	H	L	L	L	X	X	X	X	H	X	X	X	X	X	X	X	X	X	X	X	L
L	H	L	H	L	X	X	X	X	X	L	X	X	X	X	X	X	X	X	X	X	H
L	H	L	H	L	X	X	X	X	X	H	X	X	X	X	X	X	X	X	X	X	L
L	H	H	L	L	X	X	X	X	X	X	L	X	X	X	X	X	X	X	X	X	H
L	H	H	L	L	X	X	X	X	X	X	H	X	X	X	X	X	X	X	X	X	L
L	H	H	H	L	X	X	X	X	X	X	X	L	X	X	X	X	X	X	X	X	H
L	H	H	H	L	X	X	X	X	X	X	X	H	X	X	X	X	X	X	X	X	L
H	L	L	L	L	X	X	X	X	X	X	X	X	L	X	X	X	X	X	X	X	H
H	L	L	L	L	X	X	X	X	X	X	X	X	H	X	X	X	X	X	X	X	L
H	L	L	H	L	X	X	X	X	X	X	X	X	X	L	X	X	X	X	X	X	H
H	L	L	H	L	X	X	X	X	X	X	X	X	X	H	X	X	X	X	X	X	L
H	L	H	L	L	X	X	X	X	X	X	X	X	X	X	L	X	X	X	X	X	H
H	L	H	L	L	X	X	X	X	X	X	X	X	X	X	H	X	X	X	X	X	L
H	L	H	H	L	X	X	X	X	X	X	X	X	X	X	X	L	X	X	X	X	H
H	L	H	H	L	X	X	X	X	X	X	X	X	X	X	X	H	X	X	X	X	L
H	H	L	L	L	X	X	X	X	X	X	X	X	X	X	X	X	L	X	X	X	H
H	H	L	L	L	X	X	X	X	X	X	X	X	X	X	X	X	H	X	X	X	L
H	H	L	H	L	X	X	X	X	X	X	X	X	X	X	X	X	X	L	X	X	H
H	H	L	H	L	X	X	X	X	X	X	X	X	X	X	X	X	X	H	X	X	L
H	H	H	L	L	X	X	X	X	X	X	X	X	X	X	X	X	X	X	L	X	H
H	H	H	L	L	X	X	X	X	X	X	X	X	X	X	X	X	X	X	H	X	H
H	H	H	H	L	X	X	X	X	X	X	X	X	X	X	X	X	X	X	X	L	H
H	H	H	H	L	X	X	X	X	X	X	X	X	X	X	X	X	X	X	X	H	L

H = High voltage level
L = Low voltage level
X = Don't care

74151 8-input multiplexer

The '151 is a logical implementation of a single-pole, eight-position switch with the switch position controlled by the state of three select inputs: S_0, S_1, and S_2. True (Y) and complement ($\overline{Y}$) outputs are both provided. The enable input ($\overline{E}$) is active LOW when E is HIGH and Y output is LOW, regardless of all other inputs.

In one package, the '151 provides the ability to select from eight sources of data or control information. The device can provide any logic function of four variables and its negation with correct manipulation.

Type of Output	74S Standard	74 Standard	74LS Standard
Typ. Delay, Data to Inverting Output (ns)	4.5	8	11
Typ. Delay, Data to Noninverting Output (ns)	8	16	18
Typ. Delay Time, From Enable (ns)	9	22	27
Typ. Total Power (mW)	225	145	30

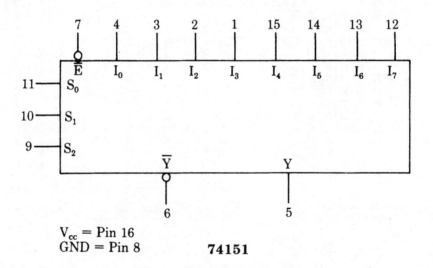

V_{cc} = Pin 16
GND = Pin 8

74151

Truth Table

$\overline{E}$	S_2	S_1	S_0	I_0	I_1	I_2	I_3	I_4	I_5	I_6	I_7	$\overline{Y}$	Y
			INPUTS									OUTPUTS	
H	X	X	X	X	X	X	X	X	X	X	X	H	L
L	L	L	L	L	X	X	X	X	X	X	X	H	L
L	L	L	L	H	X	X	X	X	X	X	X	L	H
L	L	L	H	X	L	X	X	X	X	X	X	H	L
L	L	L	H	X	H	X	X	X	X	X	X	L	H
L	L	H	L	X	X	L	X	X	X	X	X	H	L
L	L	H	L	X	X	H	X	X	X	X	X	L	H
L	L	H	H	X	X	X	L	X	X	X	X	H	L
L	L	H	H	X	X	X	H	X	X	X	X	L	H
L	H	L	L	X	X	X	X	L	X	X	X	H	L
L	H	L	L	X	X	X	X	H	X	X	X	L	H
L	H	L	H	X	X	X	X	X	L	X	X	H	L
L	H	L	H	X	X	X	X	X	H	X	X	L	H
L	H	H	L	X	X	X	X	X	X	L	X	H	L
L	H	H	L	X	X	X	X	X	X	H	X	L	H
L	H	H	H	X	X	X	X	X	X	X	L	H	L
L	H	H	H	X	X	X	X	X	X	X	H	L	H

H = High voltage level
L = Low voltage level
X = Don't care

74153 Dual 4-line to 1-line multiplexer

The '153 is a dual 4-input multiplexer that can select two bits of data from up to four sources under control of the common select inputs (S_0 and S_1). The two 4-input multiplexer circuits have individual active LOW enables ($\overline{E}_a$, $\overline{E}_b$) that can be used to strobe the outputs independently. Outputs (Y_a and Y_b) are forced LOW when the corresponding enables (E_a and E_b) are HIGH.

The device is the logical implementation of a 2-pole, 4-position switch, where the position of the switch is determined by the logic levels supplied to the two select inputs. The '153 can be used to move data to a common output bus from a group of registers. The state of the select inputs would determine the particular register from which the data came. An alternative application is as a function generator. The device can generate two functions of three variables, which is useful for implementing highly irregular random logic.

	74S Standard	74 Standard	74LS Standard
Type of Output			
Typ. Delay, Data to Noninverting Output (ns)	6	14	14
Typ. Delay Time, From Enable (ns)	9.5	17	17
Typ. Total Power (mW)	225	180	31

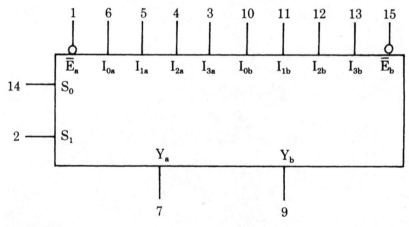

V_{cc} = Pin 16
GND = Pin 8

74153

Truth Table

SELECT INPUTS		INPUTS (a or b)					OUTPUT
S_0	S_1	$\bar{E}$	I_0	I_1	I_2	I_3	Y
X	X	H	X	X	X	X	L
L	L	L	L	X	X	X	L
L	L	L	H	X	X	X	H
H	L	L	X	L	X	X	L
H	L	L	X	H	X	X	H
L	H	L	X	X	L	X	L
L	H	L	X	X	H	X	H
H	H	L	X	X	X	L	L
H	H	L	X	X	X	H	H

H = High voltage level
L = Low voltage level
X = Don't care

74154 1-of-16 decoder/demultiplexer

The '154 accepts four active HIGH binary address inputs and provides 16 mutually exclusive active LOW outputs. The two-input enable gate can be used to strobe the decoder to eliminate the normal decoding glitches on the outputs, or it can be used to expand the decoder. The enable gate has two ANDed inputs that must be LOW to enable the outputs.

The '154 can be used as a 1-of-16 demultiplexer by using one of the enable inputs as the multiplexed data input. When the other enable is LOW, the addressed output will follow the state of the applied data.

Type of Output	74 Totem Pole	74LS Totem Pole	74L Totem Pole
Typ. Select Time (ns)	19.5	23	55
Typ. Enable Time (ns)	17.5	19	45
Typ. Total Power (mW)	170	45	24

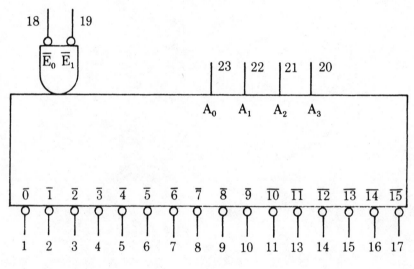

V_{cc} = Pin 24
GND = Pin 12

74154

Truth Table

INPUTS						OUTPUTS															
$\bar{E_0}$	$\bar{E_1}$	A_3	A_2	A_1	A_0	$\bar{0}$	$\bar{1}$	$\bar{2}$	$\bar{3}$	$\bar{4}$	$\bar{5}$	$\bar{6}$	$\bar{7}$	$\bar{8}$	$\bar{9}$	$\bar{10}$	$\bar{11}$	$\bar{12}$	$\bar{13}$	$\bar{14}$	$\bar{15}$
L	H	X	X	X	X	H	H	H	H	H	H	H	H	H	H	H	H	H	H	H	H
H	L	X	X	X	X	H	H	H	H	H	H	H	H	H	H	H	H	H	H	H	H
H	H	X	X	X	X	H	H	H	H	H	H	H	H	H	H	H	H	H	H	H	H
L	L	L	L	L	L	L	H	H	H	H	H	H	H	H	H	H	H	H	H	H	H
L	L	L	L	L	H	H	L	H	H	H	H	H	H	H	H	H	H	H	H	H	H
L	L	L	L	H	L	H	H	L	H	H	H	H	H	H	H	H	H	H	H	H	H
L	L	L	L	H	H	H	H	H	L	H	H	H	H	H	H	H	H	H	H	H	H
L	L	L	H	L	L	H	H	H	H	L	H	H	H	H	H	H	H	H	H	H	H
L	L	L	H	L	H	H	H	H	H	H	L	H	H	H	H	H	H	H	H	H	H
L	L	L	H	H	L	H	H	H	H	H	H	L	H	H	H	H	H	H	H	H	H
L	L	L	H	H	H	H	H	H	H	H	H	H	L	H	H	H	H	H	H	H	H
L	L	H	L	L	L	H	H	H	H	H	H	H	H	L	H	H	H	H	H	H	H
L	L	H	L	L	H	H	H	H	H	H	H	H	H	H	L	H	H	H	H	H	H
L	L	H	L	H	L	H	H	H	H	H	H	H	H	H	H	L	H	H	H	H	H
L	L	H	L	H	H	H	H	H	H	H	H	H	H	H	H	H	L	H	H	H	H
L	L	H	H	L	L	H	H	H	H	H	H	H	H	H	H	H	H	L	H	H	H
L	L	H	H	L	H	H	H	H	H	H	H	H	H	H	H	H	H	H	L	H	H
L	L	H	H	H	L	H	H	H	H	H	H	H	H	H	H	H	H	H	H	L	H
L	L	H	H	H	H	H	H	H	H	H	H	H	H	H	H	H	H	H	H	H	L

H = High voltage level
L = Low voltage level
X = Don't care

74155 Dual 2-line to 4-line decoder/demultiplexer

The '155 is a dual of 1-of-4 decoder/demultiplexer with common-address inputs and separate gated enable inputs. Each decoder section, when enabled, will accept the binary weighted address input (A_0 and A_1) and provide four mutually exclusive active LOW outputs (0 through 3). When the enable requirements of each decoder are not met, all outputs of that decoder are HIGH.

Both decoder sections have a two-input enable gate. For decoder a, the enable gate requires one active HIGH input and one active LOW input ($E_a \bullet \overline{E}_a$). Decoder a can accept either true or complemented data in demultiplexing applications, by using the $\overline{E}_a$ or E_a inputs respectively. The decoder b enable gate requires two active LOW inputs ($E_b \bullet \overline{E}_b$). The device can be used as a 1-of-8 decoder/demultiplexer by tying E_a or $\overline{E}_b$ and relabeling the common connection address as A2, forming the common enable by connecting the remaining $\overline{E}_b$ and $\overline{E}_a$.

	74LS	**74**
Type of Output	Totem Pole	Totem Pole
Typ. Select Time (ns)	18	21
Typ. Enable Time (ns)	15	16
Type. Total Power (mW)	30	250

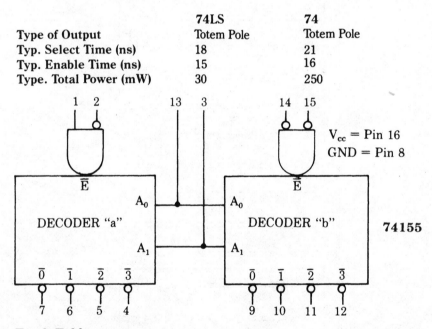

V_{cc} = Pin 16
GND = Pin 8

74155

Truth Table

ADDRESS		ENABLE "a"		OUTPUT "a"				ENABLE "b"		OUTPUT "b"			
A_0	A_1	E_a	$\overline{E}_a$	$\overline{0}$	$\overline{1}$	$\overline{2}$	$\overline{3}$	$\overline{E}_b$	$\overline{E}_b$	$\overline{0}$	$\overline{1}$	$\overline{2}$	$\overline{3}$
X	X	L	X	H	H	H	H	H	X	H	H	H	H
X	X	X	H	H	H	H	H	X	H	H	H	H	H
L	L	H	L	L	H	H	H	L	L	L	H	H	H
H	L	H	L	H	L	H	H	L	L	H	L	H	H
L	H	H	L	H	H	L	H	L	L	H	H	L	H
H	H	H	L	H	H	H	L	L	L	H	H	H	L

H = High voltage level
L = Low voltage level
X = Don't care

74156 Dual 2-line to 4-line decoder/demultiplexer (O.C.)

The '156 is a dual 1-of-4 decoder/demultiplexer with common address inputs and gated enable inputs. Each decoder section, when enabled, will accept the binary weighted address inputs (A_0 and A_1) and provide four mutually exclusive action LOW outputs (0 through $\bar{3}$). When the enable requirements of each decoder are not met, all outputs of that decoder are HIGH.

Both decoder sections have a two-input enable gate. For decoder a, the enable gate requires one active HIGH input and one active LOW input ($E_a \bullet \bar{E}_a$). Decoder a can accept either true or complemented data in demultiplexing applications by using the $\bar{E}_a$ or E_a inputs, respectively. The decoder b enable gate requires two active LOW inputs ($\bar{E}_b \bullet \bar{E}_b$). The device can be used as a 1-of-8 decoder/demultiplexer by tying E_a to $\bar{E}_b$ and relabeling the common connection address as A_2, forming the common enable by connecting the remaining $\bar{E}_b$ and $\bar{E}_a$.

The '156 can be used to generate all four minterms of two variables. The four minterms are useful to replace multiple gate functions in some applications.

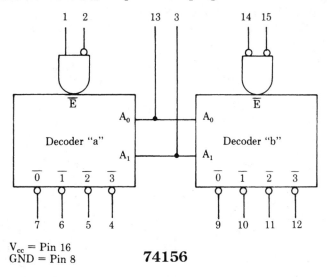

V_{cc} = Pin 16
GND = Pin 8

74156

	74	74LS
Typ. of Output	Open Collector	Open Collector
Typ. Select Time (ns)	23	33
Typ. Enable Time (ns)	18	26
Typ. Total Power (mW)	250	31

Truth Table

ADDRESS		ENABLE "a"		OUTPUT "a"				ENABLE "b"		OUTPUT "b"			
A_0	A_1	E_a	$\bar{E}_a$	$\bar{0}$	$\bar{1}$	$\bar{2}$	$\bar{3}$	$\bar{E}_b$	$\bar{E}_b$	$\bar{0}$	$\bar{1}$	$\bar{2}$	$\bar{3}$
X	X	L	X	H	H	H	H	H	X	H	H	H	H
X	X	X	H	H	H	H	H	X	H	H	H	H	H
L	L	H	L	L	H	H	H	L	L	L	H	H	H
H	L	H	L	H	L	H	H	L	L	H	L	H	H
L	H	H	L	H	H	L	H	L	L	H	H	L	H
H	H	H	L	H	H	H	L	L	L	H	H	H	L

H = High voltage level
L = Low voltage level
X = Don't care

74157 Quad 2-input data selector/multiplexer (noninverted)

The 74157 is a quad two-input multiplexer that selects four bits of data from two sources under the control of a common select input (S). The enable input ($\overline{E}$)is active LOW. When $\overline{E}$ is HIGH, all of the outputs (Y_a through Y_d) are forced LOW, regardless of all other input conditions.

Moving data from two groups of registers to four common output busses is a common use of the 74157. The state of the select input (S) determines the particular register from which the data comes. It can also be used as a function generator. The device is useful for implementing highly irregular logic by generating any four of the 16 different functions of two variables with one variable common. The device is the logic implementation of a four-pole, two-position switch, where the "position" of the switch is determined by the logic levels supplied to the select input.

	74S	74LS	74	74L
Type of Output	Standard	Standard	Standard	Standard
Type Delay, Data to Noninverting Output (ns)	5	9	9	40
Typ. Delay, from Enable (ns)	8	14	4	60
Typ. Total Power (mW)	250	49	150	15

V_{cc} = Pin 16
GND = Pin 8

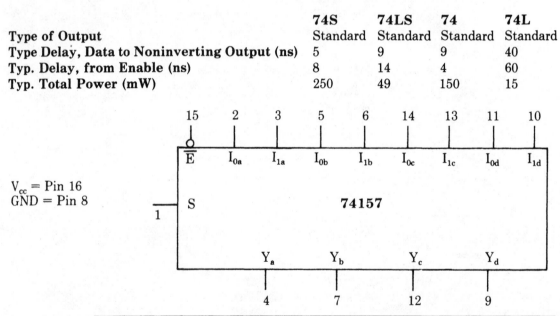

Truth Table

ENABLE	SELECT INPUT	DATA INPUTS		OUTPUT
$\overline{E}$	S	I_0	I_1	Y
H	X	X	X	L
L	H	X	L	L
L	H	X	H	H
L	L	L	X	L
L	L	H	X	H

H = High voltage level
L = Low voltage level
X = Don't care

74158 Quad 2-input data selector/multiplexer (inverted)

The '158 is a quad two-input multiplexer that selects four bits of data from two sources under the control of a common select input (S), presenting the data in inverted form at the four outputs $(\overline{Y})$. The enable input $(\overline{E})$ is active LOW. When $\overline{E}$ is HIGH, all of the outputs $(\overline{Y})$ are forced HIGH, regardless of all other input conditions.

Moving data from two groups of registers to four common output busses is a common use of the 74158. The state of the select input determines the particular register from which the data comes. It can also be used as a function generator. The device is useful for implementing gating functions by generating any four functions of two variables with one variable common. The device is the logic implementation of a four-pole, two-position switch, where the position of the switch is determined by the logic levels supplied to the select input.

Type of Output	74S Standard	74LS Standard
Typ. Delay, Data to Inverting Output (ns)	4	7
Typ. Delay, From Enable (ns)	7	12
Typ. Total Power (mW)	195	24

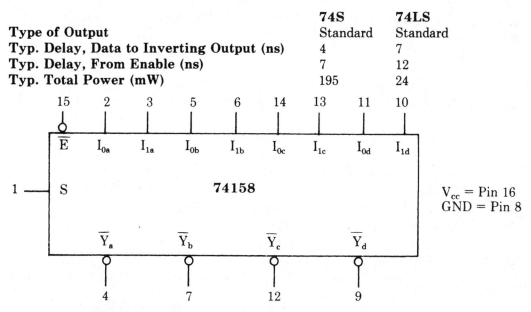

V_{cc} = Pin 16
GND = Pin 8

Truth Table

ENABLE	SELECT INPUT	DATA INPUTS		OUTPUTS
$\overline{E}$	S	I_0	I_1	$\overline{Y}$
H	X	X	X	H
L	L	L	X	H
L	L	H	X	L
L	H	X	L	H
L	H	X	H	L

H = High voltage level
L = Low voltage level
X = Don't care

113

74160 DEC decade counter

The '160 is a high-speed BCD decade counter. The counters are positive edge-triggered, synchronously presettable, and are easily cascaded to n-bit synchronous applications without additional gating. A terminal count output is provided, which detects a count of HLLH. The master reset asynchronously clears all flip-flops.

The '160 is a synchronous presettable BCD decade counter featuring an internal carry lookahead for applications in high-speed counting designs. Synchronous operation is provided by having all flip-flops clocked simultaneously so that the outputs change coincident with each other when so instructed by the count-enable inputs and internal gating. This mode of operation eliminates the output counting that is normally associated with asynchronous (ripple clock) counters. A buffered clock input triggers the four flip-flops on the positive-going edge of the clock. The clock input on the LS 160 features about 400 mV of hysteresis to reduce false triggering caused by noise on the clock line or by slowly rising clock edges.

The counter is fully programmable; that is, the outputs can be preset to either level. Presetting is synchronous with the clock, and takes place regardless of the levels of the count-enable inputs. A LOW level on the parallel-enable ($\overline{PE}$) input disables the counter and causes the data at the D_n inputs to be loaded into the counter on the next LOW-to-HIGH transition of the clock. The reset (clear) function for the '160 is asynchronous. A LOW level on the master reset ($\overline{MR}$) input sets all four of the flip-flop outputs LOW, regardless of the levels of the CP, $\overline{PE}$, CE_T, and CE_P inputs.

The carry lookahead circuitry provides for cascading counters for n-bit synchronous applications without additional gating. Instrumental in accomplishing this function are two count-enable inputs ($CE_T \bullet CE_P$) and a terminal count (TC) output. Both count-enable inputs must be HIGH to count. The CE_T input is fed forward to enable the TC output. The TC output thus enabled will produce a HIGH output pulse with a duration approximately equal to the HIGH-level portion of the Q_0 output. This HIGH-level TC pulse is used to enable successive cascaded stages. The fast synchronous multistage counting connections are shown here.

All changes of the Q outputs (except due to the asynchronous master reset) occur as a result of, and synchronous with, the LOW-to-HIGH transition of the clock input (CP). As long as the setup time requirements are met, there are no special timing or activity constraints on any of the mode control or data inputs. However, for conventional operation of the 74160 the following transitions should be avoided:

- HIGH-to-LOW transition on the CE_P or CE_T input if the clock is LOW.
- LOW-to-HIGH transition on the parallel enable input when the CP is LOW, if the count enables are high at or before the transition.

For some applications, the designer might want to change those inputs while the clock is LOW. In this case, the 74160 will behave in a predictable manner. For example, if $\overline{PE}$ goes HIGH while the clock is LOW, and the count enable is not active during the remaining clock LOW period (i.e., CE_P or CE_T are LOW), the subsequent LOW-to-HIGH clock transition will change Q_0 through Q_3 to the D_0 through D_3 data that existed at the setup time before the rising edge of $\overline{PE}$. If $\overline{PE}$ goes HIGH while the clock is LOW, and the count enable is active (CE_P and CE_T are HIGH) during some portion of the remaining clock LOW period, the 74160 will perform a mixture of counting and loading. On the LOW-to-HIGH clock transition, outputs Q_0 through Q_3 will change as the count sequence or the loading requires. Only the outputs that would not change in the count sequence and that are also reloaded with their present value stay constant.

If the count enable is active (i.e., CE_P and CE_T are high) during some portion of the clock LOW period, and $\overline{PE}$ is HIGH (inactive) during the entire clock LOW period, the subsequent LOW-to-HIGH clock transition will change Q_0 through Q_3 to the next count value.

	74LS	74
Count Frequency (MHz)	25	25
Parallel Load	Synchronous	Synchronous
Clear	Asynchronous-Low	Asynchronous-Low
Typ. Total Power (mW)	93	305

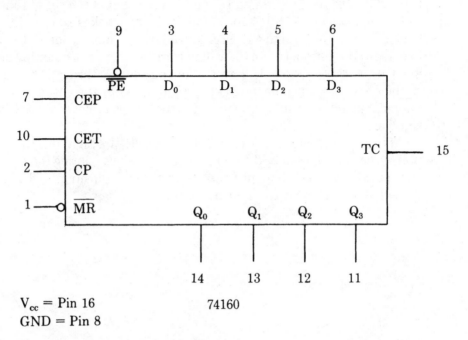

V_{cc} = Pin 16
GND = Pin 8

74160

MODE SELECT-FUNCTION TABLE

OPERATING MODE	INPUTS						OUTPUTS	
	$\overline{MR}$	CP	CEP	CET	$\overline{PE}$	D_n	Q_n	TC
Reset (Clear)	L	X	X	X	X	X	L	L
Parallel Load	H	↑	X	X	l	l	L	L
	H	↑	X	X	l	h	H	(b)
Count	H	↑	h	h	h(d)	X	count	(b)
Hold (do nothing)	H	X	l(c)	X	h(d)	X	q_n	(b)
	H	X	X	l(c)	h(d)	X	q_n	L

H = High voltage level steady state.
L = Low voltage level steady state.
h = High voltage level one setup time prior to the low-to-high clock transition.
l = Low voltage level one setup time prior to the low-to-high clock transition.
x = Don't care.
q = Lower case letters indicate the state of the referenced output prior to the low-to-high clock transition.
↑ = Low-to-high clock transition.

NOTES
(b) The TC output is high when CET is high and the counter is at Terminal Count (HLLH for "160").
(c) The high-to-low transition of CEP or CET on the 54/74160 should only occur while CP is high for conventional operation.
(d) The low-to-high transition of $\overline{PE}$ on the 54/74160 should only occur while CP is high for conventional operation.

Synchronous Multistage Counting Scheme

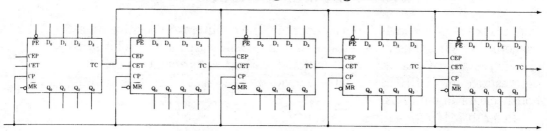

74161 4-Bit binary counter

The '161 is a high-speed 4-bit binary counter. The counters are positive edge-triggered, synchronously presettable and are easily cascaded to n-bit synchronous applications without additional gating. A terminal count that detects a count of HHHH output is provided. The master reset asynchronously clears all flip-flops.

Refer to 74160 for a further explanation of binary counters.

	74LS	74
Count Frequency (MHz)	25	25
Parallel Load	Synchronous	Synchronous
Clear	Asynchronous-Low	Asynchronous-Low
Typ. Total Power (mW)	93	305

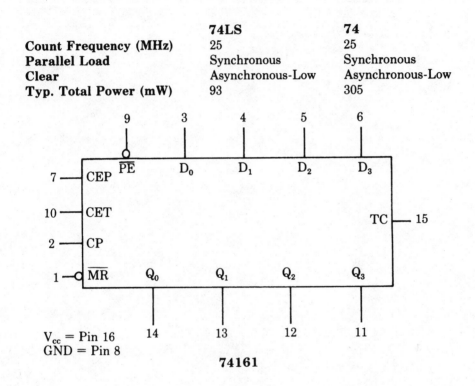

V_{cc} = Pin 16
GND = Pin 8

74161

MODE SELECT-FUNCTION TABLE

OPERATING MODE	INPUTS						OUTPUTS	
	$\overline{MR}$	CP	CEP	CET	$\overline{PE}$	D_n	Q_n	TC
Reset (Clear)	L	X	X	X	X	X	L	L
Parallel Load	H	↑	X	X	l	l	L	L
	H	↑	X	X	l	h	H	(b)
Count	H	↑	h	h	h(d)	X	count	(b)
Hold (do nothing)	H	X	l(c)	X	h(d)	X	q_n	(b)
	H	X	X	l(c)	h(d)	X	q_n	L

H = High voltage level steady state.
L = Low voltage level steady state.
h = High voltage level one setup time prior to the low-to-high clock transition.
l = Low voltage level one setup time prior to the low-to-high clock transition.
X = Don't care.
q = Lower case letters indicate the state of the referenced output prior to the low-to-high clock transition.
↑ = Low-to-high clock transition.

NOTES
(b) The TC output is high when CET is high and the counter is at Terminal Count (HHHH for "161").
(c) The high-to-low transition of $\overline{CEP}$ or CET on the 54/74161 should only occur while CP is high for conventional operation.
(d) The low-to-high transition of $\overline{PE}$ on the 54/74161 should only occur while CP is high for conventional operation.

74162 BCD decade counter

The '162 is a high-speed BCD decade counter. The counters are positive edge-triggered, synchronously presettable and are easily cascaded to n-bit synchronous applications without additional gating. A terminal count output that detects a count of HLLH is provided. The synchronous reset is edge-triggered. It overrides all control inputs, but is active only during the rising clock edge.

Refer to 74160 for a further explanation of binary counters.

	74LS	74
Count Frequency (MHz)	25	25
Parallel Load	Synchronous	Synchronous
Clear	Synchronous-Low	Synchronous-Low
Typ. Total Power (mW)	93	305

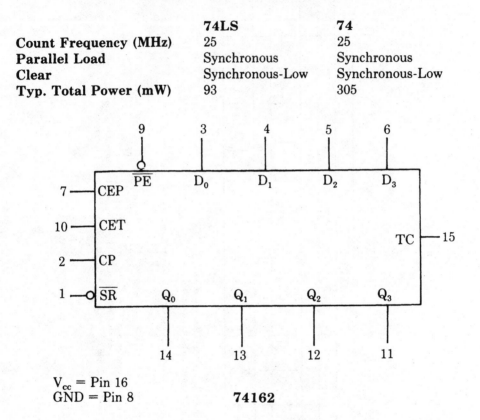

V_{cc} = Pin 16
GND = Pin 8

74162

MODE SELECT-FUNCTION TABLE

OPERATING MODE	INPUTS						OUTPUTS	
	$\overline{SR}$	CP	CEP	CET	$\overline{PE}$	D_n	Q_n	TC
Reset (Clear)	l	↑	X	X	X	X	L	L
Parallel Load	h(d)	↑	X	X	l	l	L	L
	h(d)	↑	X	X	l	h	H	(b)
Count	h(d)	↑	h	h	h(d)	X	count	(b)
Hold (do nothing)	h(d)	X	l(c)	X	h(d)	X	q_n	(b)
	h(d)	X	X	l(c)	h(d)	X	q_n	L

H = High voltage level steady state.
L = Low voltage level steady state.
h = High voltage level one setup time prior to the low-to-high clock transition.
l = Low voltage level one setup time prior to the low-to-high clock transition.
X = Don't care.
q = Lower case letters indicate the state of the referenced output prior to the low-to-high clock transition.
↑ = Low-to-high clock transition.

NOTES
(b) The TC output is high when CET is high and the counter is at Terminal Count (HLLH for "162").
(c) The high-to-low transition of CEP or CET on the 54/74162 should only occur while CP is high for conventional operation.
(d) The low-to-high transition of $\overline{PE}$ or $\overline{SR}$ on the 54/74162 should only occur while CP is high for conventional operation.

74163 4-Bit binary counter

The '163 is a high-speed 4-bit binary counter. The counters are positive edge-triggered, synchronously presettable and are easily cascaded to n-bit synchronous applications without additional gating. A terminal count output that detects a count of HHHH is provided. The synchronous reset is edge-triggered. It overrides all other control inputs, but is active only during the rising clock edge.

Refer to 74160 for a further explanation of binary counters.

	74LS	**74**
Count Frequency (MHz)	25	25
Parallel Load	Synchronous	Synchronous
Clear	Synchronous-Low	Synchronous-Low
Typ. Total Power (mW)	93	305

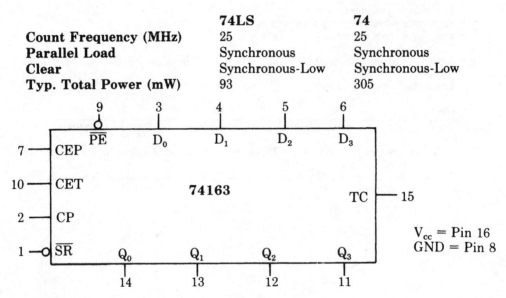

V_{cc} = Pin 16
GND = Pin 8

MODE SELECT-FUNCTION TABLE

OPERATING MODE	INPUTS						OUTPUTS	
	$\overline{SR}$	CP	CEP	CET	$\overline{PE}$	D_n	Q_n	TC
Reset (Clear)	l	↑	X	X	X	X	L	L
Parallel Load	h(d)	↑	X	X	l	l	L	L
	h(d)	↑	X	X	l	h	H	(b)
Count	h(d)	↑	h	h	h(d)	X	count	(b)
Hold (do nothing)	h(d)	X	l(c)	X	h(d)	X	q_n	(b)
	h(d)	X	X	l(c)	h(d)	X	q_n	L

H = High voltage level steady state.
L = Low voltage level steady state.
h = High voltage level one setup time prior to the low-to-high clock transition.
l = Low voltage level one setup time prior to the low-to-high clock transition.
X = Don't care.
q = Lower case letters indicate the state of the referenced output prior to the low-to-high clock transition.
↑ = Low-to-high clock transition.

NOTES
(b) The TC output is high when CET is high and the counter is at Terminal Count (HHHH for "163").
(c) The high-to-low transition of CEP or CET on the 54/74163 should only occur while CP is high for conventional operation.
(d) The low-to-high transition of $\overline{PE}$ or $\overline{SR}$ on the 54/74163 should only occur while CP is high for conventional operation.

122

74164 8-bit serial-in, parallel-out shift register

The 74168 is an eight-bit edge-triggered shift register with serial data entry and a parallel output from each of the eight stages. Data are entered serially through one of two inputs (D_{sa} or D_{sb}). Either input can be used as an active HIGH enable for data entry through the other input. If one input is not used in a given application, both inputs must be connected together, or the unused input must be tied HIGH.

Data shift one place to the right on each LOW-to-HIGH transition of the clock (CP) input, and enter into Q_0 the logical AND of the two data inputs (D_{sa} and D_{sb}) that existed one setup time before the rising clock edge. A LOW level on the Master Reset ($\overline{MR}$) input overrides all other inputs and clears the register asynchronously, forcing all outputs LOW.

	74LS	**74**	**74L**
Shift Frequency (MHz)	25	25	6
Serial Data Input	Gated D	Gated D	Gated D
Asynchronous Clear	Low	Low	Low
Shift-Right Mode	Yes	Yes	Yes
Shift-Left Mode	No	No	No
Load	No	No	No
Hold	No	No	No
Typ. Total Power (mW)	80	175	30

V_{cc} = Pin 14
GND = Pin 7

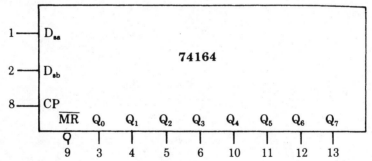

Truth Table

OPERATING MODE	INPUTS				OUTPUTS			
	$\overline{MR}$	CP	D_{sa}	D_{sb}	Q_0	Q_1	–	Q_7
Reset (Clear)	L	X	X	X	L	L	–	L
Shift	H	↑	l	l	L	q_0	–	q_6
	H	↑	l	h	L	q_0	–	q_6
	H	↑	h	l	L	q_0	–	q_6
	H	↑	h	h	H	q_0	–	q_6

H = High voltage level
h = High voltage level one setup time prior to the low-to-high clock transition
L = Low voltage level
l = Low voltage level one setup time prior to the low-to-high clock transition
q = Lower case letters indicate the state of the referenced input (or output) one setup time prior to the low-to-high clock transition
X = Don't care
↑ = Low-to-high clock transition

74165 8-Bit serial/parallel-in serial-out shift register

The '165 is an 8-bit parallel load or serial-in shift register with complementary serial outputs (Q_7 and $\overline{Q_7}$) available from the last stage. When the parallel load ($\overline{PL}$) input is LOW, parallel data from the D_0 through D_7 inputs are loaded into the register asynchronously. When the $\overline{PL}$ input is HIGH, data enter the register serially at the D_S input and shift one place to the right ($Q_0 \rightarrow Q_1 \rightarrow Q_2$, etc.) with each positive-going clock transition. This feature allows parallel to serial converter expansion by tying the Q_7 output to the D_S input of the succeeding stage.

The clock input is a gated OR structure that allows one input to be used as an active LOW clock enable ($\overline{CE}$) input. The pin assignment for the CP and $\overline{CE}$ inputs is arbitrary and can be reversed for layout convenience. The low-to-high transition of the $\overline{CE}$ input should only take place while the CP is high for predictable operation. Also, the CP and $\overline{CE}$ inputs should be LOW before the LOW-to-HIGH transition of $\overline{PL}$ to prevent shifting the data when $\overline{PL}$ is released.

	74
Shift Frequency (MHz)	25
Serial Data Input	D
Asynchronous Clear	None
Shift-Right Mode	Yes
Shift-Left Mode	No
Load	Yes
Hold	Yes
Typ. Total Power (mW)	200

MODE SELECT-FUNCTION TABLE

OPERATING MODES	INPUTS					Q_n REGISTER		OUTPUTS	
	$\overline{PL}$	$\overline{CE}$	CP	D_S	$D_0 - D_7$	Q_0	$Q_1 - Q_6$	Q_7	$\overline{Q_7}$
Parallel Load	L	X	X	X	L	L	L - L	L	H
	L	X	X	X	H	H	H - H	H	L
Serial Shift	H	L	↑	l	X	L	$q_0 - q_5$	q_6	$\overline{q_6}$
	H	L	↑	h	X	H	$q_0 - q_5$	q_6	$\overline{q_6}$
Hold "Do Nothing"	H	H	X	X	X	q_0	$q_1 - q_6$	q_7	$\overline{q_7}$

H = High voltage level.
h = High voltage level one setup time prior to the low-to-high clock transition.
L = Low voltage level.
l = Low voltage level one setup time prior to the low-to-high clock transition.
q_n = Lower case letters indicate the state of the referenced output one setup time prior to the low-to-high clock transition.
X = Don't care.
↑ = Low-to-high clock transition.

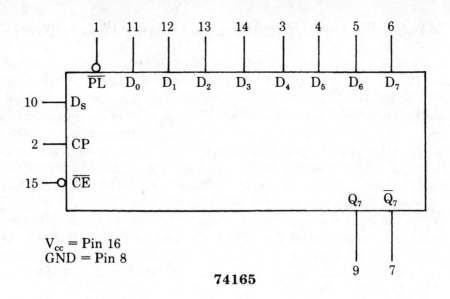

V_{cc} = Pin 16
GND = Pin 8

74165

74166 8-Bit serial/parallel-in serial out shift register

The '166 is an 8-bit shift register that has fully synchronous serial or parallel data entry selected by an active LOW parallel enable ($\overline{PE}$) input. When the $\overline{PE}$ is LOW one setup time before the LOW-to-HIGH clock transition, parallel data are entered into the register. When $\overline{PE}$ is HIGH, data is entered into the register. When $\overline{PE}$ is HIGH, data is entered into internal bit position Q_0 from serial data input (D_s), and the remaining bits are shifted one place to the right ($Q_0 \rightarrow Q_1 \rightarrow Q_2$) with each positive-going clock transition. To expand the register in parallel to serial converters, the Q_7 output is connected to the D_s input of the succeeding stage.

The clock input is a gated OR structure that allows one input to be used as an active LOW clock enable ($\overline{CE}$) input. The pin assignment for the CP and $\overline{CE}$ inputs is arbitrary and can be reversed for layout convenience. The LOW-to-HIGH transition of $\overline{CE}$ input should only occur while the CP is HIGH for predictable operation.

A LOW on the master reset ($\overline{MR}$) input overrides all other inputs and clears the register asynchronously, forcing all bit positions to a LOW state.

	74
Shift Frequency (MHz)	20
Serial Data Input	D
Asynchronous Clear	Low
Shift-Right Mode	Yes
Shift-Left Mode	No
Load	Yes
Hold	Yes
Typ. Total Power (mW)	360

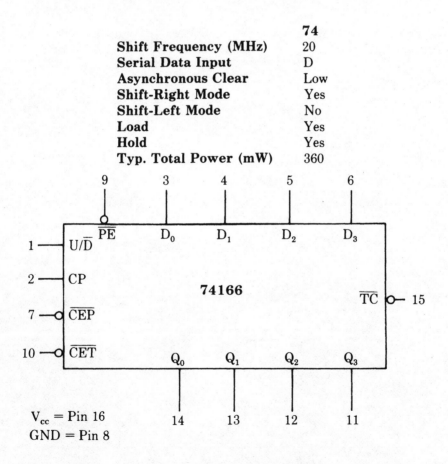

V_{cc} = Pin 16
GND = Pin 8

MODE SELECT-FUNCTION TABLE

OPERATING MODE	INPUTS						OUTPUTS	
	CP	U/$\overline{\text{D}}$	$\overline{\text{CEP}}$	$\overline{\text{CET}}$	$\overline{\text{PE}}$	D_n	Q_n	$\overline{\text{TC}}$
Parallel Load	↑	X	X	X	l	l	L	(b)
	↑	X	X	X	l	h	H	(b)
Count Up	↑	h	l	l	h	X	Count Up	(b)
Count Down	↑	l	l	l	h	X	Count Down	(b)
Hold (do nothing)	↑	X	h	X	h	X	q_n	(b)
	↑	X	X	h	h	X	q_n	H

H = High voltage level steady state.
L = Low voltage level steady state.
h = High voltage level one setup time prior to the low-to-high clock transition.
l = Low voltage level one setup time prior to the low-to-high clock transition.
X = Don't care.
q = Lower case letters indicate the state of the referenced output prior to the low-to-high clock transition.
↑ = Low-to-high clock transition.

NOTE
(b) The $\overline{\text{TC}}$ is low when $\overline{\text{CET}}$ is low and the counter is at Terminal Count. Terminal Count Up is (HLLH), and Terminal Count Down is (LLLL).

74168 4-Bit up/down synchronous counter

The '168 is a synchronous presettable BCD decade up/down counter featuring an internal carry lookahead for applications in high-speed counting designs. Synchronous operation is provided by having all flip-flops clocked simultaneously so that the outputs change coincident with each other when so instructed by the count-enable inputs and internal gating. This mode of operation eliminates the output spikes which are normally associated with asynchronous (ripple clock) counters. A buffered clock input triggers the flip-flops on the LOW-to-HIGH transition of the clock.

The counter is fully programmable; that is, the outputs can be present to either level. Presetting is synchronous with the clock and occurs regardless of the levels of the count-enable inputs. A low level on the parallel enable ($\overline{PE}$) input disables the counter and causes the data at the D_n inputs to be loaded into the counter on the next LOW-to-HIGH transition of the clock. The direction of counting is controlled by the up/down ($U/\overline{D}$) input; a HIGH will cause the count to increase, and a LOW will cause the count to decrease.

The carry lookahead circuitry provides for cascading counters for n-bit synchronous applications without additional gating. Instrumental in accomplishing this function are two count-enable inputs ($\overline{CET} \cdot \overline{CEP}$) and a terminal count ($\overline{TC}$) output. Both count-enable inputs must be LOW to count. The $\overline{CET}$ input is fed forward to enable the $\overline{TC}$ output. The $\overline{TC}$ output, thus enabled, will produce a LOW output pulse with a duration approximately equal to the HIGH-level portion of the Q_0 output. This LOW-level $\overline{TC}$ pulse is used to enable successive cascaded stages. See 74168 for the fast synchronous multistage counting connections.

	74LS
Count Frequency (MHz)	25
Parallel Load	Synchronous
Clear	None
Typ. Total Power (mW)	100

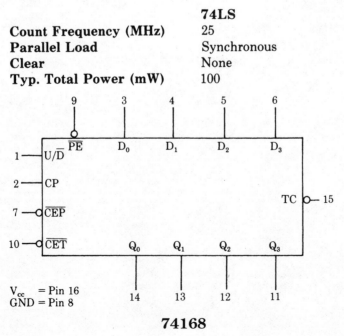

V_{cc} = Pin 16
GND = Pin 8

74168

MODE SELECT-FUNCTION TABLE

OPERATING MODE	INPUTS						OUTPUTS	
	CP	U/$\overline{D}$	$\overline{CEP}$	$\overline{CET}$	$\overline{PE}$	D_a	Q_a	$\overline{TC}$
Parallel Load	↑	X	X	X	l	l	L	(b)
	↑	X	X	X	l	h	H	(b)
Count Up	↑	h	l	l	h	X	Count Up	(b)
Count Down	↑	l	l	l	h	X	Count Down	(b)
Hold (do nothing)	↑	X	h	X	h	X	q_n	(b)
	↑	X	X	h	h	X	q_n	H

H = High voltage level steady state.
L = Low voltage level steady state.
h = High voltage level one setup time prior to the low-to-high clock transition.
l = Low voltage level one setup time prior to the low-to-high clock transition.
X = Don't care.
q = Lower case letters indicate the state of the referenced output prior to the low-to-high clock transition.
↑ = Low-to-high clock transition.

NOTE
(b) The $\overline{TC}$ is low when $\overline{CET}$ is low and the counter is at Terminal Count. Terminal Count Up is (HLLH), and Terminal Count Down is (LLLL).

74169 4-Bit up/down synchronous counter

The '169 is a synchronous presettable module 16 binary up/down counter featuring an internal carry lookahead for applications in high-speed counting designs. Synchronous operation is provided by having all flip-flops clocked simultaneously so that the outputs change coincident with each other when so instructed by the count-enable inputs and internal gating. This mode of operation eliminates the output spikes that are normally associated with asynchronous (ripple clock) counters. A buffered clock input triggers the flip-flops on the LOW-to-HIGH transition of the clock.

The counter is fully programmable; that is, the outputs can be preset to either level. Presetting is synchronous with the clock and occurs regardless of the levels of the count enable inputs. A LOW level on the parallel enable ($\overline{PE}$) input disables the counter and causes the data at the D_n inputs to be loaded into the counter on the next LOW-to-HIGH transition of the clock. The direction of counting is controlled by the up/down ($U/\overline{D}$) input. A HIGH will cause the count to increase, and a LOW will cause the count to decrease.

The carry lookahead circuitry provides for cascading counters for n-bit synchronous applications without additional gating. Instrumental in accomplishing this function are two count-enable inputs ($\overline{CET}$ • $\overline{CEP}$) and a terminal count ($\overline{TC}$) output. Both count-enable inputs must be LOW to count. The $\overline{CET}$ input is fed forward to enable the $\overline{TC}$ output. The $\overline{TC}$ output, thus enabled, will produce a low output pulse with a duration approximately equal to the HIGH-level portion of the Q_0 output. This LOW-level $\overline{TC}$ pulse is used to enable successive cascaded stages. See 74168 for the fast synchronous multistage counting connections.

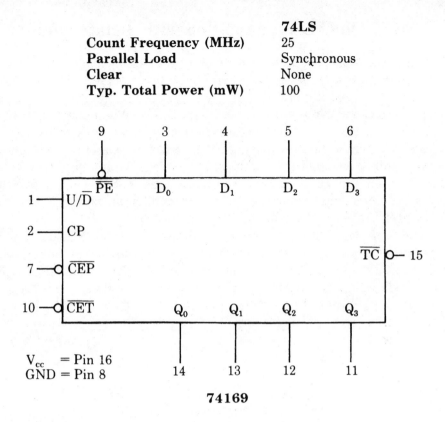

	74LS
Count Frequency (MHz)	25
Parallel Load	Synchronous
Clear	None
Typ. Total Power (mW)	100

V_{cc} = Pin 16
GND = Pin 8

74169

MODE SELECT — FUNCTION TABLE

OPERATING MODE	INPUTS						OUTPUTS	
	CP	U/$\overline{D}$	$\overline{CEP}$	$\overline{CET}$	$\overline{PE}$	D_n	Q_n	$\overline{TC}$
Parallel Load	↑	X	X	X	l	l	L	(b)
	↑	X	X	X	l	h	H	(b)
Count Up	↑	h	l	l	h	X	Count Up	(b)
Count Down	↑	l	l	l	h	X	Count Down	(b)
Hold (do nothing)	↑	X	h	X	h	X	q_n	(b)
	↑	X	X	h	h	X	q_n	H

H = High voltage level steady state.
L = Low voltage level steady state.
h = High voltage level one setup time prior to the low-to-high clock transition.
l = Low voltage level one setup time prior to the low-to-high clock transition.
X = Don't care.
q = Lower case letters indicate the state of the referenced output prior to the low-to-high clock transition.
↑ = Low-to-high clock transition.

NOTE
(b) The $\overline{TC}$ is low when $\overline{CET}$ is low and the counter is at Terminal Count. Terminal Count up is (HHHH) and Terminal Count Down is (LLLL).

74173 Quad D-type flip-flop with 3-state outputs

The '173 is a 4-bit parallel load register with clock enable control, 3-state buffered outputs and master reset. When the two clock-enable ($\overline{E}_1$ and $\overline{E}_2$) inputs are LOW, data on the D inputs are loaded into the register synchronously with the LOW-to-HIGH clock (CP) transition. When one or both $\overline{E}$ inputs are HIGH one setup time before the LOW-to-HIGH clock transition, the register will retain the previous data. The data inputs and clock-enable inputs are fully edge-triggered and must be stable only one setup time before the LOW-to-HIGH clock transition. The master reset (MR) is an active HIGH asynchronous input. When the MR is HIGH, all four flip-flops are reset (cleared) independently of any other input condition.

The 3-state output buffers are controlled by a 2-input NOR gate. When both output enable ($\overline{OE}_1$ and $\overline{OE}_2$) inputs are LOW, the data in the register is presented at the Q outputs. When one or both $\overline{OE}$ inputs is HIGH, the outputs are forced to a high-impedance off state. The 3-state output buffers are completely independent of the register operation. The OE transitions do not affect the clock and reset operations.

	74
Frequency (MHz)	25
Asynchronous Clear	High
Typ. Total Power (mW)	250

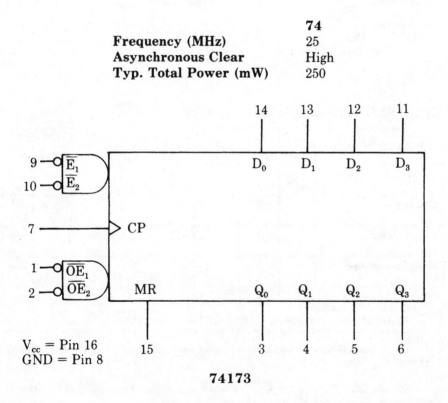

74173

MODE SELECT — FUNCTION TABLE

REGISTER OPERATING MODES	INPUTS					OUTPUTS
	MR	CP	$\overline{E}_1$	$\overline{E}_2$	D_n	Q_n (Register)
Reset (clear)	H	X	X	X	X	L
Parallel Load	L	↑	l	l	l	L
	L	↑	l	l	h	H
Hold (No change)	L	X	h	X	X	q_n
	L	X	X	h	X	q_n

3-STATE BUFFER OPERATING MODES	INPUTS			OUTPUTS
	Q_n (Register)	$\overline{OE}_1$	$\overline{OE}_2$	Q_0, Q_1, Q_2, Q_3
Read	L	L	L	L
	H	L	L	H
Disabled	X	H	X	(Z)
	X	X	H	(Z)

NOTES

H = High voltage level

h = High voltage level one setup time prior to the low-to-high clock transition

L = Low voltage level

l = Low voltage level one setup time prior to the low-to-high clock transition

q_n = Lower case letters indicate the state of the referenced input (or output) one setup time prior to the low-to-high clock transition

X = Don't care

(Z) = High impedance "off" state

↑ = Low-to-high transition

74174 Hex D flip-flop

The '174 has six edge-triggered D-type flip-flops with individual D inputs and Q outputs. The common buffered clock (CP) and master reset ($\overline{MR}$) inputs load and reset (clear) all flip-flops simultaneously. The register is fully edge-triggered. The state of each D input, one setup time before the LOW-to-HIGH clock transition, is transferred to the corresponding Q output of the flip-flop.

All outputs will be forced LOW independently of clock or data inputs by a low voltage level on the $\overline{MR}$ input. The device is useful for applications where the true output only is required and the clock and master reset are common to all storage elements.

	74S	74LS	74
Frequency (MHz)	75	30	25
Asynchronous Clear	Low	Low	Low
Typ. Total Power (mW)	450	80	225

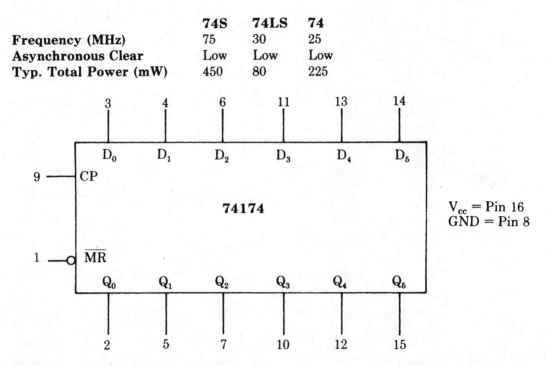

V_{cc} = Pin 16
GND = Pin 8

MODE SELECT-FUNCTION TABLE

OPERATING MODE	INPUTS			OUTPUTS
	$\overline{MR}$	CP	D_n	Q_n
Reset (clear)	L	X	X	L
Load "1"	H	↑	h	H
Load "0"	H	↑	l	L

H = High voltage level steady state
h = High voltage level one setup time prior to the low-to-high clock transition
L = Low voltage level steady state
l = Low voltage level one setup time prior to the low-to-high clock transition
X = Don't care
↑ = Low-to-high clock transition

74175 Quad D flip-flop

The 74175 is a quad edge-triggered D-type flip-flop with individual D inputs and both Q and $\overline{Q}$ outputs. The common buffered clock (CP) and master reset ($\overline{MR}$) inputs load and reset (clear) all flip-flops simultaneously.

The register is fully edge-triggered. The state of each D input, one setup time before the LOW-to-HIGH clock transition, is transferred to the corresponding Q output of the flip-flop.

All Q outputs will be forced LOW independently of clock or data inputs by a LOW voltage level on the $\overline{MR}$ input. This device is useful for applications, where both true and complement outputs are required, and the clock and master reset functions are common to all storage elements.

	74S	74LS	74
Frequency (MHz)	75	30	25
Asynchronous Clear	Low	Low	Low
Typ. Total Power (mW)	300	55	150

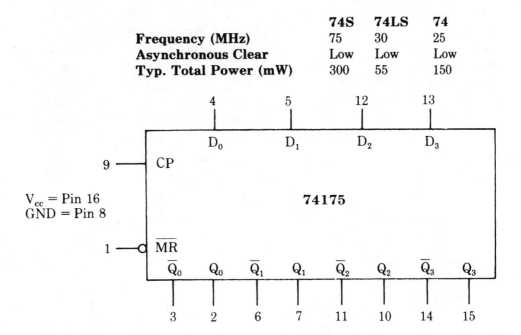

V_{cc} = Pin 16
GND = Pin 8

MODE SELECT—FUNCTION TABLE

OPERATING MODE	INPUTS			OUTPUTS	
	$\overline{MR}$	CP	D_n	Q_n	$\overline{Q}_n$
Reset (clear)	L	X	X	L	H
Load "1"	H	↑	h	H	L
Load "0"	H	↑	l	L	H

H = High voltage level steady state.
h = High voltage level one setup time prior to the low-to-high clock transition.
L = Low voltage level steady state.
l = Low voltage level one setup time prior to the low-to-high clock transition.
X = Don't care.
↑ = Low-to-high clock transition.

74180 9-Bit odd/even parity generator/checker

The '180 is a 9-bit parity generator or checker commonly used to detect errors in high-speed data transmissions or data-retrieval systems. Both even and odd parity enable inputs and parity outputs are available for generating or checking parity on 9-bits.

True active HIGH or true active LOW parity can be generated at both the even and odd outputs. True active HIGH parity is established with even parity enable input (P_E) set HIGH and the odd parity enable input (P_O) set LOW. True active LOW parity is established when P_E is LOW and P_O is HIGH. When both enable inputs are at the same logic level, both outputs will be forced to the opposite logic level.

Parity checking of a 9-bit word (8-bit plus parity) is possible by using the two enable inputs plus an inverter as the ninth data input. To check for true active HIGH parity, the ninth data input is tied to the P_O input and an inverter is connected between the P_O and P_E inputs. To check for true active LOW parity, the ninth data input is tied to the P_E input and an inverter is connected between the P_E and P_O inputs.

Expansion to larger word sizes is accomplished by serially cascading the '180 in 8-bit increments. The even and odd parity outputs of the first stage are connected to the corresponding P_E and P_O inputs, respectively, of the succeeding stage.

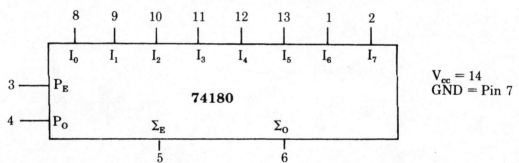

$V_{cc} = 14$
GND = Pin 7

Truth Table

INPUTS			OUTPUTS	
Number of HIGH Data Inputs (I_0–I_7)	P_E	P_O	Σ_E	Σ_O
Even	H	L	H	L
Odd	H	L	L	H
Even	L	H	L	H
Odd	L	H	H	L
X	H	H	L	L
X	L	L	H	H

H = High voltage level
L = Low voltage level
X = Don't care

	74
Typ. Delay Time (ns)	35
Typ. Total Power (mW)	170

136

74181 4-Bit arithmetic logic unit

The '181 is a 4-bit high-speed parallel arithmetic logic unit (ALU). Controlled by the four function select inputs ($S_0 \ldots S_3$) and the mode control input (M), it can perform all the 16 possible logic operations or 16 different arithmetic operations on active HIGH or active LOW operands. The function table lists these operations.

When the mode control input (M) is HIGH, all internal carries are inhibited and the device performs logic operations on the individual bits as listed. When the mode control input is LOW, the carries are enabled and the device performs arithmetic operations on the two 4-bit words. The device incorporates full internal carry look ahead and provides for either ripple carry between devices using the C_{n+4} output, or for carry look ahead between packages using the signals $\overline{P}$ (Carry Propagate) and $\overline{G}$ (Carry Generate). $\overline{P}$ and $\overline{G}$ are not affected by carry in. When speed requirements are not stringent, it can be used in a simple ripple-carry mode by connecting the carry output ($C_n + 4$) signal to the carry input (C_n) of the next unit. For high-speed operation, the device is used in conjunction with the '182 carry-look-ahead circuit. One carry-look-ahead package is required for each group of four '181 devices. Carry look ahead can be provided at various levels and offers high-speed capability over extremely long word lengths.

The A=B output from the device goes HIGH when all four $\overline{F}$ outputs are HIGH and can be used to indicate logic equivalence over four bits when the unit is in the subtract mode. The A=B output is an open collector and can be wired-AND with other A=B outputs to give a comparison for more than four bits. The A=B signal can be used with the C_{n+4} signal to indicate A>B and A<B.

The function table lists the arithmetic operations that are performed without a carry-in. An incoming carry adds a one to each operation. Thus, select code LHHL generates A minus B minus 1 (2s complement notation) without a carry-in and generates A minus B when a carry is applied. Because subtraction is actually performed by complementary addition (1s complement), a carry-out means borrow; thus, a carry is generated when there is no underflow and no carry is generated when there is underflow.

This device can be used with either active LOW inputs producing active LOW outputs or with active HIGH inputs producing active HIGH outputs. For either case, the table lists the operations that are performed to the operands labeled inside the logic symbol.

	74
Typ. Carry Time (ns)	12.5
Typ. Add Time (ns)	24
Typ. Total Power (mW)	455

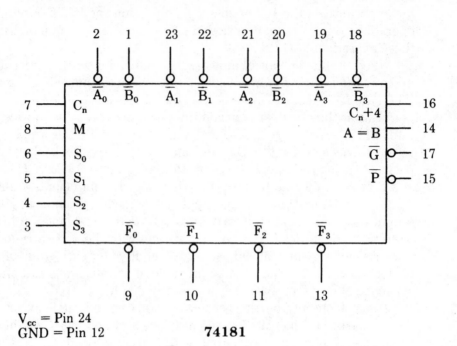

V_{cc} = Pin 24
GND = Pin 12

74181

MODE SELECT - FUNCTION TABLE

MODE SELECT INPUTS				ACTIVE HIGH INPUTS & OUTPUTS	
S_3	S_2	S_1	S_0	LOGIC (M = H)	ARITHMETIC** (M = L) (C_n = H)
L	L	L	L	$\overline{A}$	A
L	L	L	H	$\overline{A + B}$	$A + B$
L	L	H	L	$\overline{A}B$	$A + \overline{B}$
L	L	H	H	Logical 0	minus 1
L	H	L	L	$\overline{AB}$	A plus $A\overline{B}$
L	H	L	H	$\overline{B}$	(A + B) plus $A\overline{B}$
L	H	H	L	$A \oplus B$	A minus B minus 1
L	H	H	H	$A\overline{B}$	AB minus 1
H	L	L	L	$\overline{A} + B$	A plus AB
H	L	L	H	$\overline{A \oplus B}$	A plus B
H	L	H	L	B	$(A + \overline{B})$ plus AB
H	L	H	H	AB	AB minus 1
H	H	L	L	Logical 1	A plus A*
H	H	L	H	$A + \overline{B}$	$(A + \overline{B})$ plus A
H	H	H	L	$A + B$	$(A + \overline{B})$ plus A
H	H	H	H	A	A minus 1

*Each bit is shifted to the next more significant position
**Arithmetic operations expressed in 2s complement notation

L = Low voltage level
H = High voltage level

MODE SELECT INPUTS				ACTIVE LOW INPUTS & OUTPUTS	
S_3	S_2	S_1	S_0	LOGIC (M = H)	ARITHMETIC** (M = L) (C_n = L)
L	L	L	L	$\overline{A}$	A minus 1
L	L	L	H	$\overline{AB}$	$A\overline{B}$ minus 1
L	L	H	L	$\overline{A} + B$	$A\overline{B}$ minus 1
L	L	H	H	Logical 1	minus 1
L	H	L	L	$\overline{A} + B$	A plus $(A + \overline{B})$
L	H	L	H	$\overline{B}$	AB plus $(A + \overline{B})$
L	H	H	L	$\overline{A \oplus B}$	A minus B minus 1
L	H	H	H	$A + \overline{B}$	$A + \overline{B}$
H	L	L	L	$\overline{A}B$	A plus (A + B)
H	L	L	H	$A \oplus B$	A plus B
H	L	H	L	B	$A\overline{B}$ plus (A + B)
H	L	H	H	A + B	A + B
H	H	L	L	Logical 0	A plus A*
H	H	L	H	$A\overline{B}$	$A\overline{B}$ plus A
H	H	H	L	AB	$A\overline{B}$ plus A
H	H	H	H	A	A

*Each bit is shifted to the next more significant position
**Arithmetic operations expressed in 2s complement notation

L = Low voltage level
H = High voltage level

74182 Carry-look-ahead generator

The 74182 carry-look-ahead generator accepts up to four pairs of active low carry propogate ($\overline{P}_0$, $\overline{P}_1$, $\overline{P}_2$, and $\overline{P}_3$) and carry generate ($\overline{G}_0$, $\overline{G}_1$, $\overline{G}_2$, and $\overline{G}_3$) signals and an active HIGH carry input (C_n). From these inputs, the 74182 provides anticipated active HIGH carries (C_{n+x}, C_{n+y}, and C_{n+z}) across four groups of binary adders. The 74182 also has active LOW carry propogate ($\overline{P}$) and carry generate ($\overline{G}$) outputs that can be used for further levels of look-ahead.

Also, the 74182 can be used with binary ALUs in an active LOW or active HIGH input operand mode. The connections to and from the ALU to the carry-look-ahead generator are identical in both cases.

Truth Table

INPUTS									OUTPUTS				
C_n	$\overline{G}_0$	$\overline{P}_0$	$\overline{G}_1$	$\overline{P}_1$	$\overline{G}_2$	$\overline{P}_2$	$\overline{G}_3$	$\overline{P}_3$	C_{n+x}	C_{n+y}	C_{n+z}	$\overline{G}$	$\overline{P}$
X	H	H							L				
L	H	X							L				
X	L	X							H				
H	X	L							H				
X	X	X	H	H						L			
X	H	H	H	X						L			
L	H	X	H	X						L			
X	X	X	L	X						H			
X	L	X	X	L						H			
H	X	L	X	L						H			
X	X	X	X	X	H	H					L		
X	X	X	H	H	H	X					L		
X	H	H	H	X	H	X					L		
L	H	X	H	X	H	X					L		
X	X	X	X	X	L	X					H		
X	X	X	L	X	X	L					H		
X	L	X	X	L	X	L					H		
H	X	L	X	L	X	L					H		
	X	X	X	X	X	X	H	H				H	
	X	X	X	X	H	H	H	X				H	
	X	X	H	H	H	X	H	X				H	
	H	H	H	X	H	X	H	X				H	
	X	X	X	X	X	X	L	X				L	
	X	X	X	X	L	X	X	L				L	
	X	X	L	X	X	L	X	L				L	
	L	X	X	L	X	L	X	L				L	
		H		X		X		X					H
		X		H		X		X					H
		X		X		H		X					H
		X		X		X		H					H
		L		L		L		L					L

H = High voltage level
L = Low voltage level
X = Don't care

	74S	74
Typ. Carry Time (ns)	7	13
Typ. Total Power (mW)	260	180

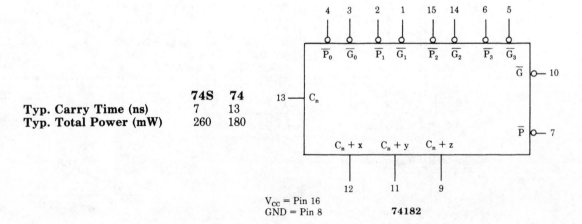

V_{CC} = Pin 16
GND = Pin 8

74182

74190 Presettable BCD/decade up/down counter

The '190 is an asynchronously presettable up/down BCD decade counter. It contains four master-slave flip-flops with internal gating and steering logic to provide asynchronous preset and synchronous countup and countdown operations.

Asynchronous parallel load capability permits the counter to be preset to any desired number. Information present on the parallel-data inputs (D_0 through D_3) is loaded into the counter and appears on the outputs when the parallel-load ($\overline{PL}$) input is LOW. As indicated in the mode-select table, this operation overrides the counting function.

Counting is inhibited by a HIGH level on the count-enable ($\overline{CE}$) input. When $\overline{CE}$ is LOW, internal state changes are initiated synchronously by the LOW-to-HIGH transition of the clock input. The up/down ($\overline{U}/D$) input signal determines the direction of counting, as indicated in the mode-select table. The $\overline{CE}$ input can go LOW when the clock is in either state; however, the LOW-to-HIGH $\overline{CE}$ transition must occur only while the clock is HIGH. Also, the $\overline{U}/D$ input should be changed only when either $\overline{CE}$ or CP is HIGH.

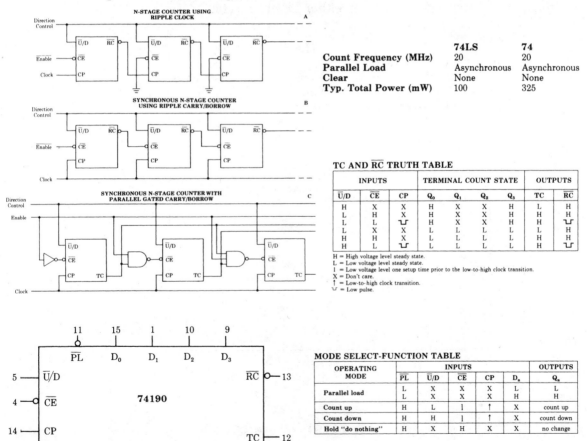

	74LS	74
Count Frequency (MHz)	20	20
Parallel Load	Asynchronous	Asynchronous
Clear	None	None
Typ. Total Power (mW)	100	325

TC AND $\overline{RC}$ TRUTH TABLE

INPUTS			TERMINAL COUNT STATE				OUTPUTS	
$\overline{U}/D$	$\overline{CE}$	CP	Q_0	Q_1	Q_2	Q_3	TC	$\overline{RC}$
H	X	X	H	X	X	H	L	H
L	H	X	H	X	X	H	H	H
L	L	⊔	H	X	X	H	H	⊔
L	X	X	L	L	L	L	L	H
H	H	X	L	L	L	L	H	H
H	L	⊔	L	L	L	L	H	⊔

H = High voltage level steady state.
L = Low voltage level steady state.
l = Low voltage level one setup time prior to the low-to-high clock transition.
X = Don't care.
↑ = Low-to-high clock transition.
⊔ = Low pulse.

MODE SELECT-FUNCTION TABLE

OPERATING MODE	INPUTS					OUTPUTS
	$\overline{PL}$	$\overline{U}/D$	$\overline{CE}$	CP	D_n	Q_n
Parallel load	L	X	X	L	L	L
	L	X	X	X	H	H
Count up	H	L	l	↑	X	count up
Count down	H	H	l	↑	X	count down
Hold "do nothing"	H	X	H	X	X	no change

V_{CC} = Pin 16
GND = Pin 8

74191 Presettable 4-bit binary up/down counter

The '191 is an asynchronously presettable up/down 4-bit binary counter. It contains four master-slave flip-flops with internal gating and steering logic to provide asynchronous preset and synchronous countup and countdown operation.

Asynchronous parallel load capability permits the counter to be preset to any desired number. Information present on the parallel data inputs (D_0 through D_3) is loaded into the counter and appears on the outputs when the parallel-load ($\overline{PL}$) input is LOW. As indicated in the mode-select table, this operation overrides the counting function.

Counting is inhibited by a HIGH level on the count-enable ($\overline{CE}$) input. When $\overline{CE}$ is LOW, internal state changes are initiated synchronously by the LOW-to-HIGH transition of the clock input. The up/down ($\overline{U}/D$) input signal determines the direction of counting as indicated in the mode-select table. The $\overline{CE}$ input can go LOW when the clock is in either state; however, the LOW-to-HIGH $\overline{CE}$ transition must occur only while the clock is HIGH. Also, the $\overline{U}/D$ input should be changed only when either $\overline{CE}$ or CP is HIGH.

	74LS	**74**
Count Frequency (MHz)	20	20
Parallel Load	Asynchronous	Asynchronous
Clear	None	None
Typ. Total Power (mW)	90	325

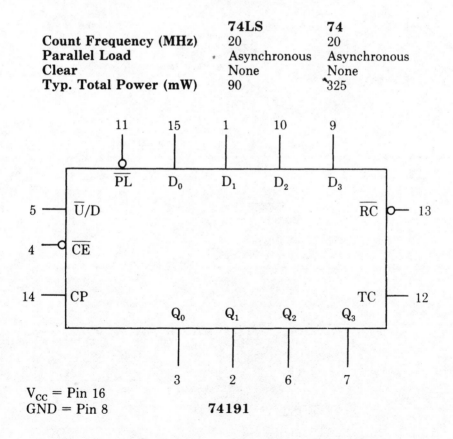

V_{CC} = Pin 16
GND = Pin 8

74191

MODE SELECT-FUNCTION TABLE

OPERATING MODE	INPUTS					OUTPUTS
	$\overline{PL}$	$\overline{U}/D$	$\overline{CE}$	CP	D_n	Q_n
Parallel load	L	X	X	X	L	L
	L	X	X	X	H	H
Count up	H	L	l	↑	X	count up
Count down	H	H	l	↑	X	count down
Hold "do nothing"	H	X	H	X	X	no change

TC AND $\overline{RC}$ TRUTH TABLE

INPUTS			TERMINAL COUNT STATE				OUTPUTS	
$\overline{U}/D$	$\overline{CE}$	CP	Q_0	Q_1	Q_2	Q_3	TC	$\overline{RC}$
H	X	X	H	H	H	H	L	H
L	H	X	H	H	H	H	H	H
L	L	⊔	H	H	H	H	H	⊔
L	X	X	L	L	L	L	L	H
H	H	X	L	L	L	L	H	H
H	L	⊔	L	L	L	L	H	⊔

H = High voltage level steady state.
L = Low voltage level steady state.
l = Low voltage level one setup time prior to the low-to-high clock transition.
X = Don't care.
↑ = Low-to-high clock transition.
⊔ = Low pulse.

74192 Presettable BCD decade up/down counter

The '192 is an asynchronously presettable up/down (reversible) internal gating and steering logic to provide asynchronous master reset (clear), parallel load, and synchronous countup and countdown operations.

Each flip-flop contains JK feedback from slave to master such that a LOW-to-HIGH transition on the clock inputs causes the Q outputs to change state synchronously. A LOW-to-HIGH transition on the countdown-clock-pulse (CP_D) input will decrease the count by one, while a similar transition on the countup-clock-pulse (CP_U) input will advance the count by one. One clock should be held HIGH while counting with the other because the circuit will either count by twos or not at all, depending on the state of the first flip-flop, which cannot toggle as long as either clock input is LOW. Applications requiring reversible operation must make the reversing decision while the activating clock is HIGH to avoid erroneous counts.

The terminal countup ($\overline{TC_U}$) and terminal countdown ($\overline{TC_D}$) outputs are normally HIGH. When the circuit has reached the maximum count state of nine, the next HIGH-to-LOW transition of CP_U will cause $\overline{TC_U}$ to go LOW. $\overline{TC_U}$ will stay LOW until CP_U goes HIGH again, duplicating the countup clock, although delayed by two gate delays. Likewise, the $\overline{TC_D}$ output will go LOW when the circuit is in the zero state and the CP_D goes LOW. The $\overline{TC}$ outputs can be used as the clock-input signals to the next higher-order circuit in a multistage counter, because they duplicate the clock waveforms. Multistage counters will not be fully synchronous, because there is a two-gate delay time difference added for each stage that is added.

The counter can be preset by the asynchronous parallel load capability of the circuit. Information present on the parallel-data inputs (D_0 through D_3) is loaded into the counter and appears on the outputs, regardless of the conditions of the clock inputs when the parallel-load ($\overline{PL}$) input is LOW. A HIGH level on the master-reset (MR) input will disable the parallel-load gates, override both clock inputs and set all Q outputs LOW. If one of the clock inputs is LOW during and after a reset or load operation, the next LOW-to-HIGH transition of that clock will be interpreted as a legitimate signal and will be counted.

$$\overline{TC}_U = Q_0 \cdot Q_3 \cdot CP_U$$
$$\overline{TC}_D = Q_0 \cdot Q_1 \cdot Q_2 \cdot Q_3 \cdot CP_D$$

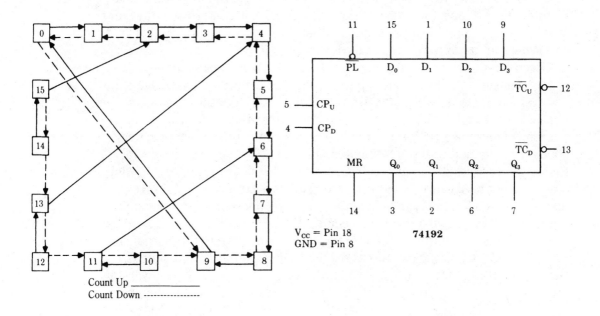

Count Up _____
Count Down ------------------

V_{CC} = Pin 18
GND = Pin 8

74192

MODE SELECT-FUNCTION TABLE

OPERATING MODE	INPUTS								OUTPUTS					
	MR	$\overline{PL}$	CP_U	CP_D	D_0,	D_1,	D_2,	D_3	Q_0,	Q_1,	Q_2,	Q_3	$\overline{TC}_U$	$\overline{TC}_D$
Reset (clear)	H	X	X	L	X	X	X	X	L	L	L	L	H	L
	H	X	X	H	X	X	X	X	L	L	L	L	H	H
Parallel load	L	L	X	L	L	L	L	L	L	L	L	L	H	L
	L	L	X	H	L	L	L	L	L	L	L	L	H	H
	L	L	L	X	H	X	X	H	$Q_a = D_a$				L	H
	L	L	H	X	H	X	X	H	$Q_a = D_a$				H	H
Count up	L	H	↑	H	X	X	X	X	Count up				H[(b)]	H
Count down	L	H	H	↑	X	X	X	X	Count down				H	H[(c)]

H = High voltage level
L = Low voltage level
X = Don't care
↑ = Low-to-high clock transition

NOTES
(b) $\overline{TC}_U$ = CP_U at terminal count up (HLLH)
(c) $\overline{TC}_D$ = CP_D at terminal count down (LLLL)

	74LS	**74**	**74L**
Count Frequency (MHz)	25	20	6
Parallel Load	Asynchronous	Asynchronous	Asynchronous
Clear	Asynchronous- High	Asynchronous- High	Asynchronous- High
Typ. Total Power (mW)	85	325	40

74193 Presettable 4-bit binary up/down counter

The '193 is an asynchronously presettable, up/down (reversible) 4-bit binary counter. It contains four master-slave flip-flops with internal gating and steering logic to provide asynchronous master reset (dear) and parallel load, and synchronous countup and countdown operations.

Each flip-flop contains JK feedback from slave to master such that a LOW-to-HIGH transition on the clock inputs causes the Q outputs to change state synchronously. A LOW-to-HIGH transition on the countdown-clock-pulse (CP_D) input will decrease the count by one, while similar transition on the countup-clock-pulse (CP_U) input will advance the count by one. One clock should be held HIGH while counting with the other, because the circuit will either count by twos or not at all, depending on the state of the first flip-flop, which cannot toggle as long as either clock input is LOW. Applications requiring reversible operation must make the reversing decision while the activating clock is HIGH to avoid erroneous counts.

The terminal countup ($\overline{TC_U}$) and terminal countdown ($\overline{TC_D}$) outputs are normally HIGH. When the circuit has reached the maximum count state of 15, the next HIGH-to-LOW transition of CP_U will cause $\overline{TC_U}$ to go LOW. $\overline{TC_U}$ will stay LOW until CP_U goes HIGH again, duplicating the countup clock, although delayed by two gate delays. Likewise, the $\overline{TC_D}$ output will go LOW when the circuit is in the zero state and the CP_D goes LOW. The $\overline{TC}$ outputs can be used as the clock-input signals to the next higher order circuit in a multistage counter, because they duplicate the clock waveforms. Multistage counters will not be fully synchronous for there is a two-gate delay time difference added for each stage that is added.

The counter can be preset by the asynchronous parallel load capability of the circuit. Information present on the parallel-data inputs (D_0 through D_3) is loaded into the counter and appears on the outputs, regardless of the conditions of the clock inputs when the parallel-load ($\overline{PL}$) input is LOW. A HIGH level on the master-reset (MR) input will disable the parallel-load gates, override both clock inputs and set all Q outputs LOW. If one of the clock inputs is LOW during and after a reset or load operation, the next LOW-to-HIGH transition of that clock will be interpreted as a legitimate signal and will be counted.

LOGIC EQUATIONS FOR TERMINAL COUNT

$$\overline{TC}_U = Q_0 \cdot Q_1 \cdot Q_2 \cdot Q_3 \cdot CP_U$$
$$\overline{TC}_D = Q_0 \cdot Q_1 \cdot Q_2 \cdot Q_3 \cdot CP_D$$

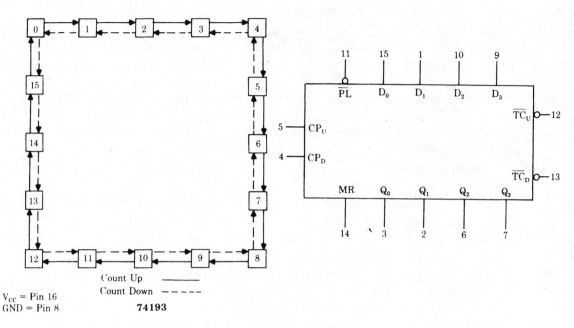

Count Up ————
Count Down — — — —

V_{CC} = Pin 16
GND = Pin 8

74193

MODE SELECT-FUNCTION TABLE

OPERATING MODE	INPUTS								OUTPUTS					
	MR	$\overline{PL}$	CP_U	CP_D	D_0,	D_1,	D_2,	D_3	Q_0,	Q_1,	Q_2,	Q_3	$\overline{TC}_U$	$\overline{TC}_D$
Reset (clear)	H	X	X	L	X	X	X	X	L	L	L	L	H	L
	H	X	X	H	X	X	X	X	L	L	L	L	H	H
Parallel load	L	L	X	L	L	L	L	L	L	L	L	L	H	L
	L	L	X	H	L	L	L	L	L	L	L	L	H	H
	L	L	L	X	H	H	H	H	H	H	H	H	L	H
	L	L	H	X	H	H	H	H	H	H	H	H	H	H
Count up	L	H	↑	H	X	X	X	X	Count up				H[b]	H
Count down	L	H	H	↑	X	X	X	X	Count down				H	H[c]

H = High voltage level
L = Low voltage level
X = Don't care
↑ = Low-to-high clock transition

NOTES
(b) $\overline{TC}_U = CP_U$ at terminal count up (HHHH)
(c) $\overline{TC}_D = CP_D$ at terminal count down (LLLL)

	74LS	**74**	**74L**
Count Frequency (MHz)	25	20	6
Parallel Load	Asynchronous	Asynchronous	Asynchronous
Clear	Asynchronous-High	Asynchronous-High	Asynchronous-High
Typ. Total Power (mW)	85	325	40

74194 4-Bit directional universal shift register

The functional characteristics of the '194 4-bit bidirectional shift register are indicated in the logic diagram and truth table. The register is full-synchronous, with all operations taking place in less than 20 nanoseconds (typical) for the 74 and 74LS, and 12 ns (typical) for 74S, making the device especially useful for implementing very high-speed CPUs, or for memory-buffer registers.

The '194 design has special logic features that increase the application range. The synchronous operation of the device is determined by two mode-select inputs, S_0 and S_1. As shown in the mode-select table, data can be entered and shifted from left to right (shift right, $Q_0 \rightarrow Q_1$, etc.) or right to left (shift left, $Q_3 \rightarrow Q_2$, etc.) or parallel data can be entered, loading all four bits of the register simultaneously. When both S_0 and S_1 are LOW, existing data is retained in a hold (do nothing) mode. The interfering with parallel-load operation.

Mode-select and data inputs on the 74S194 and 74LS194A are edge-triggered, responding only to the LOW-to-HIGH transition of the clock (CP). Therefore, the only timing restriction is that the mode-control and selected data inputs must be stable one setup time prior to the positive transition of the clock pulse. The mode select inputs of the 74194 are gated with the clock and should be changed from HIGH-to-LOW only while the clock input is HIGH.

The four parallel data inputs (D_0 through D_3) are D-type inputs. Data appearing on D_0 through D_3 inputs when S_0 and S_1 are HIGH are transferred to the Q_0 through Q_3 outputs respectively, following the next LOW-to-HIGH transition of the clock. When LOW, the asynchronous master reset ($\overline{MR}$) overrides all other input conditions and forces the Q outputs LOW.

	74S	74LS	74
Shift Frequency (MHz)	70	25	25
Serial Data Input	D	D	D
Asynchronous Clear	Low	Low	Low
Shift-Right Mode	Yes	Yes	Yes
Shift-Left Mode	Yes	Yes	Yes
Load	Yes	Yes	Yes
Hold	Yes	Yes	Yes
Typ. Total Power (mW)	450	75	195

V_{CC} = Pin 16
GND = Pin 8

74194

MODE SELECT-FUNCTION TABLE

OPERATING MODE	INPUTS							OUTPUTS			
	CP	$\overline{MR}$	S_1	S_0	D_{SR}	D_{SL}	D_n	Q_0	Q_1	Q_2	Q_3
Reset (clear)	X	L	X	X	X	X	X	L	L	L	L
Hold (do nothing)	X	H	l(b)	l(b)	X	X	X	q_0	q_1	q_2	q_3
Shift Left	↑	H	h	l(b)	X	l	X	q_1	q_2	q_3	L
	↑	H	h	l(b)	X	h	X	q_1	q_2	q_3	H
Shift Right	↑	H	l(b)	h	l	X	X	L	q_0	q_1	q_2
	↑	H	l(b)	h	h	X	X	H	q_0	q_1	q_2
Parallel Load	↑	H	h	h	X	X	d_n	d_0	d_1	d_2	d_3

H = High voltage level
h = High voltage level one setup time prior to the low-to-high clock transition
L = Low voltage level
l = Low voltage level one setup time prior to the low-to-high clock transition
d_n (q_n) = Lower case letters indicate the state of the referenced input (or output) one setup time prior to the low-to-high clock transition
X = Don't care
↑ = Low-to-high clock transition

NOTES
(b) The high-to-low transition of the S_0 and S_1 inputs on the 54/74194 should only take place while CP is high for conventional operation.

149

74195 4-Bit parallel access shift register

The functional characteristics of the '195 four-bit parallel-access register are indicated in the logic diagram and function table. The device is useful in a wide variety of shifting, counting and storage applications. It performs serial, parallel, serial-to-parallel, or parallel-to-serial data transfers at very high speeds.

The '195 operates on two primary modes: shift right ($Q_0 \rightarrow Q_1$) and parallel load, which are both controlled by the state of the parallel enable ($\overline{PE}$) input. Serial data enters the first flip-flop (Q_0) via the J and $\overline{K}$ inputs when the $\overline{PE}$ input is HIGH, and is shifted one bit in the direction $Q_0 \rightarrow Q_1 \rightarrow Q_2 \rightarrow Q_3$ following each LOW-to-HIGH clock transition. The J and $\overline{K}$ inputs provide the flexibility of the JK-type input for special applications and, by tying the two pins together, the simple D-type input for general applications. The device appears as four common clocked D flip-flops when the $\overline{PE}$ input is LOW. After the LOW-to-HIGH clock transition, data on the parallel inputs (D_0 through D_3) is transferred to the respective Q_0 through Q_3 outputs. Shift left operation ($Q_3 \rightarrow Q_2$) can be achieved by tying the Q_n outputs to the D_{n-1} inputs and holding the $\overline{PE}$ input LOW.

All parallel and serial data transfers are synchronous, occurring after each LOW-to-HIGH clock transition. The '195 utilizes edge-triggering; therefore, there is no restriction on the activity of the J, $\overline{K}$, D_n, and $\overline{PE}$ inputs for logic operation, other than the setup and release time requirements.

A LOW on the asynchronous master reset ($\overline{MR}$) input sets all Q outputs LOW, independent of any other input condition. The $\overline{MR}$ on the 54/74195 is gated with the clock. Therefore, the LOW-to-HIGH $\overline{MR}$ transition should only occur while the clock is LOW to avoid false clocking of the 74195.

	74S	74	74LS
Shift Frequency (MHz)	70	30	30
Serial Data Input	J-$\overline{K}$	J-$\overline{K}$	J-$\overline{K}$
Asynchronous Clear	Low	Low	Low
Shift-Right Mode	Yes	Yes	Yes
Shift-Left Mode	No	No	No
Load	Yes	Yes	Yes
Hold	No	No	No
Typ. Total Power (mW)	375	195	70

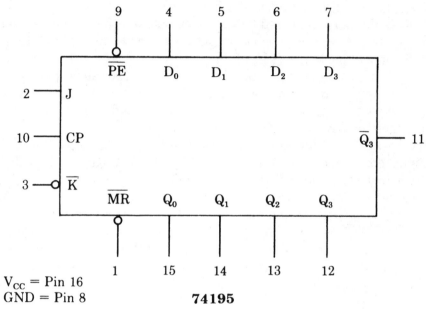

V_{CC} = Pin 16
GND = Pin 8

74195

MODE SELECT-FUNCTION TABLE

OPERATING MODES	INPUTS						OUTPUTS				
	$\overline{MR}$	CP	$\overline{PE}$	J	$\overline{K}$	D_n	Q_0	Q_1	Q_2	Q_3	$\overline{Q_3}$
Asynchronous Reset	L	X	X	X	X	X	L	L	L	L	H
Shift, Set First Stage	H	↑	h	h	h	X	H	q_0	q_1	q_2	$\overline{q_2}$
Shift, Reset First Stage	H	↑	h	l	l	X	L	q_0	q_1	q_2	$\overline{q_2}$
Shift, Toggle First Stage	H	↑	h	h	l	X	$\overline{q_0}$	q_0	q_1	q_2	$\overline{q_2}$
Shift, Retain First Stage	H	↑	h	l	h	X	q_0	q_0	q_1	q_2	$\overline{q_2}$
Parallel Load	H	↑	l	X	X	d_n	d_0	d_1	d_2	d_3	$\overline{d_3}$

H = High voltage level.
L = Low voltage level.
X = Don't care.
l = Low voltage level one setup time prior to the low-to-high clock transition.
h = High voltage level one setup time prior to the low-to-high clock transition.
d_n (q_n) = Lower case letters indicate the state of the referenced input (or output) one setup time prior to the low-to-high clock transition.
↑ = Low-to-high clock transition.

74196 Presettable decade ripple counter

The 74196 is an asynchronously presettable decade ripple counter that is partitioned into divide-by-two and divide-by-five sections, with each section having a separate clock input. State changes are initiated in the counting modes by the HIGH-to-LOW transition of the clock inputs; however, state changes of the Q outputs do not occur simultaneously because of the internal ripple delays. When using external logic to decode the Q outputs, circuit designers should remember that unequal delays can lead to spikes. Thus, a decoded signal should not be used as a strobe or clock.

The Q_0 flip-flip is triggered by the $\overline{CP_0}$ input, and the $\overline{CP_1}$ input triggers the divide-by-five section. The Q_0 output is designed and specified to drive the rated fanout plus the CP_1 input.

As indicated in the count sequence tables, the 74196 can be connected to operate in two different count sequences. The device counts in the BCD (8, 4, 2, 1) sequence with the input connected to CP_0 and with Q_0 driving CP_1. Q_0 becomes the low frequency output and has a 50% duty cycle waveform (quarter wave) with the input connected to $\overline{CP_1}$ and Q_3 driving $\overline{CP_0}$. The maximum counting rate is reduced in the biquinary configuration because of the interstage gating delay within the divide-by-five section.

This device has an asynchronous active LOW master reset input ($\overline{MR}$) which overrides all other inputs and forces all outputs LOW. The counter is also asynchronously presettable. A LOW on the parallel-load input ($\overline{PL}$) overrides the clock inputs and loads the data from the parallel-data (D_0 through D_3) inputs into the flip-flops. The counter acts as a transparent latch while the PL is LOW, and any change in the D_n inputs will be reflected in the outputs.

	74	74LS
Count Frequency (MHz)	50	30
Parallel Load	Yes	Yes
Clear	Low	Low
Typ. Total Power (mW)	240	60

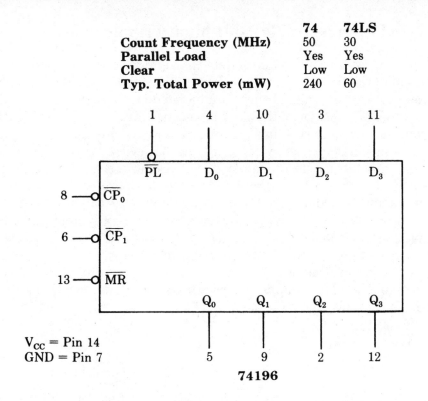

V_{CC} = Pin 14
GND = Pin 7

74196

MODE SELECT-FUNCTION TABLE

OPERATING MODE	INPUTS				OUTPUTS
	$\overline{MR}$	$\overline{PL}$	$\overline{CP}$	D_n	Q_n
Reset (Clear)	L	X	X	X	L
Parallel Load	H	L	X	L	L
	H	L	X	H	H
Count	H	H	↓	X	count

H = High voltage level
L = Low voltage level
X = Don't care
↓ = High-to-low clock transition

COUNT SEQUENCES

BCD DECADE[b]					BIQUINARY[c]				
COUNT	Q_3	Q_2	Q_1	Q_0	COUNT	Q_0	Q_3	Q_2	Q_1
0	L	L	L	L	0	L	L	L	L
1	L	L	L	H	1	L	L	L	H
2	L	L	H	L	2	L	L	H	L
3	L	L	H	H	3	L	L	H	H
4	L	H	L	L	4	L	H	L	L
5	L	H	L	H	5	H	L	L	L
6	L	H	H	L	6	H	L	L	H
7	L	H	H	H	7	H	L	H	L
8	H	L	L	L	8	H	L	H	H
9	H	L	L	H	9	H	H	L	L

NOTES

(b) Input applied to CP_0; Q_0 connected to CP_1
(c) Input applied to CP_1; Q_3 connected to CP_0

74197 Presettable 4-bit binary ripple counter

The '197 is an asynchronously presettable binary ripple counter that is partitioned into divide-by-2 and divide-by-8 sections, with each section having a separate clock input. State changes are initiated in the counting modes by the HIGH-to-LOW transition of the clock inputs; however, stage changes of the Q outputs do not occur simultaneously because of the internal ripple delays. Designers should remember, when using external logic to decode the Q outputs, that the unequal delays can lead to decoding spikes, and thus a decoded signal should not be used as a strobe or clock. The Q_0 output is designed and specified to drive the rated fanout plus the CP_1 input.

The device has an asynchronous active LOW master-reset input ($\overline{MR}$) that overrides all other inputs and forces all outputs LOW. The counter is also asynchronously presettable. A LOW on the parallel-load input ($\overline{PL}$) overrides the clock inputs and loads the data from parallel-data (D_0 through D_3) inputs into the flip-flops. The counter acts as a transparent latch while the $\overline{PL}$ is LOW, and any change in the D_n inputs will be reflected in the outputs.

LOGIC SYMBOL

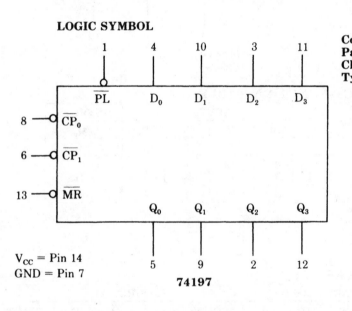

V_{CC} = Pin 14
GND = Pin 7

74197

	74	74LS
Count Frequency (MHz)	50	30
Parallel Load	Yes	Yes
Clear	Low	Low
Typ. Total Power (mW)	240	60

COUNT SEQUENCE

COUNT	4-Bit Binary[b]			
	Q_3	Q_2	Q_1	Q_0
0	L	L	L	L
1	L	L	L	H
2	L	L	H	L
3	L	L	H	H
4	L	H	L	L
5	L	H	L	H
6	L	H	H	L
7	L	H	H	H
8	H	L	L	L
9	H	L	L	H
10	H	L	H	L
11	H	L	H	H
12	H	H	L	L
13	H	H	L	H
14	H	H	H	L
15	H	H	H	H

NOTE
(b) Q_0 connected to $\overline{CP_1}$; input applied to $\overline{CP_0}$

MODE SELECT-FUNCTION TABLE

OPERATING MODE	INPUTS				OUTPUTS
	$\overline{MR}$	$\overline{PL}$	$\overline{CP}$	D_n	Q_n
Reset (Clear)	L	X	X	X	L
Parallel Load	H	L	X	L	L
	H	L	X	H	H
Count	H	H	↓	X	count

H = High voltage level
L = Low voltage level
X = Don't care
↓ = High-to-low clock transition

154

74240 Octal inverter/line driver with tri-state outputs

The 74240 contains eight tri-state inverter/buffers or line drivers. This chip was designed for such applications as bus-oriented transmitters/receivers, clock drivers, and memory address drivers. Although the chip's functions are fairly simple and could be easily duplicated with other available devices, the use of the 74240 will usually result in a lower overall parts count.

The eight inverter stages on the 74240 are combined into two sets of four each. Each group can be independently controlled for the tri-state function. All four inverters in a group can be placed in the high-impedance output state via a HIGH signal on the appropriate $\overline{O_e}$ input. The inverters in the group are functionally enabled by a LOW signal at the $\overline{O_e}$ input.

The input circuitry includes clamp diodes to limit high-speed termination effects. The outputs can sink up to 64 mA, and the source current is rated for 15 mA.

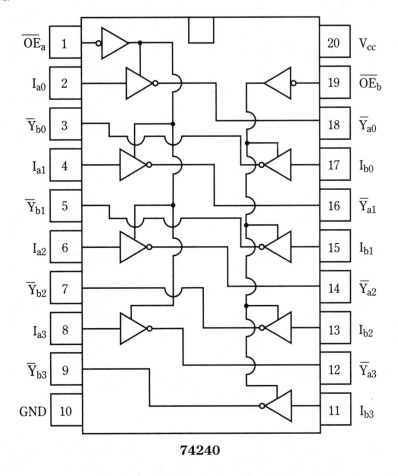

74240

74240 Truth Table

O_{ea}	I_a	O_{eb}	I_b	Y_a	Y_b
0	0	0	0	1	1
0	0	0	1	1	0
0	0	1	x	1	Z
0	1	1	x	0	Z
0	1	0	0	0	1
1	x	0	0	Z	1
1	x	0	1	Z	0
1	x	1	x	Z	Z

x = Don't care
Z = High impedance

74241 Octal inverter/line driver with tri-state outputs

The 74241 contains eight tri-state buffers or line drivers. This chip was designed for applications, such as bus-oriented transmitters/receivers, clock drivers, and memory address drivers. Although the chip's functions are fairly simple and could be easily duplicated with other available devices, the use of the 74240 will usually result in a lower overall parts count.

The eight buffer stages on the 74241 are combined into two sets of four each. Each group can be independently controlled for the tri-state function. All four buffers in a group can be placed in the high-impedance output state via a HIGH signal on the appropriate $\overline{O_e}$ input. The buffers in the group are functionally enabled by a LOW signal at the $\overline{O_e}$ input.

The input circuitry includes clamp diodes to limit high-speed termination effects. The outputs can sink up to 64 mA and the source current is rated for 15 mA. The difference between the 74241 and the 74240 is that the 74240 does not invert the date signals fed through it.

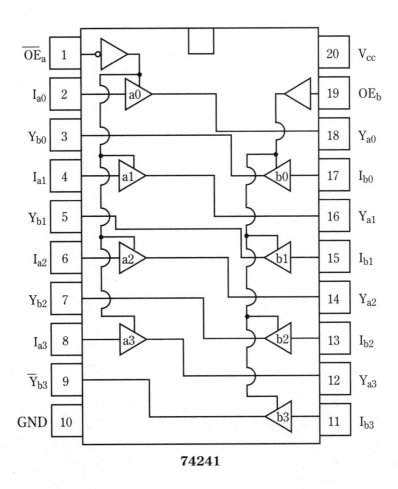

74241

74241 Truth Table

O_{ea}	I_a	O_{eb}	Ib	Y_a	Y_b
0	x	0	x	Z	Z
0	x	1	0	Z	0
0	x	1	1	Z	1
1	0	0	x	0	Z
1	1	0	x	1	Z
1	0	1	0	0	0
1	0	1	1	0	1
1	1	1	0	1	0
1	1	1	1	1	1

x = Don't care
Z = High impedance

74242 Quad bus transceiver with tri-state outputs

The 74242 is an inverting quad bus transceiver intended to provide four-line asynchronous two-way data communication between data buses. The input circuitry includes clamp diodes to limit high-speed termination effects.

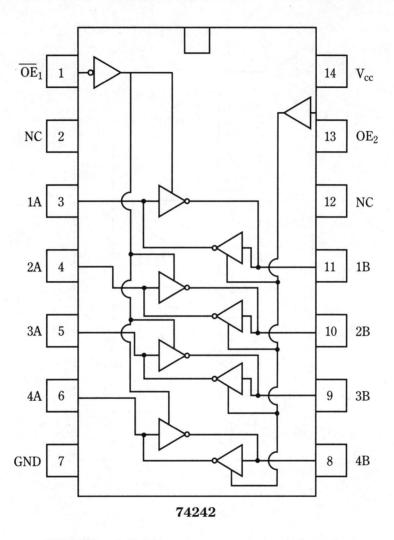

74242

74242 Truth Table

INPUTS		OUTPUT	INPUTS		OUTPUT
$\overline{Oe_1}$	D		Oe_2	D	
0	0	1	0	0	Z
0	1	0	0	1	Z
1	0	Z	1	0	1
1	1	Z	1	1	0

x = Don't care
Z = High impedance

74243 Quad bus transceiver with tri-state outputs

The 74243 is a noninverting quad bus transceiver intended to provide four-line asynchronous two-way data communication between data buses. The input circuitry includes clamp diodes to limit high-speed termination effects.

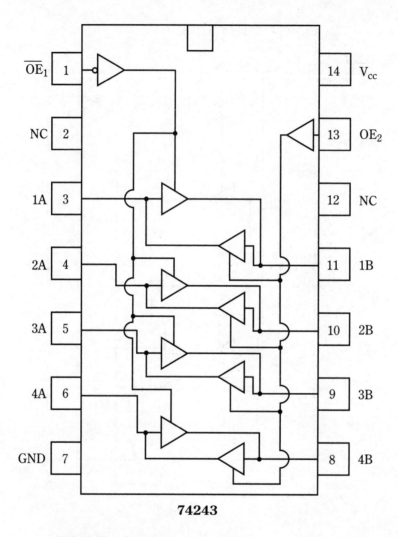

74243

74243 Truth table

INPUTS		OUTPUT	INPUTS		OUTPUT
O_{e1}	D		O_{e2}	D	
0	0	0	0	0	Z
0	1	1	0	1	Z
1	0	Z	1	0	0
1	1	Z	1	1	1

x = Don't care
Z = High impedance

74244 Octal inverter/line driver with tri-state outputs

The 74244 contains eight tri-state buffers or line drivers. This chip was designed for applications such as bus-oriented transmitters/receivers, clock drivers, and memory address drivers. While the functioning of this chip is fairly simple, and could be easily duplicated with other available devices, the use of the 74240 will usually result in a lower overall parts count.

The eight buffer stages on the 74244 are combined into two sets of four each. Each group can be independently controlled for the tri-state function. All four buffers in a group can be placed in the high-impedance output state via a HIGH signal on the appropriate $\overline{O_e}$ input. The buffers in the group are functionally enabled by a LOW signal at the O_e input.

The input circuitry includes clamp diodes to limit high-speed termination effects. The outputs can sink up to 64 mA and the source current is rated for 15 mA.

The 74244 is functionally quite similar to the 74241. The only difference is how the O_e inputs respond to control signals.

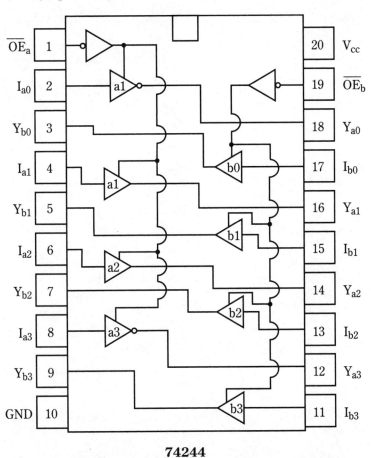

74244

74244 Truth Table

O_{ea}	I_a	O_{eb}	I_b	Y_a	Y_b
0	0	0	0	0	0
0	0	0	1	0	1
0	0	1	x	0	Z
0	1	1	x	1	Z
0	1	0	0	1	0
1	x	0	0	Z	0
1	x	0	1	Z	1
1	x	1	x	Z	Z

x = Don't care
Z = High impedance

74245 Octal bidirectional transceiver with tri-state inputs/outputs

The 74245 contains eight pairs of noninverting buffers with tri-state inputs and outputs. The pairs of buffers are arranged for bidirectional operation. That is, data can be routed along an eight-line bus in either direction.

The 74245 can sink up to 24 mA of current at its A ports, and up to 64 mA at its B ports. Pin 1 controls the direction of data flow. It is labelled T/R for "Transmit/not Receive." The device(s) connected to the A ports is considered the system "transmitter," and the device(s) connected to the B ports are defined as the system "receiver." In many applications, this distinction will be more or less arbitrary and negligible, but the labels help clarify the operation of the 74245. A HIGH signal at the T/R input activates the "Transmit" function, enabling data to flow from the A ports to the B ports. The "Receive" mode, predictably, works in just the opposite way. It is enabled by a LOW signal at the T/R input, and it permits data to flow from the B ports to the A ports.

Pin 19 ($\overline{Q}$) controls the tri-state function for both sets of ports simultaneously. A LOW signal on this pin enables the normal data flow through the ports, and a HIGH signal on this pin activates the high-impedance state for all of the buffer ports.

The buffers in the 74245 are all of the noninverting type, so the data itself is unaffected by flowing through the ports. For each port, the output is the same as the input.

74245 Function Table

CONTROL	INPUTS	OUTPUT FUNCTION
O_e	T/$\overline{R}$	
0	0	Bus B data to Bus A
0	1	Bus A data to Bus B
1	x	High-impedance state (ports disabled)

x = Don't care

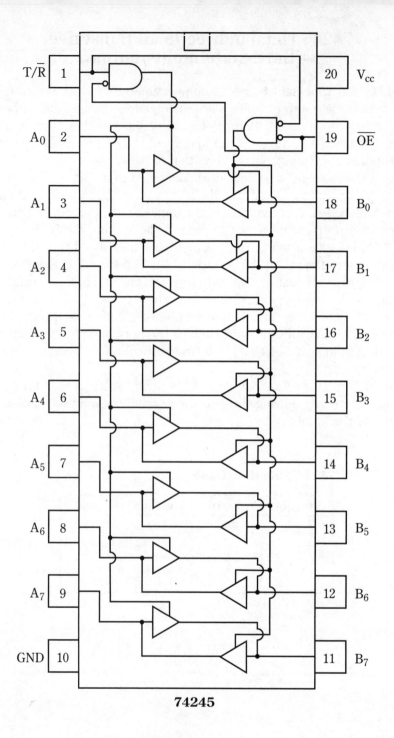

74245

74251 8-Input multiplexer (3-state)

The '251 is a logical implementation of a single-pole, 8-position switch with the state of the three select inputs (S_0, S_1, and S_2) controlling the switch position. Assertion (Y) and negation ($\overline{Y}$) outputs are both provided. The output enable input ($\overline{OE}$) is active LOW. The logic function provided at the output, when activated is:

$$Y = \overline{OE} \ (I_0 \bullet \overline{S}_0 \bullet \overline{S}_2 + I_1 \bullet S_0 \bullet S_1 \bullet \overline{S}_2 + I_2 \bullet \overline{S}_0$$
$$\bullet S_1 \bullet \overline{S}_2 + I_3 \bullet S_0 \bullet S_1 \bullet \overline{S}_2 + I_4 \bullet \overline{S}_0 \bullet \overline{S}_1$$
$$\bullet S_2 + I_5 \bullet S_0 \bullet \overline{S}_1 \bullet S_2 + I_6 \bullet \overline{S}_0 \bullet S_1 \bullet S_2 + I_7$$
$$\bullet S_0 \bullet S_1 \bullet S_2).$$

Both outputs are in the high-impedance (high-Z) state when the output enable is HIGH, allowing multiplexer expansion by tying the outputs of up to 128 devices together. All but one device must be in the high-impedance state to avoid HIGH currents that would exceed the maximum ratings when the outputs of the 3-state devices are tied together. Design of the output enable signals must ensure there is no overlap in the active LOW portion of the enable voltages.

	74S	74	74LS
Type of Output	3-state	3-state	3-state
Typ. Delay, Data to Inverting Output (ns)	4.5	11	17
Typ. Delay, Data to Noninverting Output (ns)	8	18	21
Typ. Delay, From Enable (ns)	14	17	21
Typ. Total Power (mW)	275	155	35

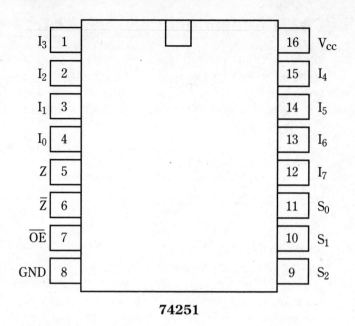

74251

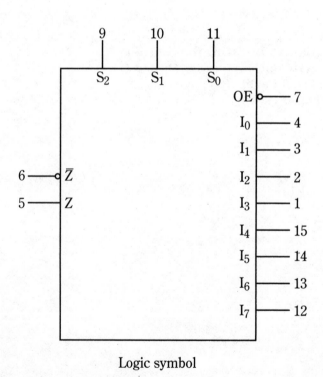

Logic symbol

Truth Table

INPUTS												OUTPUTS	
$\overline{OE}$	S_2	S_1	S_0	I_0	I_1	I_2	I_3	I_4	I_5	I_6	I_7	Y	$\overline{Y}$
H	X	X	X	X	X	X	X	X	X	X	X	(Z)	(Z)
L	L	L	L	L	X	X	X	X	X	X	X	H	L
L	L	L	L	H	X	X	X	X	X	X	X	L	H
L	L	L	H	X	L	X	X	X	X	X	X	H	L
L	L	L	H	X	H	X	X	X	X	X	X	L	H
L	L	H	L	X	X	L	X	X	X	X	X	H	L
L	L	H	L	X	X	H	X	X	X	X	X	L	H
L	L	H	H	X	X	X	L	X	X	X	X	H	L
L	L	H	H	X	X	X	H	X	X	X	X	L	H
L	H	L	L	X	X	X	X	L	X	X	X	H	L
L	H	L	L	X	X	X	X	H	X	X	X	L	H
L	H	L	H	X	X	X	X	X	L	X	X	H	L
L	H	L	H	X	X	X	X	X	H	X	X	L	H
L	H	H	L	X	X	X	X	X	X	L	X	H	L
L	H	H	L	X	X	X	X	X	X	H	X	L	H
L	H	H	H	X	X	X	X	X	X	X	L	H	L
L	H	H	H	X	X	X	X	X	X	X	H	L	H

H = High voltage level
L = Low voltage level
X = Don't care
(Z) = High impedance (off)

74253 Dual 4-input multiplexer with tri-state outputs

The 74253 has two identical four-input multiplexers with tri-state outputs, which select two bits from four sources determined by the common select inputs (S_0 and S_1). When the individual output enable ($\overline{E_0}$ and $\overline{E_{0b}}$) inputs of the four-input multiplexers are HIGH, the outputs are forced to a high-impedance (high-Z) state, regardless of any other input conditions.

The 74253 is the logic implementation of a two-pole, four-position switch. The effective "position" of the switch is determined by the logic levels supplied to the two select inputs. Logic equations for the outputs are:

$$Ya = \overline{OE_a} \, (\overline{I_{oa}} \bullet S_1 \bullet S_0 + I_{1a} \bullet S_1$$
$$\bullet \, S_0 + I_{2a} \bullet S_1 \bullet S_0 + I_{3a} \bullet S_1 \bullet S_0)$$
$$Y_b = \overline{OE} \bullet (I_{ob} \bullet S_1 \bullet S_0 + I_{1b} \bullet S_1$$
$$\bullet \, S_0 + I_{2b} \bullet S_1 \bullet S_0 + I_{3b} \bullet S_1 \bullet S_0)$$

All but one device must be in the high-impedance state to avoid high currents exceeding the maximum ratings, if the outputs of tri-state devices are tied together. The output enable signal design must ensure that there is no overlap.

	74LS
Type of Output	3-State
Typ. Delay, Data to Noninverting Output (ns)	12
Typ. Delay, From Enable (ns)	16
Typ. Total Power (mW)	35

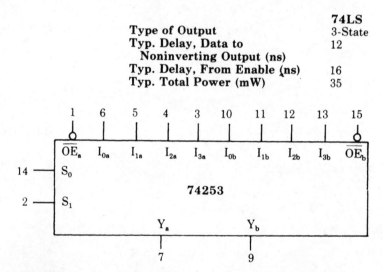

Truth Table

SELECT INPUTS		DATA INPUTS				OUTPUT ENABLE	OUTPUT
S_0	S_1	I_0	I_1	I_2	I_3	$\overline{OE}$	Y
X	X	X	X	X	X	H	(Z)
L	L	L	X	X	X	L	L
L	L	H	X	X	X	L	H
H	L	X	L	X	X	L	L
H	L	X	H	X	X	L	H
L	H	X	X	L	X	L	L
L	H	X	X	H	X	L	H
H	H	X	X	X	L	L	L
H	H	X	X	X	H	L	H

V_{CC} = Pin 16
GND = Pin 8

H = High voltage level
L = Low voltage level
X = Don't care
(Z) = High-impedance (off) state

168

74256 Dual four-bit addressable latch

The 74256 combines an eight-bit latch and dual demultiplexer with series to parallel capability. It offers random (i.e., addressable) data entry, and output from each storage bit is available. This device is easily expandable for larger data buses. Its primary intended application is as a dual 1-of-4 active HIGH decoder.

The Clear ("Master Reset," $\overline{\text{MR}}$, pin 15) and Enable ($\overline{\text{E}}$, pin 14) inputs are used to select one of four operational modes, as summarized. If both of these control inputs are LOW, the chip operates in its demultiplex mode. If the D input (D_a, pin 3) or (D_b, pin 13) is held HIGH, the 74256 will function as an active HIGH decoder in this operating mode.

If both control inputs are HIGH, the 74256 is in its store mode, and does nothing. The outputs hold their preceding values.

When the $\overline{\text{MR}}$ pin is LOW and the $\overline{\text{E}}$ pin is HIGH, the master reset function is activated. In effect, this acts as a common clear input.

Finally, the addressable latch mode is accessed when $\overline{\text{MR}}$ is HIGH and $\overline{\text{E}}$ is LOW.

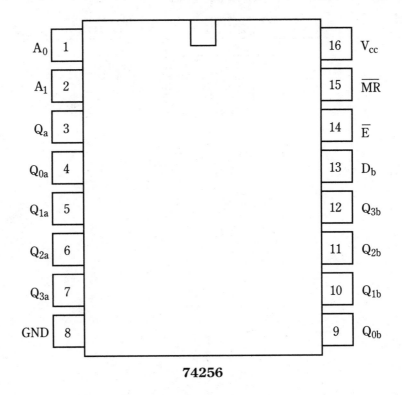

74256

74256 Truth Table

INPUTS					OUTPUTS			
$\overline{MR}$	$\overline{E}$	D	A_1	A_0	Q_0	Q_1	Q_2	Q_3
0	0	d	0	0	d	0	0	0
0	0	d	0	1	0	d	0	0
0	0	d	1	0	0	0	d	0
0	0	d	1	1	0	0	0	d
0	1	x	x	x	0	0	0	0
1	0	d	0	0	d	q_1	q_2	q_3
1	0	d	0	1	q_0	d	q_2	q_3
1	0	d	1	0	q_0	q_1	d	q_3
1	0	d	1	1	q_0	q_1	q_2	d
1	1	x	x	x	q_0	q_1	q_2	q_3

x   =   don't care
d   =   value currently at D input
q_n =   previous value of that Q output

74256 Function Table

CONTROL INPUTS		FUNCTION
$\overline{MR}$	$\overline{E}$	
0	0	Demultiplex
0	1	Master Reset
1	0	Addressable Latch
1	1	Store (Do Nothing)

74257 Quad 2-line-to-1-line data selector/multiplexer (3-state)

The '257 has four identical 2-input multiplexers with 3-state outputs that select four bits of data from two sources under control of a common-data select input (S). The I_0 inputs are selected when the select input is LOW, and the I_1 inputs are selected when the select input is HIGH. Data appears at the outputs in true (noninverted) form from the selected inputs.

The '257 is the logic implementation of a 4-pole, 2-position switch where the position of the switch is determined by the logic levels supplied to the select input. Outputs are forced to a high-impedance OFF-state when the output enable input ($\overline{OE}$) is HIGH. All but one device must be in the high-impedance state to avoid currents exceeding the maximum ratings if outputs are tied together. Design of the output enable signals must ensure that there is no overlap when outputs of 3-state devices are tied together.

	74S	74LS
Type of Output	3-state	3-state
Typ. Delay, Data to Inverting Output (ns)	–	–
Typ. Delay, Data to Noninverting Output (ns)	5	12
Typ. Delay, From Enable (ns)	14	20
Typ. Total Power (mW)	320	50

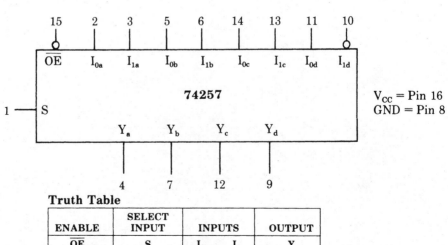

V_{CC} = Pin 16
GND = Pin 8

Truth Table

ENABLE	SELECT INPUT	INPUTS		OUTPUT
$\overline{OE}$	S	I_0	I_1	Y
H	X	X	X	(Z)
L	H	X	L	L
L	H	X	H	H
L	L	L	X	L
L	L	H	X	H

H = High voltage level   X = Don't care
L = Low voltage level   (Z) = High-impedance (off) state

74258 Quad 2-line-to-1-line data selector/multiplexer with tri-state outputs

The 74258 has four identical two-input multiplexers with tri-state outputs. These multiplexers select four bits of data from two sources under control of a common data select input (S). The I_0 inputs are selected when the select input is LOW, and the I_1 inputs are selected when the select input is HIGH. Data appear at the outputs in inverted (complimentary) form.

The 74258 is the logic implementation of a four-pole, two-position switch, where the effective "position" of the switch is determined by the logic level supplied to the select input. All outputs are forced to the third high-impedance OFF state when the output enable input ($\overline{OE}$) is HIGH. All but one device must be in the high-impedance state to avoid currents exceeding the maximum ratings if the outputs of the tri-state devices are tied together. Design of the output enable signals must ensure that there is no overlap when outputs of tri-state devices are tied together.

	74S	74LS
Type of Output	3-State	3-State
Typ. Delay, Data to Inverting Output (ns)	4	12
Typ. Delay, Data to Noninverting Output (ns)	–	–
Typ. Delay, From Enable (ns)	4	20
Typ. Total Power (mW)	280	35

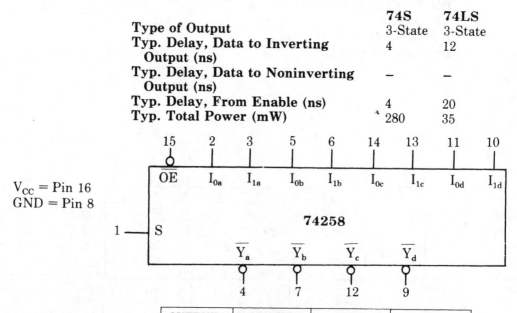

V_{CC} = Pin 16
GND = Pin 8

OUTPUT ENABLE	SELECT INPUT	DATA INPUTS		OUTPUTS
$\overline{OE}$	S	I_0	I_1	$\overline{Y}$
H	X	X	X	(Z)
L	H	X	L	H
L	H	X	H	L
L	L	L	X	H
L	L	H	X	L

Truth Table

H = High voltage level
L = Low voltage level
X = Don't care
(Z) = High-impedance (off) state

172

74259 8-Bit addressable latch

The '259 addressable latch has four distinct modes of operation that are selectable by controlling the clear and enable inputs. In the addressable latch mode, data at the data (D) inputs are written into the addressable latches. The addressed latches will follow the data input with all unaddressed latches remaining in their previous states. In the memory mode, all latches remain in their previous states and are unaffected by the data or address inputs. To eliminate the possibility of entering erroneous data in the latches, the enable should be held HIGH (inactive) while the address lines are changing. In the one-of-eight decoding or demultiplexing mode, ($\overline{\text{CLR}} = \overline{\text{E}} = $ LOW) addressed outputs will follow the level of the D inputs with all other outputs LOW. In the clear mode, all outputs are LOW and unaffected by the address and data inputs.

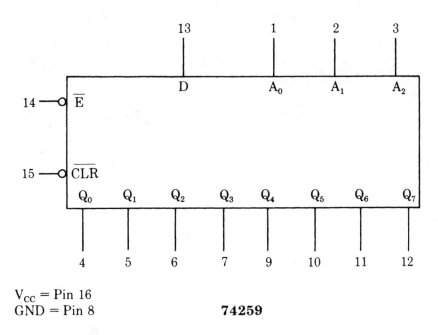

V_{CC} = Pin 16
GND = Pin 8

74259

173

MODE SELECT-FUNCTION TABLE

OPERATING MODE	INPUTS						OUTPUTS							
	$\overline{\text{CLR}}$	$\overline{\text{E}}$	D	A_0	A_1	A_2	Q_0	Q_1	Q_2	Q_3	Q_4	Q_5	Q_6	Q_7
Clear	L	H	X	X	X	X	L	L	L	L	L	L	L	L
Demultiplex (active High decoder when D = H)	L	L	d	L	L	L	Q = d	L	L	L	L	L	L	L
	L	L	d	H	L	L	L	Q = d	L	L	L	L	L	L
	L	L	d	L	H	L	L	L	Q = d	L	L	L	L	L
	•	•	•	•	•	•	•	•	•	•	•	•	•	•
	•	•	•	•	•	•	•	•	•	•	•	•	•	•
	•	•	•	•	•	•	•	•	•	•	•	•	•	•
	L	L	d	H	H	H	L	L	L	L	L	L	L	Q = d
Store (do nothing)	H	H	X	X	X	X	q_0	q_1	q_2	q_3	q_4	q_5	q_6	q_7
Addressable latch	H	L	d	L	L	L	Q = d	q_1	q_2	q_3	q_4	q_5	q_6	q_7
	H	L	d	H	L	L	q_0	Q = d	q_2	q_3	q_4	q_5	q_6	q_7
	H	L	d	L	H	L	q_0	q_1	Q = d	q_3	q_4	q_5	q_6	q_7
	•	•	•	•	•	•	•	•	•	•	•	•	•	•
	•	•	•	•	•	•	•	•	•	•	•	•	•	•
	•	•	•	•	•	•	•	•	•	•	•	•	•	•
	H	L	d	H	H	H	q_0	q_1	q_2	q_3	q_4	q_5	q_6	Q = d

H = High voltage level steady state
L = Low voltage level steady state
X = Don't care
d = High or low data one setup time prior to low-to-high enable transition
q = Lower case letters indicate the state of the referenced output established during the last cycle in which it was addressed or cleared.

74266 Quad 2-input Exclusive-NOR gate with open collector outputs

The Exclusive-NOR gate is rather uncommon, but it is essentially just the opposite of an Exclusive-OR gate (refer to the description of the 7486).

In an Exclusive-OR gate, the output is HIGH if one, but not both, of the inputs is HIGH. To create an Exclusive-NOR gate, the output of an Exclusive-OR gate is inverted. Therefore, in an Exclusive-NOR gate, the output is HIGH if both outputs are LOW, or if both outputs are HIGH.

Either an exclusive-OR gate or an Exclusive-NOR gate can be considered a one-bit digital comparator. An Exclusive-OR functions as a "difference detector," but an Exclusive-NOR gate operates as an "equality detector."

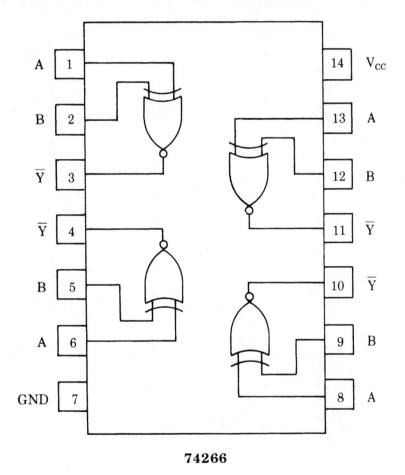

74266

175

74269 8-bit bidirectional binary counter

The 74269 is a fully synchronous up/down binary counter. Its built-in features include an U/$\overline{\text{D}}$ input to control the direction of the count, preset capability for programmable operation, and carry look-ahead for cascading.

All state changes, both for parallel loading and counting, occur on the rising edge of the clock signal.

Supply Current	95 mA (typical)
Output Current (LOW)	20 mA (maximum)
Output Current (HIGH)	−1.0 mA (maximum)
Count Frequency	115 MHz (typical)

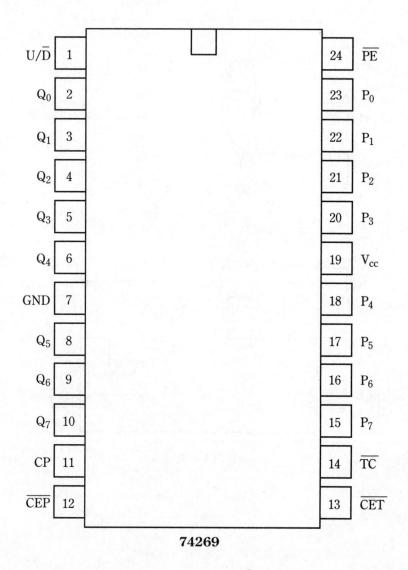

74269

74269 Function Table

INPUTS					OUTPUTS		OPERATING MODE
U/D	CEP	CET	PE	Pn	Qn	TC	
x	x	x	0	0	0	*	Parallel Load
x	x	x	0	1	1	*	Parallel Load
0	0	0	1	x	C	*	Count Down
1	0	0	1	x	C	*	Count Up
x	1	x	1	x	q_n	*	Hold
x	x	1	1	1	q_n	1	Do Nothing

x = Don't care
C = Count value
q_n = Previous count value
* = TC is LOW when CET is LOW and the counter is at Terminal Count. Count Up is when all Qn outputs are HIGH. Count Down is when all Qn outputs are LOW.

74273 Octal D-type flip-flop

The 74273 has eight edge-triggered D-type flip-flops with individual inputs and Q outputs. To accommodate the number of pins available, there are no complementary ($\overline{Q}$) outputs. The common buffered clock (CP) and master reset ($\overline{MR}$) inputs load and reset (clear) all flip-flops simultaneously.

The register is fully edge-triggered. The state of each D input, one setup time before the LOW-to-HIGH clock transition, is transferred to the corresponding flip-flop Q output.

All outputs will be forced LOW independently of clock or data inputs by a LOW voltage level on the $\overline{MR}$ input.

This device is useful for applications where only the true (uninverted) output is required, and clock and master reset functions are common to all storage elements.

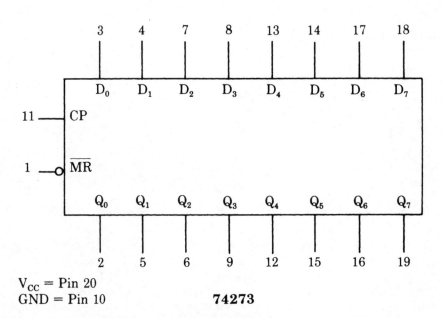

V_{CC} = Pin 20
GND = Pin 10

74273

MODE SELECT-FUNCTION TABLE

OPERATING MODE	INPUTS			OUTPUTS
	$\overline{MR}$	CP	D_n	Q_n
Reset (clear)	L	X	X	L
Load "1"	H	↑	h	H
Load "0"	H	↑	l	L

H = High voltage level steady state.
h = High voltage level one setup time prior to the low-to-high clock transition.
L = Low voltage level steady state.
l = Low voltage level one setup time prior to the low-to-high clock transition.
X = Don't care.
↑ = Low-to-high clock transition.

178

74279A Quad SR latch

A latch is basically a simplified flip-flop. It is used to hold a digital value (a 1 or a 0)—even after the original signal is removed.

The $\overline{S}$ inputs set the latch, and the $\overline{R}$ input resets it. Both the $\overline{S}$ and $\overline{R}$ inputs are inverted in the 279A.

Notice that two of the four latches in the 279A have two $\overline{S}$ inputs and two have only one. All four latches have just a single $\overline{R}$ input.

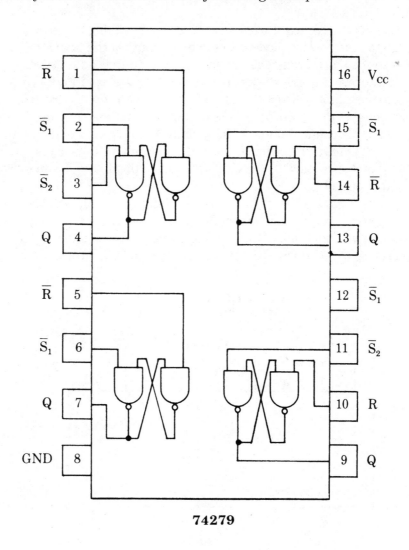

74279

74280 Nine-bit odd/even parity generator/checker

The 74280 is a 9-bit parity generator or checker, commonly used to detect errors in high-speed data transmission or data retrieval systems. Both even and odd parity outputs are available for generating or checking even or odd parity on up to nine bits.

The even parity output (ΣE) is HIGH when an even number of data inputs (I_0 through I_8) is HIGH. The odd parity output (Σ_O) is HIGH when an odd number of data inputs is HIGH.

Expansion to larger word sizes is accomplished by tying the even outputs (ΣE) of up to nine parallel devices to the data inputs of the final stage.

Because the output from the divide-by-two section is not internally connected to the succeeding stages, the device can be operated in various counting modes. In a BCD (8-4-2-1) counter, the CP_1 input must be externally connected to the Q_0 output. The CP_0 input receives the incoming count, producing a BCD count sequence. In a symmetrical biquinary divide-by-10 counter, the Q_3 output must be connected externally to the CP_0 input. The input count is then applied to the CP_1 input, and a divide-by-10 square wave is obtained at output Q_0.

To operate as a divide-by-two or divide-by-five counter, no external interconnections are required. The first flip-flop is used as a binary element for the divide-by-two function (CP_0 as the input and Q_0 as the output). The CP_1 input is used to obtain divide-by-five operation at the Q_3 output.

	74S
Typ. Delay Time (ns)	13
Typ. Total Power (mW)	335

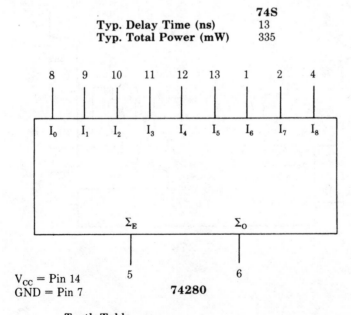

V_{CC} = Pin 14
GND = Pin 7

74280

Truth Table

INPUTS	OUTPUTS	
Number of HIGH data inputs ($I_0 - I_8$)	Σ_E	Σ_O
Even	H	L
Odd	L	H

74283 4-Bit full adder with fast carry

The '283 adds two 4-bit binary words (A_n plus B_n) plus the incoming carry. The binary sum appears on the sum outputs [Σ_1–Σ_4 and the outgoing carry (C_{out})] according to the equation:

$$C_{IN} + (A_1 + B_1) + 2(A_2 + B_2) + 4(A_3 + B_3) + 8(A_4 + B_4) =$$
$$\Sigma_1 + 2\,\Sigma_2 + 4\,\Sigma_3 + 8\,\Sigma_4 + 16 C_{OUT},$$

Where (+) = plus.

Due in the symmetry of the binary add function, the '283 can be used with either all active high operands (positive logic) or with all active LOW operands (negative logic). With active HIGH inputs, C_{IN} cannot be left open; it must be held LOW when no carry in is intended. Interchanging inputs of equal weight does not affect the operation; thus, C_{IN}, A_1, and B_1 can arbitrarily be assigned to pins 5, 6, 7, etc.

	74LS
Typ. Carry Time (ns)	10
Typ. Add Time (ns)	15
Typ. Power Per Bit (mW)	24

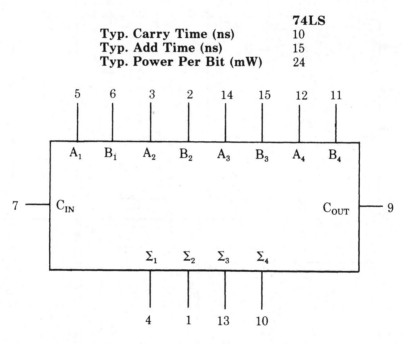

V_{CC} = Pin 16
GND = Pin 8   **74283**

PINS	C_{IN}	A_1	A_2	A_3	A_4	B_1	B_2	B_3	B_4	Σ_1	Σ_2	Σ_3	Σ_4	C_{OUT}
Logic Levels	L	L	H	L	H	H	L	L	H	H	H	L	L	H
Active high	0	0	1	0	1	1	0	0	1	1	1	0	0	1
Active low	1	1	0	1	0	0	1	1	0	0	0	1	1	0

74289 65-Bit random access memory (O.C)

The '289 is a high-speed array of 64 memory cells organized as 16 words of four bits each. A one-of-16 address decoder selects a single word that is specified by the four address inputs (A_0 through A_3). A READ operation is initiated.

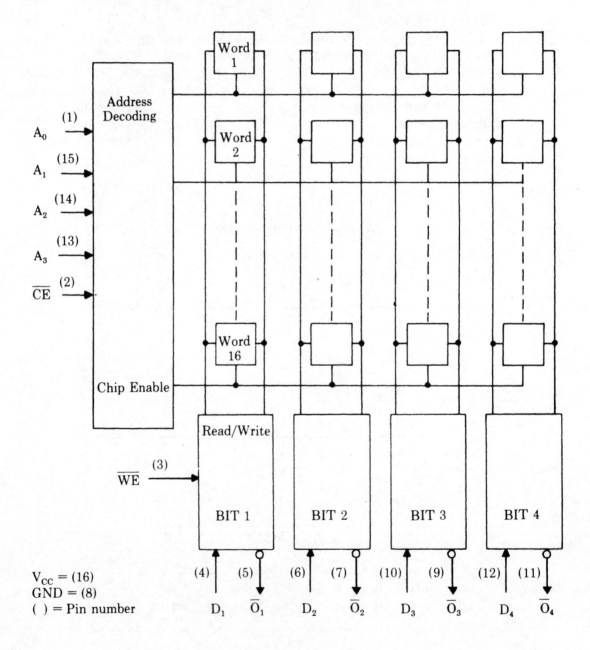

MODE SELECT-FUNCTION TABLE

OPERATING MODE	INPUTS			OUTPUTS
	$\overline{CS}$	$\overline{WE}$	D_n	$\overline{O}_n$
Write — Disable Outputs	L	L	L	H
	L	L	H	H
Read	L	H	X	$\overline{Data}$
Store-Disable Outputs	H	X	X	H

H = High voltage level
L = Low voltage level
X = Don't care
$\overline{Data}$ = Read complement of data from addressed word location

74290 Decade counter

The '290 is a 4-bit ripple-type decade counter. The device consists of four master-slave flip-flops internally connected to provide a divide-by-2 section and a divide-by-5 section. Each section has a separate clock input to initiate state changes of the counter on the HIGH-to-LOW clock transition. State changes of the Q outputs do not occur simultaneously because of internal ripple delays. Therefore, decoded output signals are subject to decoding spikes and should not be used for clocks or strobes. The Q_0 output is designed and specified to drive the rate fallout plus the $\overline{CP_1}$ input of the device.

A gated AND asynchronous master reset ($MR_1 \bullet MR_2$) is provided, which overrides both clocks and resets (clears) all the flip-flops. Also provided is a gated AND asynchronous master set ($MS_1 \bullet MS_2$), which overrides the clocks and the MR inputs, setting the outputs to nine (HLLH).

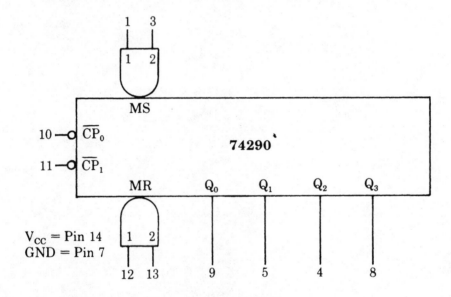

MODE SELECTION-TRUTH TABLE

RESET/SET INPUTS				OUTPUTS			
MR_1	MR_2	MS_1	MS_2	Q_0	Q_1	Q_2	Q_3
H	H	L	X	L	L	L	L
H	H	X	L	L	L	L	L
X	X	H	H	H	L	L	H
L	X	L	X		Count		
X	L	X	L		Count		
L	X	X	L		Count		
X	L	L	X		Count		

H = High voltage level
L = Low voltage level
X = Don't care

BCD COUNT SEQUENCE — TRUTH TABLE

COUNT	OUTPUT			
	Q_0	Q_1	Q_2	Q_3
0	L	L	L	L
1	H	L	L	L
2	L	H	L	L
3	H	H	L	L
4	L	L	H	L
5	H	L	H	L
6	L	H	H	L
7	H	H	H	L
8	L	L	L	H
9	H	L	L	H

NOTE: Output Q_0 connected to input $\overline{CP}_1$

74293 Four-bit binary ripple counter

The 74293 is a four-bit ripple-type binary counter. This device consists of four master/slave flip-flops internally connected to provide a divide-by-two section and a divide-by-eight section. Each section has a separate clock input to initiate state changes of the counter on the HIGH-to-LOW clock transition. State changes of the Q outputs do not occur simultaneously because of internal ripple delays. Therefore, decoded output signals are subject to decoding spikes and should not be used for clocks or strobes or similar critical applications.

The Q_0 output is designed and specified to drive the rated fanout, plus the $\overline{CP_1}$ input of the device. A gated AND asynchronous master rest (MR_1 and MR_2) is provided, which overrides both clocks and resets (clears) all the flip-flops.

Because the output from the divide-by-two section is not internally connected to the succeeding stages, the device can be operated in various counting modes. In a four-bit ripple counter, output Q_0 must be connected externally to input $\overline{CP_1}$. The input count pulses are applied to input $\overline{CP_0}$. Simultaneous divisions of 2, 4, 8, and 16 are performed at the Q_0, Q_1, Q_2, and Q_3 outputs, as shown in the Truth Table.

As a three-bit ripple counter, the input count pulses are applied to input CP_1. Simultaneous frequency divisions of 2, 4, and 8 are available at the Q_1, Q_2, and Q_3 outputs. Independent use of the first flip-flop is available if the reset function coincides with the reset of the three-bit ripple-through counter.

MODE SELECTION

RESET INPUTS		OUTPUTS			
MR_1	MR_2	Q_0	Q_1	Q_2	Q_3
H	H	L	L	L	L
L	H		Count		
H	L		Count		
L	L		Count		

H = High voltage level
L = Low voltage level
X = Don't care

Truth Table

COUNT	OUTPUT			
	Q_0	Q_1	Q_2	Q_3
0	L	L	L	L
1	H	L	L	L
2	L	H	L	L
3	H	H	L	L
4	L	L	H	L
5	H	L	H	L
6	L	H	H	L
7	H	H	H	L
8	L	L	L	H
9	H	L	L	H
10	L	H	L	H
11	H	H	L	H
12	L	L	H	H
13	H	L	H	H
14	L	H	H	H
15	H	H	H	H

NOTE: Output Q_0 connected to input $\overline{CP_1}$

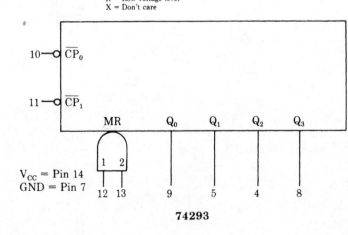

V_{CC} = Pin 14
GND = Pin 7

74293

74299 8-Bit universal shift/storage register

The '229 is an 8-bit general-purpose shift-storage register useful in a wide variety of shifting and 3-state bus interface applications. The register has four synchronous operating modes controlled by the two select inputs, as shown in the mode-select function table. The mode select (S_0 and S_1) inputs, the serial-data (D_{S0} and D_{S7}) inputs and the parallel-data (I/O_0 through I/O_7) inputs are edge-triggered, responding only to the LOW-to-HIGH transition of the clock ($\overline{CP}$) input. Therefore, the only timing restriction is that the S_0, S_1, and selected data inputs must be stable one setup time prior to the positive transition of the clock pulse. The master reset ($\overline{MR}$) is an asynchronous active LOW input. When LOW, the $\overline{MR}$ overrides the clock and all other inputs and clears the register.

Serial mode expansion of the register is accomplished by tying the Q_0 serial output to the D_{S7} input of the preceding register, and tying the $Q7$ serial output to the D_{S0} input of the following register. Recirculating the $(n \times 8)$ bit words is accomplished by tying the Q_7 output of the last stage to the D_{S0} input of the first stage.

The 3-state bidirectional input output port has three modes of operation. When the two output-enable ($\overline{OE_1}$ and $\overline{OE_2}$) inputs are low, and one or both of the select inputs are LOW, data in the register are presented at the eight outputs. When both select inputs are HIGH, the 3-state outputs are forced to the high-impedance OFF state and the register is prepared to load data from the 3-state bus coincident with the next LOW-to-HIGH clock transition. In this parallel load mode, the select inputs disable the outputs even if $\overline{OE_1}$ and $\overline{OE_2}$ are both LOW. A HIGH level on one of the output-enable inputs will force the outputs to the high-impedance OFF-state. When disabled, the 3-state I/O ports present one unit load to the bus, because an input is tied to the I/O node. The enabled 3-state output is designed to drive heavy capacitive loads or heavily loaded 3-state buses.

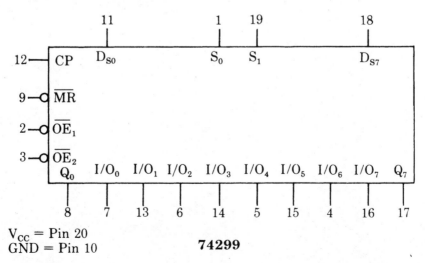

V_{CC} = Pin 20
GND = Pin 10

74299

MODE SELECT-FUNCTION TABLES

REGISTER OPERATING MODES	INPUTS							REGISTER OUTPUTS		
	$\overline{MR}$	CP	S_0	S_1	D_{80}	D_{87}	I/O_n	Q_0	Q_1---Q_6	Q_7
Reset (clear)	L	X	X	X	X	X	X	L	L---L	L
Shift right	H	↑	h	l	l	X	X	L	q_0---q_5	q_6
	H	↑	h	l	h	X	X	H	q_0---q_5	Q_6
Shift left	H	↑	l	h	X	l	X	q_1	q_2---q_7	L
	H	↑	l	h	X	h	X	q_1	q_2---q_7	H
Hold (do nothing)	H	↑	l	l	X	X	X	q_0	q_1---q_6	q_7
Parallel load	H	↑	h	h	X	X	l	L	L---L	L
	H	↑	h	h	X	X	h	H	H---H	H

3-STATE I/O PORT OPERATING MODE	INPUTS					INPUTS/OUTPUTS
	$\overline{OE_1}$	$\overline{OE_2}$	S_0	S_1	Q_n (Register)	I/O_0----I/O_7
Read register	L	L	L	X	L	L
	L	L	L	X	H	H
	L	L	X	L	L	L
	L	L	X	L	H	H
Load register	X	X	H	H	$Q_n = I/O_n$	I/O_n = inputs
Disable I/O	H	X	X	X	X	(Z)
	X	H	X	X	X	(Z)

H = High voltage level
h = High voltage level one setup time prior to low-to-high clock transition
L = Low voltage level
l = Low voltage level one setup time prior to the low-to-high clock transition
q_n = Lower case letters indicate the state of the referenced output one setup prior to the low-to-high clock transition
X = Don't care
(Z) = High-impedance "off" state
↑ = Low-to-high clock transition.

74323 8-Bit universal shift/storage register

The '323 is an 8-bit general-purpose shift-storage register useful in a wide variety of shifting and 3-state bus interface applications. The register has five synchronous operating modes controlled by the two select inputs and the synchronous reset as shown in the mode-select function table. The mode-select (S_0 and S_1) inputs, the synchronous reset ($\overline{SR}$) input, the serial-data (D_{S0} and D_{S7}) inputs and the parallel data I/O_0 through I/O_1 inputs are edge-triggered, responding only to the LOW-to-HIGH transition of the clock (CP) input. Therefore, the only timing restriction is that the $\overline{SR}$, S_0, S_1, and selected data inputs must be stable one setup time prior to the positive transition of the clock pulse. The $\overline{SR}$ input overrides the select and data inputs when LOW and clears the register coincident with the next positive clock transition.

Serial-mode expansion of the register is accomplished by tying the Q_0 serial output to the D_{S7} input of the preceding register, and tying the Q_7 serial output to the D_{S0} input of the following register. Recirculating the (n × 8) bit words is accomplished by typing the Q_7 output of the last stage to the D_{S0} input of the first stage.

The 3-state bidirectional input/output port has three modes of operation. When the two output-enable (OE_1 and OE_2) inputs are LOW, and one or both of the selected inputs are LOW, the data in the register is presented at the eight outputs. When both select inputs are HIGH, the 3-state outputs are forced to the high-impedance OFF-state, and the register is prepared to load data from the 3-state bus coincident with the next LOW-to-HIGH clock transition. In this parallel-load mode, the select inputs disable the outputs even if $\overline{OE_1}$ and $\overline{OE_2}$ are both LOW. A HIGH level on one of the output enable inputs will force the outputs to the high-impedance of state. When disabled, the 3-state I/O ports present one unit load to the bus, because an input is tied to the I/O mode. The enabled 3-state output is designed to drive heavy capacitive loads or heavily loaded 3-state buses.

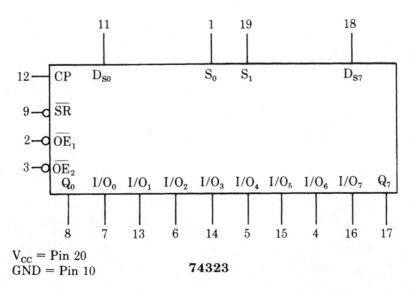

V_{CC} = Pin 20
GND = Pin 10

74323

MODE SELECT-FUNCTION TABLES

REGISTER OPERATING MODES	INPUTS							REGISTER OUTPUTS		
	$\overline{SR}$	CP	S_0	S_1	D_{80}	D_{87}	I/O_n	Q_0	Q_1---Q_6	Q_7
Reset (clear)	l	↑	X	X	X	X	X	L	L---L	L
Shift right	h	↑	h	l	l	X	X	L	q_0---q_5	q_6
	h	↑	h	l	h	X	X	H	q_0---q_5	Q_6
Shift left	h	↑	l	h	X	l	X	q_1	q_2---q_7	L
	h	↑	l	h	X	h	X	q_1	q_2---q_7	H
Hold (do nothing)	h	↑	l	l	X	X	X	q_0	q_1---q_6	q_7
Parallel load	h	↑	h	h	X	X	l	L	L---L	L
	h	↑	h	h	X	X	h	H	H---H	H

3-STATE I/O PORT OPERATING MODE	INPUTS					INPUTS/OUTPUTS
	$\overline{OE_1}$	$\overline{OE_2}$	S_0	S_1	Q_n (REGISTER)	I/O_0----I/O_7
Read register	L	L	L	X	L	L
	L	L	L	X	H	H
	L	L	X	L	L	L
	L	L	X	L	H	H
Load register	X	X	H	H	$Q_n = I/O_n$	I/O_n = inputs
Disable I/O	H	X	X	X	X	(Z)
	X	H	X	X	X	(Z)

H = High voltage level
h = High voltage level one setup time prior to low-to-high clock transition
L = Low voltage level
l = Low voltage level one setup time prior to the low-to-high clock transition
q_n = Lower case letters indicate the state of the referenced output one setup prior to the low-to-high clock transition
X = Don't care
(Z) = High-impedance "off" state
↑ = Low-to-high clock transition.

74350 4-Bit shifter with 3-state outputs

The '350 is a combination logic circuit that shifts a 4-bit word from one to three places. No clocking is required as with shift registers.

The '350 can be used to shift any number of bits to any number of places, up or down by suitable interconnection. Shifting can be:

- *Logical:* the logic zeros fill in at either end of the shifting field.
- *Arithmetic:* the sign bit is extended during a shift down.
- *End around:* the data word forms a continuous loop.

The 3-state outputs are useful for bus-interface applications or expansion to a large number of shift positions in end-around shifting. The active LOW output enable ($\overline{OE}$) input controls the state of the outputs. The outputs are in the high impedance OFF-state when $\overline{OE}$ is HIGH, and they are active when $\overline{OE}$ is LOW.

Truth Table

$\overline{OE}$	S_1	S_0	I_3	I_2	I_1	I_0	I_{-1}	I_{-2}	I_{-3}	Y_3	Y_2	Y_1	Y_0
H	X	X	X	X	X	X	X	X	X	Z	Z	Z	Z
L	L	L	D_3	D_2	D_1	D_0	X	X	X	D_3	D_2	D_1	D_0
L	L	H	X	D_2	D_1	D_0	D_{-1}	X	X	D_2	D_1	D_0	D_{-1}
L	H	L	X	X	D_1	D_0	D_{-1}	D_{-2}	X	D_1	D_0	D_{-1}	D_{-2}
L	H	H	X	X	X	D_0	D_{-1}	D_{-2}	D_{-3}	D_0	D_{-1}	D_{-2}	D_{-3}

H = High voltage level
L = Low voltage level
X = Don't care
(Z) = High-impedance (off) state
D_n = High or low state of the referenced I_n input

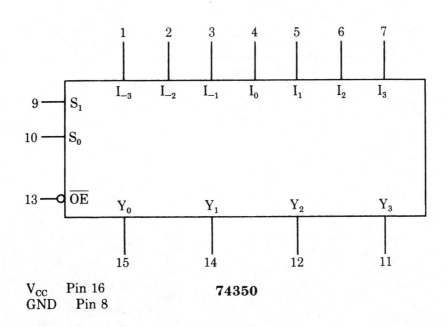

V_{CC} Pin 16
GND Pin 8

74350

74352 Dual 4-input multiplexer

The 74352 consists of a pair of multiplexer circuits—each with four inputs and buffered outputs. It is designed for very high speed operation. The two multiplexer sections can be independently enabled. The select inputs, however, are common for both halves of the chip. Two bits of data can be selected from four sources.

The outputs of the 74352 are inverted. Otherwise, it is equivalent to the 74353.

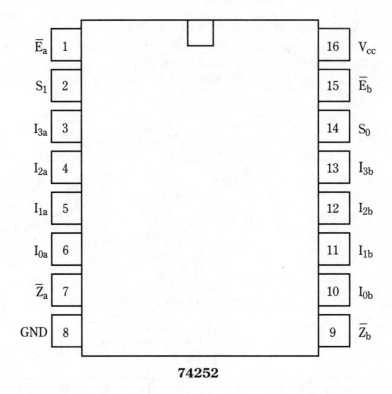

$\overline{E}_a$	1	16	V_{cc}
S_1	2	15	$\overline{E}_b$
I_{3a}	3	14	S_0
I_{2a}	4	13	I_{3b}
I_{1a}	5	12	I_{2b}
I_{0a}	6	11	I_{1b}
$\overline{Z}_a$	7	10	I_{0b}
GND	8	9	$\overline{Z}_b$

74252

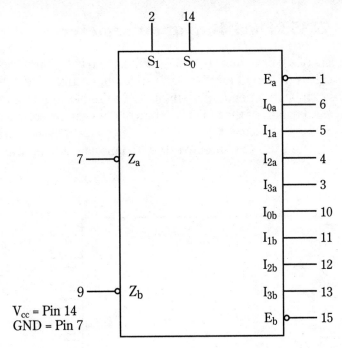

74352 Logic symbol

V_{cc} = Pin 14
GND = Pin 7

74352 Function Table

SELECT INPUTS		INPUTS (a or b)					OUTPUT
S_1	S_0	$\overline{E}$	I_0	I_2	I_3	I_4	$\overline{Z}$
0	0	0	0	x	x	x	1
0	0	0	1	x	x	x	0
0	1	0	x	0	x	x	1
0	1	0	x	1	x	x	0
1	0	0	x	x	0	x	1
1	0	0	x	x	1	x	0
1	1	0	x	x	x	0	1
1	1	0	x	x	x	1	0
x	x	1	x	x	x	x	1

x = Don't care

74353 Dual 4-input multiplexer

The 74353 consists of a pair of multiplexer circuits, each with four inputs and buffered outputs. It is designed for very high speed operation. The two multiplexer sections can be independently enabled. The select inputs, however, are common for both halves of the chip. Two bits of data can be selected from four sources.

The outputs of the 74353 are noninverted. Otherwise, it is equivalent to the 74352.

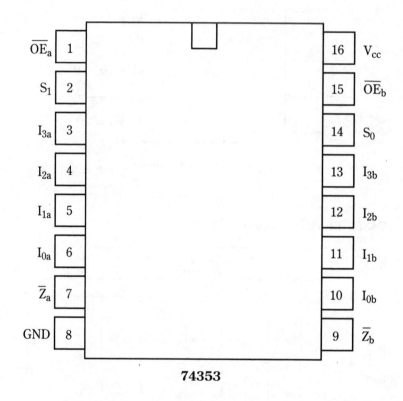

74353

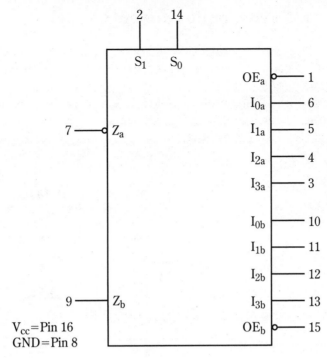

74353 Logic symbol

V_{cc}=Pin 16
GND=Pin 8

74353 Function Table

SELECT INPUTS		INPUTS (a or b)					OUTPUT
S_1	S_0	$\overline{OE}$	I_0	I_2	I_3	I_4	Z
0	0	0	0	x	x	x	0
0	0	0	1	x	x	x	1
0	1	0	x	0	x	x	0
0	1	0	x	1	x	x	1
1	0	0	x	x	0	x	0
1	0	0	x	x	1	x	1
1	1	0	x	x	x	0	0
1	1	0	x	x	x	1	1
x	x	1	x	x	x	x	1

x = Don't care

195

74365 Hex buffer/driver (3-state)

Ordinarily, each of the six buffers in the '365 operates in the usual way—the output state is the same as the input state. But this chip also features two control, or *output-enable* inputs ($\overline{OE_1}$ and $\overline{OE_2}$). When a HIGH signal is placed on either or both of the output enable pins, all six buffers are disabled. All outputs go to the third, high-impedance (Z) state. Both output enable pins must be held LOW for the buffers to recognize the signals at their inputs.

	74
Typ. Delay Time (ns)	12
Typ. Power Per Gate (mW)	54

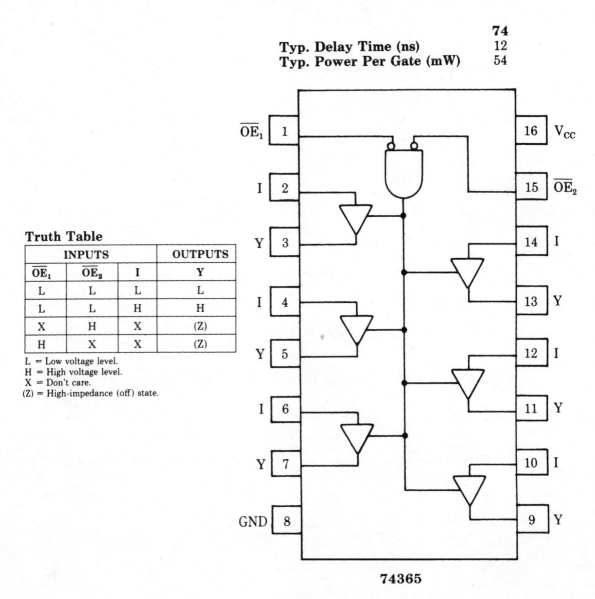

74365

Truth Table

INPUTS			OUTPUTS
$\overline{OE_1}$	$\overline{OE_2}$	I	Y
L	L	L	L
L	L	H	H
X	H	X	(Z)
H	X	X	(Z)

L = Low voltage level.
H = High voltage level.
X = Don't care.
(Z) = High-impedance (off) state.

74366 Hex inverter buffer with tri-state outputs

The 74366 is very similar to the 74365, except that six inverters are used instead of the six (noninverting) buffers in the 74365.

Ordinarily, each of the six inverters in the 74366 operates in the usual way—the output state is the opposite of the input state. That is, when the input is LOW, the output is HIGH, and vice versa. However, this chip features two control, output-enable inputs ($\overline{OE}$ and $\overline{OE}$). When a HIGH signal is placed on either or both of the output-enable pins, all six inverters are disabled. All outputs go to the third high-impedance (Z) state. Both output-enable pins must be held LOW for the inverters to recognize the signals at their inputs.

	74
Typ. Delay Time (ns)	11
Typ. Power Per Gate (mW)	49

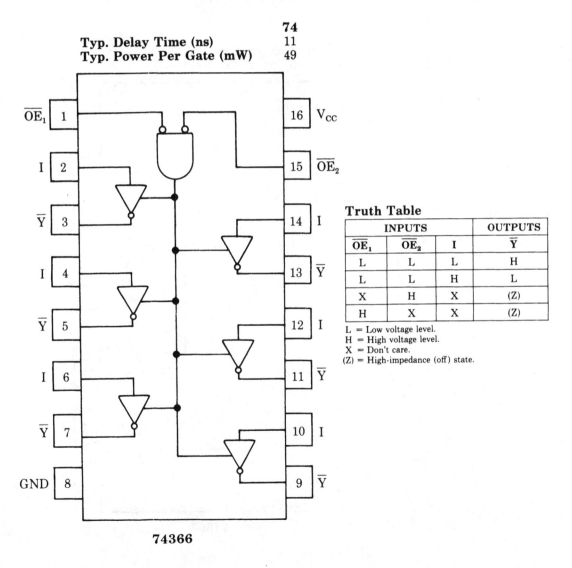

74366

Truth Table

INPUTS			OUTPUTS
$\overline{OE_1}$	$\overline{OE_2}$	I	$\overline{Y}$
L	L	L	H
L	L	H	L
X	H	X	(Z)
H	X	X	(Z)

L = Low voltage level.
H = High voltage level.
X = Don't care.
(Z) = High-impedance (off) state.

74367 Hex buffer/driver with tri-state outputs

The 74367 is quite similar to the 74365 described earlier in this section. The big difference here is that the six buffers are divided into two groups. Each group is controlled by its own independent output-enable input ($\overline{OE}$). Pin 1 is the output-enable for four of the internal buffers. The remaining two buffer stages are controlled by a second output-enable input at pin 15.

When a HIGH signal is placed on one of the output-enable pins, the appropriate buffers are disabled. The outputs of the controlled buffers go to the third high-impedance (Z) state. The appropriate output-enable pin must be held LOW for the buffers in its group to recognize the signals at their inputs. As long as the signal at the appropriate output-enable pin is LOW, each buffer's output signal will be identical to its input signal.

	74
Typ. Delay Time (ns)	12
Typ. Power Per Gate (mW)	54

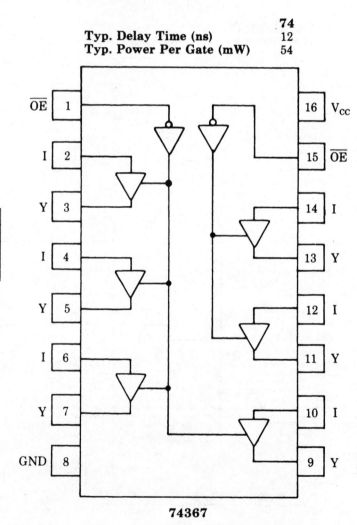

74367

Truth Table

INPUTS		OUTPUTS
$\overline{OE}$	I	Y
L	L	L
L	H	H
H	X	(Z)

L = Low voltage level
H = High voltage level
X = Don't care
(Z) = High-impedance (off) state

74368 Hex inverter buffer (3-state)

The '368 is identical to the '367 just described, except this chip contains six inverters instead of six buffers. A HIGH output-enable signal disables the inverters in its group. The affected inverter outputs go to the high-impedance (Z) state and their inputs are ignored. When the appropriate output-enable pin is held LOW, the inverters function in the normal manner. The output signal is always at the opposite state as the input signal.

	74
Typ. Delay Time (ns)	11
Typ. Power Per Gate (mW)	49

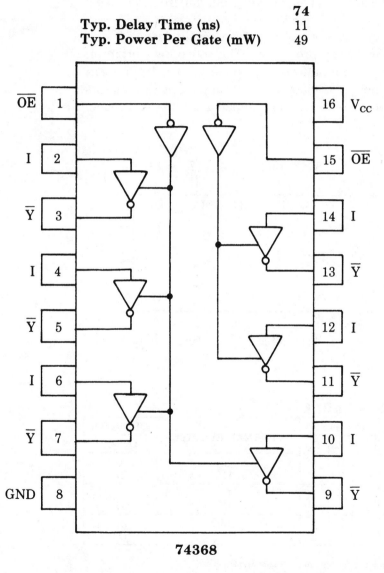

74368

Truth Table

INPUTS		OUTPUTS
$\overline{OE}$	I	$\overline{Y}$
L	L	H
L	H	L
H	X	(Z)

L = Low voltage level.
H = High voltage level.
X = Don't care.
(Z) = High-impedance (off) state.

74373 Octal transparent latch with 3-state outputs

The '373 is an octal transparent latch coupled to eight 3-state output buffers. The two sections of the device are controlled independently by latch-enable (E) and output-enable ($\overline{OE}$) control gates.

Data on the D inputs are transferred to the latch outputs when the latch-enable (E) input is HIGH. The latch remains transparent to the data inputs while E is HIGH, and stores the data present one setup time before the HIGH-to-LOW enable transition. The enable gate has about 400 mV of hysteresis built in to help minimize problems that signal and ground noise can cause in the latching operation.

The 3-state output buffers are designed to drive heavily loaded 3-state buses, MOS memories, or MOS microprocessors. The active LOW output-enable ($\overline{OE}$) controls all eight 3-state buffers independent of the latch operation. When $\overline{OE}$ is LOW, latched or transparent data appear at the outputs. When $\overline{OE}$ is HIGH, the outputs are in the high-impedance OFF-state, which means they will neither drive nor load the bus.

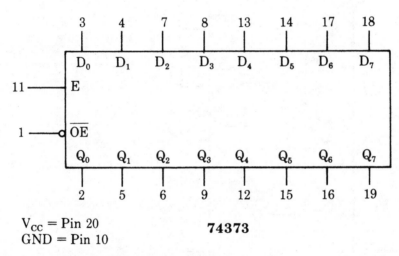

V_{CC} = Pin 20
GND = Pin 10

74373

MODE SELECT-FUNCTION TABLE

| OPERATING MODES | INPUTS | | | INTERNAL REGISTER | OUTPUTS |
	$\overline{OE}$	E	D_n		$Q_0 - Q_7$
Enable & read register	L	H	L	L	L
	L	H	H	H	H
Latch & read register	L	L	l	L	L
	L	L	h	H	H
Latch register & disable outputs	H	L	l	L	(Z)
	H	L	h	H	(Z)

H = High voltage level
h = High voltage level one setup time prior to the high-to-low enable transition
L = Low voltage level
l = Low voltage level one setup time prior to the high-to-low enable transition
(Z) = High impedance "off" state

200

74374 Octal D-type flip-flop with tri-state outputs

The 74374 is an eight-bit edge-triggered register coupled to eight tri-state output buffers. The two sections of the device are controlled independently by the clock (CP) and output-enable ($\overline{\text{OE}}$) control gates.

The register is fully edge-triggered. The state of each D input, one set-up time before the LOW-to-HIGH clock transition, is transferred to the corresponding output of the flip-flop. The clock buffer has about 400 mV of hysteresis built in to help minimize problems that signal and ground noise can cause in the clocking operation.

The tri-state output buffers are designed to drive heavily loaded tri-state buses, MOS memories, or MOS microprocessors. The active LOW output-enable ($\overline{\text{OE}}$) controls all eight tri-state buffers independently of the register operation. When $\overline{\text{OE}}$ is LOW, the data in the register appears at the outputs. When $\overline{\text{OE}}$ is HIGH, the outputs are in the high-impedance (Z) off state, which means they will neither drive nor load the bus.

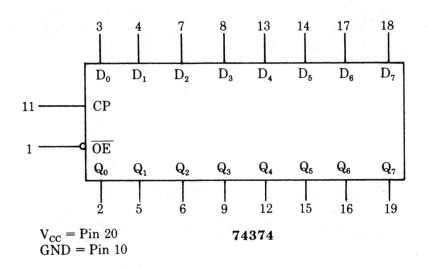

V_{CC} = Pin 20
GND = Pin 10

74374

MODE SELECT-FUNCTION TABLE

OPERATING MODES	INPUTS			INTERNAL REGISTER	OUTPUTS
	$\overline{\text{OE}}$	CP	D_n		Q_0–Q_7
Load & read register	L	↑	l	L	L
	L	↑	h	H	H
Load register & disable outputs	H	↑	l	L	(Z)
	H	↑	h	H	(Z)

H = High voltage level
h = High voltage level one setup time prior to the low-to-high clock transition
L = Low voltage level
l = Low voltage level one setup time prior to the low-to-high clock transition
(Z) = High impedance "off" state
↑ = Low-to-high clock transition

74375 Dual two-bit transparent latch

The 74375 has two independent two-bit transparent latches. Each two-bit latch is controlled by an active HIGH enable input ($\overline{E}$). When $\overline{E}$ is HIGH, data enters the latch and appears at the Q outputs, which follow the data inputs, as long as E is held HIGH. Data on the D inputs one set-up time before the HIGH-to-LOW transition of the enable input will be stored in the latch. The latched outputs will now remain stable as long as the enable input is held LOW.

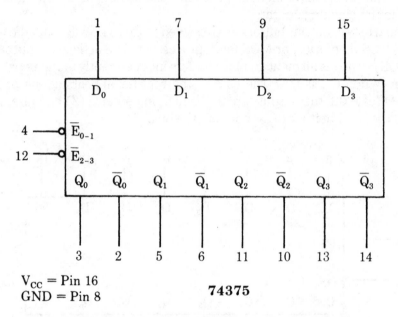

V_{CC} = Pin 16
GND = Pin 8

74375

MODE SELECT-FUNCTION TABLE

OPERATING MODE	INPUTS			OUTPUTS
	$\overline{E}$	D	Q	$\overline{Q}$
Data Enabled	H	L	L	H
	H	H	H	L
Data Latched	L	X	q	$\overline{q}$

H = High voltage level
L = Low voltage level
X = Don't care
q = Lower case letters indicate the state of referenced output one setup
   time prior to the high-to-low enable transition.

74377 Octal D flip-flop with clock enable

The '377 has eight edge-triggered D-type flip-flops with individual D inputs and Q outputs. The common buffered clock (CP) input loads all flip-flops simultaneously when the clock enable ($\overline{CE}$) is LOW.

The register is fully edge-triggered. The state of each D input, one setup time before the LOW-to-HIGH clock transition, is transferred to the corresponding Q output of the flip-flop. The $\overline{CE}$ input is also edge-triggered, and must be stable only one setup time prior to the LOW-to-HIGH clock transition for predictable operation.

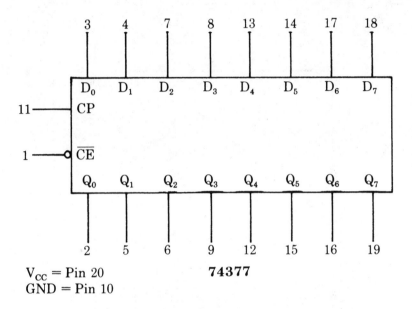

V_{CC} = Pin 20
GND = Pin 10

74377

MODE SELECT-FUNCTION TABLE

OPERATING MODE	INPUTS			OUTPUTS
	CP	$\overline{CE}$	D_n	Q_n
Load "1"	↑	l	h	H
Load "0"	↑	l	l	L
Hold (do nothing)	↑ X	h H	X X	no change no change

H = High voltage level steady state.
h = High voltage level one setup time prior to the low-to-high clock transition.
L = Low voltage level steady state.
l = Low voltage level one setup time prior to the low-to-high clock transition.
X = Don't care.
↑ = Low-to-high clock transition.

74378 Hex D flip-flop with clock enable

The '378 has six edge-triggered D-type flip-flops with individual D inputs and Q outputs. The common buffered clock (CP) input loads all flip-flops simultaneously when the clock enable ($\overline{CE}$) is LOW.

The register is fully edge-triggered. The state of each D input, one setup time before the LOW-to-HIGH clock transition, is transferred to the corresponding Q output of the flip-flop. The $\overline{CE}$ input is also edge-triggered, and must be stable only one setup time prior to the LOW-to-HIGH clock transition for predictable operation.

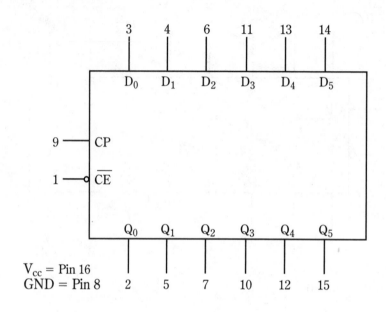

74378

V_{cc} = Pin 16
GND = Pin 8

MODE SELECT-FUNCTION TABLE

OPERATING MODE	INPUTS			OUTPUT
	CP	$\overline{CE}$	D_n	Q_n
Load "1"	↑	l	h	H
Load "0"	↑	l	l	L
Hold (do nothing)	↑	h	x	No change
	x	h	x	No change

h = High one set-up time prior to low-to-high clock transitions.
l = Low one set up time prior to low-to-high clock transition.
↑ = Low-to-high clock transition.
x = Don't care.

74379 Quad D-type flip-flop with clock enable

The 74379 is a quad edge-triggered D-type flip-flop with individual D inputs, and both Q and $\overline{Q}$ outputs. The common buffered clock (CP) input loads all flip-flops simultaneously when the clock enable (CE) is LOW.

The register is fully edge-triggered. The state of each D input, one set-up time before the LOW-to-HIGH clock transition, is transferred to the corresponding Q output of the flip-flop (and inverted at the corresponding $\overline{Q}$ output).

The CE input is also edge-triggered, and must be stable only one set-up time prior to the LOW-to-HIGH clock transition for predictable operation.

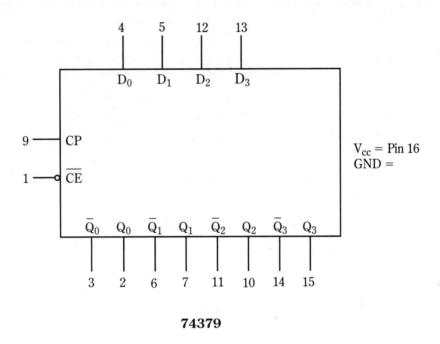

74379

V_{cc} = Pin 16
GND =

MODE SELECT-FUNCTION TABLE

OPERATING MODE	INPUTS			OUTPUTS	
	CP	$\overline{CE}$	D_n	Q_n	$\overline{Q_n}$
Load "1"	↑	l	h	H	L
Load "0"	↑	l	l	L	H
Hold (do nothing)	↑	h	x	No change	
	x	H	x	No change	

x = Don't care.
↑ = Low-to-high clock transition.
l = Low one set-up time prior to ↑.
h = High one set-up time prior to ↑.

74381 Four-bit arithmetic logic unit

An ALU, such as the 74381 is a sort of minimal, "bare bones" computer. Such a circuit is a small, but crucial element within any CPU. A separate ALU is useful in many applications that require mathematical and/or logical operations on data, but not the complexity and programmability of a full computer.

The 74381 is designed to perform three mathematical operations and three logic operations on a pair of 4-bit words, identified as A and B. Mathematically, they can be added, or subtracted in either order ($A + B$, $A - B$, and $B - A$). The two input words can also be ANDed, ORed, or X-ORed.

Clear and Preset functions can also be activated with a pair of special Select inputs to force the Function Outputs LOW or HIGH.

This ALU can be combined with another unit to expand the word length with its built-in Carry Lookahead Generator. Carry Propagate and Generate outputs are also included, to permit the 74381 to be used with the 74382.

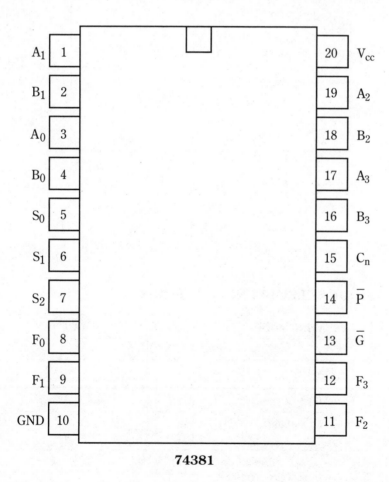

74381

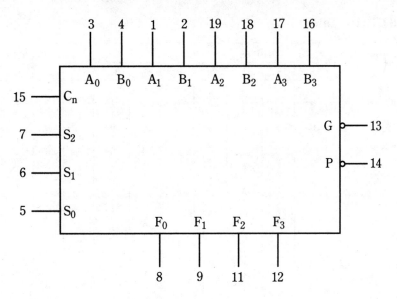

74381 Logic symbol

74381 Truth Table

| Function | S | | | | | | | | | | | |

Function	S_0	S_1	S_2	C_n	A_n	B_n	F_0	F_1	F_2	F_3	G	P
Clear	0	0	0	x	x	x	0	0	0	0	0	0
B minus A	1	0	0	0	0	0	1	1	1	1	1	0
	1	0	0	0	0	1	0	1	1	1	0	0
	1	0	0	0	1	0	0	0	0	0	1	1
	1	0	0	0	1	1	1	1	1	1	1	0
	1	0	0	1	0	0	0	0	0	0	1	0
	1	0	0	1	0	1	1	1	1	1	0	0
	1	0	0	1	1	0	1	0	0	0	1	1
	1	0	0	1	1	1	0	0	0	0	1	0
A minus B	1	0	0	0	0	0	1	1	1	1	1	0
	1	0	0	0	0	1	0	0	0	0	1	1
	1	0	0	0	1	0	0	1	1	1	0	0
	1	0	0	0	1	1	1	1	1	1	1	0
	1	0	0	1	0	0	0	0	0	0	1	0
	1	0	0	1	0	1	1	0	0	0	1	1
	1	0	0	1	1	0	1	1	1	1	0	0
	1	0	0	1	1	1	0	0	0	0	1	0
A plus B	1	1	0	0	0	0	0	0	0	0	1	1
	1	1	0	0	0	1	1	1	1	1	1	0
	1	1	0	0	1	0	1	1	1	1	1	0
	1	1	0	0	1	1	0	1	1	1	0	0
	1	1	0	1	0	0	1	0	0	0	1	1
	1	1	0	1	0	1	0	0	0	0	1	0
	1	1	0	1	1	0	0	0	0	0	1	0
	1	1	0	1	1	1	1	1	1	1	0	0
A X-OR B	0	0	1	x	0	0	0	0	0	0	0	0
	0	0	1	x	0	1	1	1	1	1	1	1
	0	0	1	x	1	0	1	1	1	1	1	0
	0	0	1	x	1	1	0	0	0	0	0	0
A OR B	1	0	1	x	0	0	0	0	0	0	0	0
	1	0	1	x	0	1	1	1	1	1	1	1
	1	0	1	x	1	0	1	1	1	1	1	1
	1	0	1	x	1	1	1	1	1	1	1	0
A AND B	0	0	1	x	0	0	0	0	0	0	0	0
	0	0	1	x	0	1	0	0	0	0	1	1
	0	0	1	x	1	0	0	0	0	0	0	0
	0	0	1	x	1	1	1	1	1	1	1	0
Preset	1	1	1	x	0	0	1	1	1	1	1	1
	1	1	1	x	0	1	1	1	1	1	1	1
	1	1	1	x	1	0	1	1	1	1	1	1

x = Don't care

74381 Function Select Table

SELECT INPUTS			SELECTED OPERATION
0	0	0	Clear
0	0	1	A X-OR B
0	1	0	A minus B
0	1	1	A AND B
1	0	0	B minus A
1	0	1	A OR B
1	1	0	A plus B
1	1	1	Preset

74382 Four-bit arithmetic logic unit

The 74382 is a companion ALU for the 74381. It performs three arithmetic operations and three logic operations. It includes a Carry output for ripple expansion, and an Overflow output to permit convenient twos compliment arithmetic. See also the data for the 74381.

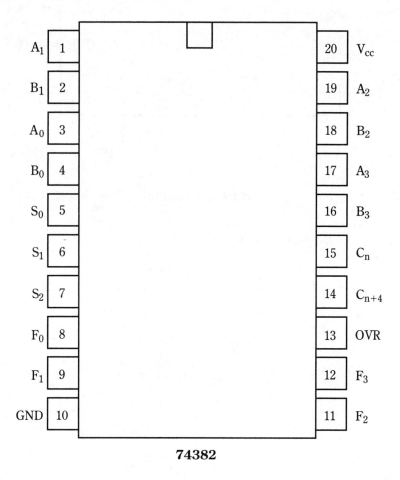

74382

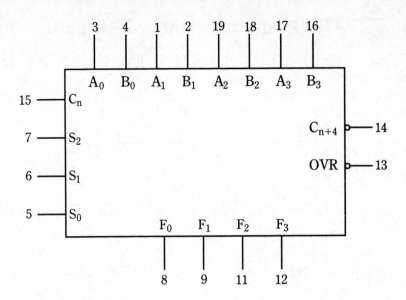

74382 Logic symbol

74382 Function Select Table

SELECT INPUTS			SELECTED OPERATION
0	0	0	Clear
0	0	1	A X-OR B
0	1	0	A minus B
0	1	1	A AND B
1	0	0	B minus A
1	0	1	A OR B
1	1	0	A plus B
1	1	1	Preset

74382 Truth Table

FUNCTION	INPUTS					OUTPUTS					
S_0	S_1	S_2	C_n	A_n	B_n	F_0	F_1	F_2	F_3	OVR	C_{n+4}
CLEAR 0	0	0	x	x	x	0	0	0	0	1	1
B minus A 1	0	0	0	0	0	1	1	1	1	0	0
1	0	0	0	0	1	0	1	1	1	0	1
1	0	0	0	1	0	0	0	0	0	0	0
1	0	0	0	1	1	1	1	1	1	0	0
1	0	0	1	0	0	0	0	0	0	0	1
1	0	0	1	0	1	1	1	1	1	0	1
1	0	0	1	1	0	1	0	0	0	0	0
1	0	0	1	1	1	0	0	0	0	0	1
A minus B 0	1	0	0	0	0	1	1	1	1	0	0
0	1	0	0	0	1	0	0	0	0	0	0
0	1	0	0	1	0	0	1	1	1	0	1
0	1	0	0	1	1	1	1	1	1	0	0
0	1	0	1	0	0	0	0	0	0	0	1
0	1	0	1	0	1	1	0	0	0	0	0
0	1	0	1	1	0	1	1	1	1	0	1
0	1	0	1	1	1	0	0	0	0	0	1
A plus B 1	1	0	0	0	0	0	0	0	0	0	0
1	1	0	0	0	1	1	1	1	1	0	0
1	1	0	0	1	0	1	1	1	1	0	0
1	1	0	0	1	1	0	1	1	1	0	1
1	1	0	1	0	0	1	0	0	0	0	0
1	1	0	1	0	1	0	0	0	0	0	1
1	1	0	1	1	0	0	0	0	0	0	1
1	1	0	1	1	1	1	1	1	1	0	1
A X-OR B 0	0	1	x	0	0	0	0	0	0	0	0
0	0	1	x	0	1	1	1	1	1	0	0
0	0	1	0	1	0	1	1	1	1	0	0
0	0	1	1	1	0	1	1	1	1	1	1
0	0	1	x	1	1	0	0	0	0	1	1
A OR B 1	0	1	x	0	0	0	0	0	0	0	0
1	0	1	x	0	1	1	1	1	1	0	0
1	0	1	x	1	0	1	1	1	1	0	0
1	0	1	0	1	1	1	1	1	1	0	0
1	0	1	1	1	1	1	1	1	1	1	1
A AND B 0	1	1	x	0	0	0	0	0	0	1	1
0	1	1	x	0	1	0	0	0	0	0	0
0	1	1	x	1	0	0	0	0	0	1	1
0	1	1	0	1	1	1	1	1	1	0	0
0	1	1	1	1	1	1	1	1	1	1	1
PRESET 1	1	1	x	0	0	1	1	1	1	0	0
1	1	1	x	0	1	1	1	1	1	0	0
1	1	1	x	1	0	1	1	1	1	0	0
1	1	1	0	1	1	1	1	1	1	0	0

74390 Dual decade ripple counter

The '390 is a dual 4-bit decade ripple counter that is divided into four separately clocked sections. The counter has two divide-by-2 sections and two divide-by-5 sections. The sections are normally used in a BCD decade or a biquinary configuration, because they share a common master reset input. If the two master resets can be used to simultaneously clear all 8 bits of the counter, a number of counting configurations are possible within one package. The separate clocks of each section allow ripple counter or frequency-division applications of divide by 2, 4, 5, 10, 20, 25, 50, or 100.

Each section is triggered by the HIGH-to-LOW transition of the clock ($\overline{CP}$) inputs. For BCD decade operation, the Q_0 output is connected to the $\overline{Cp}_1$ input of the divide-by-5 section. For biquinary decade operation (50-percent duty-cycle output), the Q_3 output is connected to the $\overline{CP}_0$ input, and Q_0 becomes the decade output.

The master resets (MR_a and MR_b) are active HIGH asynchronous inputs to each decade counter. These inputs operate on the portion of the counter identified by the a and b suffixes in the pin configuration. A HIGH level on the MR input overrides the clocks and sets the four outputs LOW.

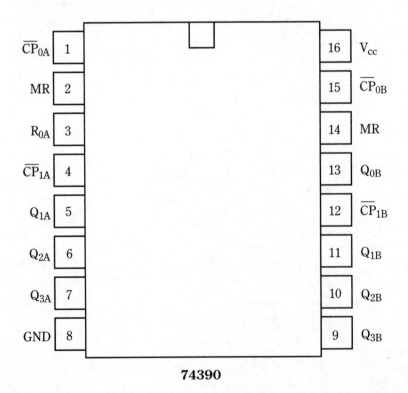

74390

COUNT SEQUENCE FOR ½ THE 74390

COUNT	BCD MODE				BIQUINARY MODE			
	Q_0	Q_1	Q_2	Q_3	Q_0	Q_1	Q_2	Q_3
0	L	L	L	L	L	L	L	L
1	H	L	L	L	L	H	L	L
2	L	H	L	L	L	L	H	L
3	H	H	L	L	L	H	H	L
4	L	L	H	L	L	L	L	H
5	H	L	H	L	H	L	L	L
6	L	H	H	L	H	H	L	L
7	H	H	H	L	H	L	H	L
8	L	L	L	H	H	H	H	L
9	H	L	L	H	H	L	L	H

74393 Dual four-bit binary ripple counter

The 74393 is a dual four-bit binary ripple counter with separate clock and master reset inputs to each counter. The operation of each half of the 74393 is the same as 7493, except that no external clock connections are required. The counters are triggered by a HIGH-to-LOW transition of the clock inputs (CP_a and CP_b).

The counter outputs are internally connected to provide clock inputs to succeeding stages. The outputs are designed to drive the internal flip-flops, plus the rated fan-out of the device. The circuit designer should be aware that this ripple counter's outputs do not change synchronously, and they should not be used for high-speed address decoding, or similarly critical applications.

The master resets (MR_a and MR_b) are active HIGH asynchronous inputs to each four-bit counter identified by the $_a$ and $_b$ suffixes in the pin-out diagram. A HIGH level on the appropriate MR input overrides the clock and sets that counter's outputs LOW.

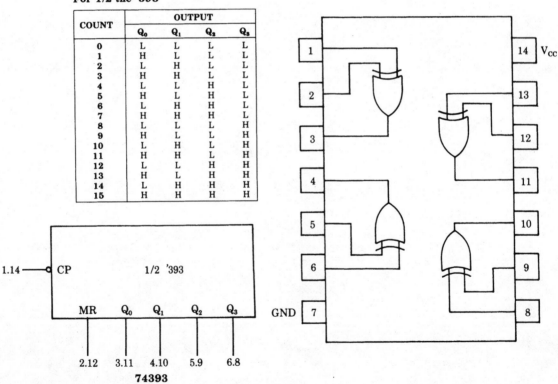

COUNT SEQUENCE
For 1/2 the '393

COUNT	OUTPUT			
	Q_0	Q_1	Q_2	Q_3
0	L	L	L	L
1	H	L	L	L
2	L	H	L	L
3	H	H	L	L
4	L	L	H	L
5	H	L	H	L
6	L	H	H	L
7	H	H	H	L
8	L	L	L	H
9	H	L	L	H
10	L	H	L	H
11	H	H	L	H
12	L	L	H	H
13	H	L	H	H
14	L	H	H	H
15	H	H	H	H

214

74398 Quad two-port register

The 74398 can select from two data sources with fully positive edge-triggered operation. It functions as four 2-input multiplexers each feeding into individual edge-triggered flip-flops. All four multiplexer units share a common select to determine which of two four-bit words is accepted. The 74398 offers both true and complement (NOT) outputs.

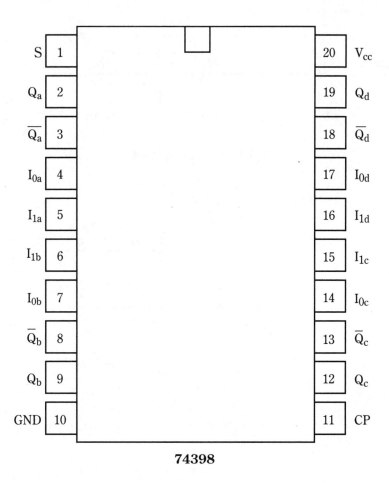

74398

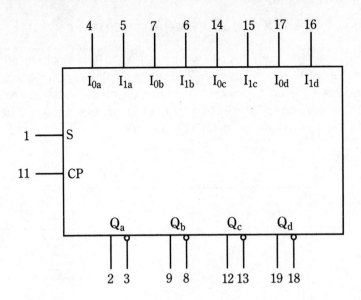

74398 Logic symbol

74398 Function Table

	INPUTS		OUTPUTS	
S	I_0	I_1	Q	Q
0	0	x	0	1
0	1	x	1	0
1	x	0	0	1
1	x	1	1	0

x = don't care

74399 Quad two-port register

The 74399 is a high-speed quad two-port register. It selects four bits of data from two sources (ports) under the control of a common select input (S). The selected data is loaded into a four-bit output register synchronous with the LOW-to-HIGH transition of the clock input (CP).

The operation of this device is fully synchronous. The data inputs (I_0 and I_1) and the select input (S) must be stable only one set-up time prior to the LOW-to-HIGH clock transition for predictable operation.

74521 Eight-bit identity comparator

The 74521 is an expandable 8-bit comparator. It looks at two words of up to eight-bits apiece. The output is LOW if and only if the two compared words are identical bit-for-bit. The comparison typically takes a mere 6.5 nS. The expansion input (A = B) functions as an active LOW enable input when it is used.

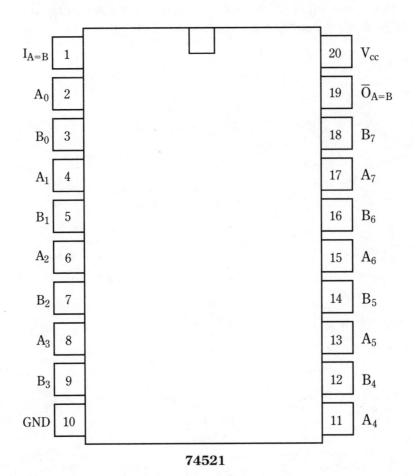

74521

74521 Function Table

i (A = B)	A = B?	O (A = B)
0	Y	0
0	N	1
1	Y	1
1	N	1

Y means all bits of input A match all bits of input B
N means one or more of the bits of A and B do not match

74533 Octal transparent latch with tri-state outputs

The 74533 contains eight tri-state latches, which effectively appear transparent to the date when the Latch Enable (LE, pin 11) is held HIGH. Applying a LOW signal to this pin causes the data that meets the setup times to be latched. The data (LOW or HIGH bits) are permitted to appear at the outputs when the Output Enable (OE, pin 1) is LOW. When the OE signal is HIGH, the bus outputs are switched into the third high-impedance state, effectively disabling the data.

The 74533 is designed for use in bus-oriented system applications.

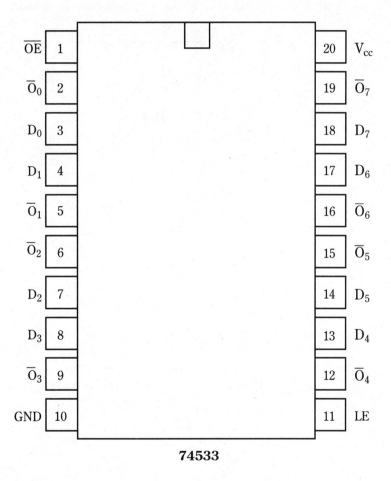

74533

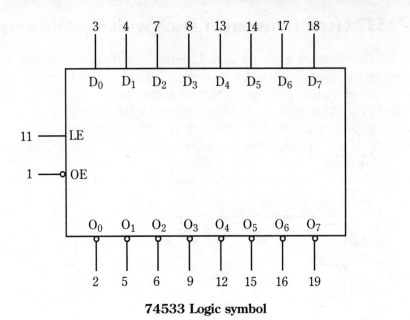

74533 Logic symbol

74534 Octal D-type flip-flop with tri-state outputs

The 74534 contains eight D-type flip-flops with tri-state outputs. Separate data inputs and outputs are provided for each of these flip-flops, but they are controlled jointly via a common buffered Clock (CP, pin 11) and Output Enable (OE, pin 1) that effect the entire chip as a whole. The inputs are edge-triggered. The flip-flops in the 74534 are designed for high-speed, low-power operation.

The 74534 is identical to the 74374, except the outputs are inverted in this device. The 74534 is well suited for bus-oriented systems applications.

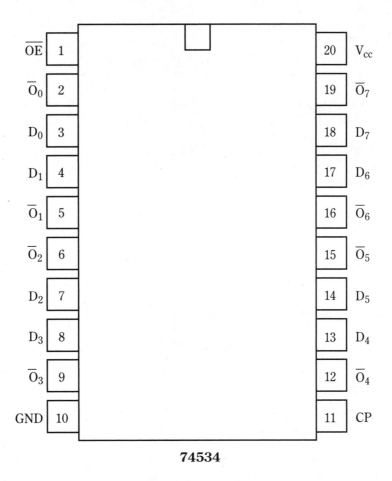

74534

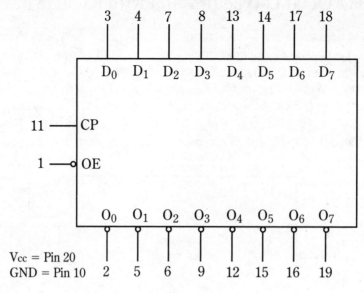

74534 Logic symbol

74568 BCD decade up/down synchronous counter (3-State)

The '568 is a synchronous presettable BCD decade up/down counter featuring an internal carry look ahead for applications in high-speed counting designs. Synchronous operation is provided by having all flip-flops clocked simultaneously so that the outputs change coincident with each other when so instructed by the count-enable inputs and internal gating. This mode of operation eliminates the output spikes that are normally associated with asynchronous (ripple clock) counters. A buffered clock input triggers the flip-flops on the LOW-to-HIGH transition of the clock.

The counter is fully programmable; that is, the outputs can be preset to either level. Presetting is synchronous with the clock and takes place regardless of the levels of the count-enable inputs. A LOW level on the parallel-enable ($\overline{PE}$) inputs disables the counter and causes the data at the D_n inputs to be loaded into the counter on the next LOW-to-HIGH transition of the clock. The synchronous reset ($\overline{SR}$), when LOW one setup time before the LOW-to-HIGH transition of the clock, overrides the $\overline{CEP}$, $\overline{CET}$, and $\overline{PE}$ inputs, and causes the flip-flops to go LOW coincident with the positive clock transition. The master reset ($\overline{MR}$) is an asynchronous overriding clear function which forces all stages to a LOW state while the $\overline{MR}$ input is LOW without regard to the clock.

The carry look ahead circuitry provides for cascading counters for n-bit synchronous applications without additional gating. Instrumental in accomplishing this function are two count-enable inputs ($\overline{CET} \bullet \overline{CEP}$) and a terminal count ($\overline{TC}$) output. Both count-enable inputs must be LOW to count. The $\overline{CET}$ input is fed forward to enable the $\overline{TC}$ output. The $\overline{TC}$ output, thus enabled, will produce a LOW output pulse with a duration approximately equal to the HIGH-level portion of the Q_0 output. This LOW-level $\overline{TC}$ pulse is used to enable successive cascaded stages. See the '168 data for the fast synchronous multistage counting connections.

The gated-clock output (GC) is a terminal-count output that provides a HIGH-LOW-HIGH pulse for a duration equal to the LOW time of the clock pulse when $\overline{TC}$ is LOW. The GC output can be used as a clock input for the next stage in a simple ripple-expansion scheme.

The direction of counting is controlled by the up/down ($U/\overline{D}$) input; a HIGH will cause the count to increase, and a LOW will cause the count to decrease.

The active LOW output enable ($\overline{OE}$) input controls the 3-state buffer outputs independent of the counter operation. When $\overline{OE}$ is LOW, the count appears at the buffer outputs. When $\overline{OE}$ is HIGH, the outputs are in the high-impedance OFF-state, which means they will neither drive nor load the bus.

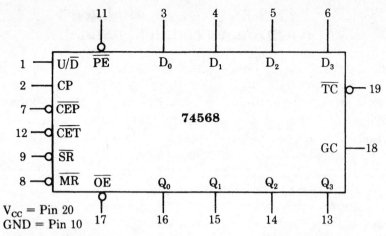

3-STATE BUFFER OPERATING MODES	INPUTS		OUTPUTS
	$\overline{OE}$	Q_n-Counter	Q_0, Q_1, Q_2, Q_3
Read counter	L	L	L
	L	H	H
Disable outputs	H	L	(Z)
	H	H	(Z)

MODE SELECT-FUNCTION TABLE

COUNTER OPERATING MODES	INPUTS								COUNTER STATES			
	$\overline{MR}$	CP	$\overline{SR}$	$U/\overline{D}$	$\overline{PE}$	$\overline{CEP}$	$\overline{CET}$	D_n	Q_0	Q_1	Q_2	Q_3
Asynchronous Reset	L	X	X	X	X	X	X	X	L	L	L	L
Synchronous Reset	H	↑	l	X	X	L	L	X	L	L	L	L
Parallel load	H	↑	h	X	l	X	X	l	L	L	L	L
	H	↑	h	X	l	X	X	h	H	H	H	H
Count up	H	↑	h	h	h	l	l	X	count up			
Count down	H	↑	h	l	h	l	l	X	count down			
Hold (do nothing)	H	↑	h	X	h	h	X	X	no change			
	H	↑	h	X	h	X	h	X	no change			

TERMINAL COUNT TRUTH TABLE

INPUTS				COUNTER STATES				OUTPUTS	
CP	$U/\overline{D}$	$\overline{CEP}$	$\overline{CET}$	Q_0	Q_1	Q_2	Q_3	$\overline{TC}$	GC
H	L	L	L	L	L	L	L	L	H
L	L	L	L	L	L	L	L	L	L
X	L	H	L	L	L	L	L	L	H
X	L	X	H	L	L	L	L	H	H
H	H	L	L	H	X	X	H	L	H
L	H	L	L	H	X	X	H	L	L
X	H	H	L	H	X	X	H	L	H
X	H	X	H	H	X	X	H	H	H

H = High voltage level
h = High voltage level one setup time prior to the low-to-high clock transition
L = Low voltage level
l = Low voltage level one setup time prior to the low-to-high clock transition
X = Don't care
(Z) = High-impedance "off" state
↑ = Low-to-high clock transition

74569 4-Bit binary up/down synchronous counter (3-state)

The '569 is a synchronous presettable module 16 binary up/down counter featuring an internal carry look ahead for applications in high-speed counting designs. Synchronous operation is provided by having all flip-flops clocked simultaneously so that the outputs change coincident with each other when so instructed by the count-enable inputs and internal gating. This mode of operation eliminates the output spikes that are normally associated with asynchronous (ripple clock) counters. A buffered clock input triggers the flip-flops on the LOW-to-HIGH transition of the clock.

The counter is fully programmable; that is, the outputs may be preset to either level. Presetting is synchronous with the clock and takes place regardless of the levels of the count enable inputs. A LOW level on the parallel-enable ($\overline{PE}$) input disables the counter and causes the data at the D_n inputs to be loaded into the counter on the next LOW-to-HIGH transition of the clock. The synchronous reset ($\overline{SR}$), when LOW one setup time before the LOW-to-HIGH transition of the clock, overrides the $\overline{CEP}$, $\overline{CET}$, and $\overline{PE}$ inputs, and cause the flip-flops to go LOW coincident with the positive clock transition. The master reset ($\overline{MR}$) is an asynchronous overriding clear function, which forces all stages to a LOW state while the $\overline{MR}$ input is LOW without regard to the clock.

The carry look ahead circuitry provides for cascading counters for n-bit synchronous applications without additional gating. Instrumental in accomplishing this function are two count enable inputs ($\overline{CET} \bullet \overline{CEP}$) and a terminal count ($\overline{TC}$) output. Both count-enable inputs must be LOW to count. The $\overline{CET}$ input is fed forward to enable the $\overline{TC}$ outputs. The $\overline{TC}$ output, thus enabled, will produce a LOW output pulse with a duration approximately equal to the HIGH-level portion of the Q_0 output. This LOW-level $\overline{TC}$ pulse is used to enable successive cascaded stages. See the '169 data for the fast synchronous multistage counting connections.

The gated-clock output (GC) is a terminal-count output that provides a HIGH-LOW-HIGH pulse for a duration equal to the LOW time of the clock pulse when $\overline{TC}$ is LOW. The GC output can be used as a clock input for the next stage in a simple ripple-expansion scheme. The direction of counting is controlled by the up/down ($U/\overline{D}$) input; a HIGH will cause the count to increase, and a LOW will cause the count to decrease.

The active LOW output enable ($\overline{OE}$) input controls the 3-state buffer outputs independent of the counter operation. When $\overline{OE}$ is LOW, the count appears at the buffer outputs. When $\overline{OE}$ is HIGH, the outputs are in the high-impedance OFF-state, which means they will neither drive nor load the bus.

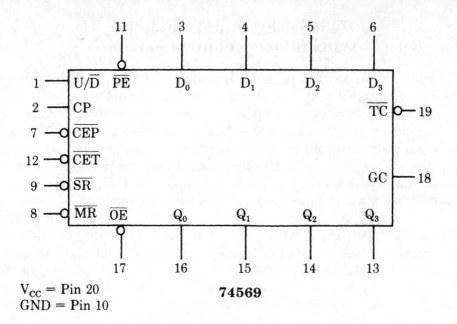

74569

V_{CC} = Pin 20
GND = Pin 10

226

2
SECTION

4xxx Digital

This section covers the 4xxx line of digital/logic devices. Devices range from simple logic gates, through flip-flops, counters, registers, and more exotic and complex special-purpose devices.

This standardized numbering scheme was created specifically for CMOS (Complementary Metal-Oxide Semiconductor) ICs, as opposed to TTL devices. CMOS chips also sometimes use the 74Cxx numbering scheme, with the same pin-outs (though not electrical compatibility) with the standard 74xx numbering scheme. There is no connection between the 4xxx and 74xx numbering schemes. Devices with similar numbers in these two series have different pin-outs, and usually completely different functions. For example, a 7406 is a hex inverter buffer/driver, but a 4006 is a 18-stage static shift register.

Many manufacturers use a "CD" prefix. For example, a CD4009 chip is the same as a 4009. You might also encounter CMOS chips numbered 14xxx. Again, there is no functional difference. A 14018 is identical to a 4018.

Although CMOS devices can perform the same logic functions as TTL devices, they are not electrically compatible. A CMOS gate can not directly drive a TTL gate, or vice versa. They can be interfaced, but extra external circuitry is required. These two logic families do not use the same voltage levels to represent the LOW and HIGH states, so confusion can result. In effect, TTL and CMOS gates don't "speak the same language."

One of the biggest advantages of CMOS over TTL is much greater flexibility in the supply voltage. Most TTL devices require a tightly regulated 5-V supply voltage. CMOS gates can be operated off anything from 3 V to 15 V. Generally speaking, supply voltages of 9 V to 12 V will be the best choice.

Power dissipation in a CMOS gate varies with the operating frequency. At dc or low frequencies, almost no power at all is dissipated. At high frequencies, however, a CMOS gate can dissipate 10 mW to 15 mW, and sometimes even more.

The propagation delay of standard CMOS gates tends to be rather slow. Delays of 90 nS are not uncommon for 4xxx gates.

4000 Dual three-input NOR gate plus inverter

The 4000 contains two independent NOR gates with three inputs each. Because some extra pins and space were left over on this device, an extra inverter stage is also included. Only the power-supply connections are common to all three logic elements.

A NOR gate produces a HIGH input if, and only if, all of its inputs are LOW. If one or more inputs goes HIGH, the output goes LOW.

An inverter reverses the state of the input signal at its output. That is, a HIGH input becomes a LOW output, and vice versa.

Supply Voltage Range	3 to 15 V
Power	10 nW typ.
Noise Immunity	0.45 V_{DD} typ.

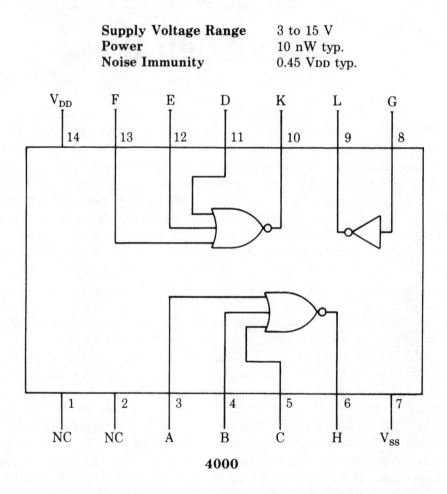

4000

4000 Truth Table (NOR Gates)

	INPUTS			OUTPUT
gate 1	a	b	c	h
gate 2	d	e	f	k
	0	0	0	1
	0	0	1	0
	0	1	0	0
	0	1	1	0
	1	0	0	0
	1	0	1	0
	1	1	0	0
	1	1	1	0

4000 Truth Table (Inverter)

INPUT	OUTPUT
L	G
0	1
1	0

4001 Quad two-input NOR gate

The 4001 contains four independent NOR gates, each with two inputs. Only the power supply connections are common to all four logic elements.

A NOR gate produces a HIGH input if, and only if, all of its inputs are LOW. If one or more inputs goes HIGH, the output goes LOW.

Supply Voltage Range	3 to 15 V
Power	10 nW (typ.)
Noise Immunity	0.45 V_{DD} (typ.)

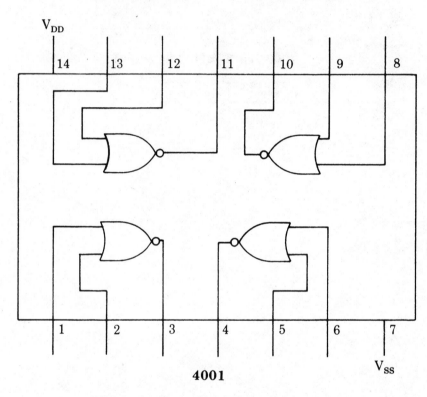

4001

4001 Truth Table

INPUTS		OUTPUT
0	0	1
0	1	0
1	0	0
1	1	0

4002 Dual four-input NOR gate

The 4002 contains a pair of NOR gates, each with four inputs. Only the power supply connections are common to both gates.

As with any NOR gate, the output is HIGH if, and only if, all inputs are LOW. If one or more inputs goes HIGH, the output will go LOW.

Supply Voltage Range	3 to 15 V
Power	10 nW (typical)
Noise Immunity	0.45 V_{DD} (typical)

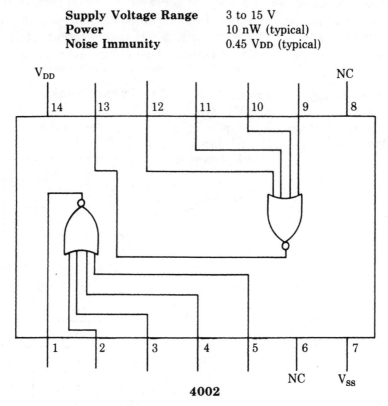

4002

4002 Truth Table

INPUTS				OUTPUT
0	0	0	0	1
0	0	0	1	0
0	0	1	0	0
0	0	1	1	0
0	1	0	0	0
0	1	0	1	0
0	1	1	0	0
0	1	1	1	0
1	0	0	0	0
1	0	0	1	0
1	0	1	0	0
1	0	1	1	0
1	1	0	0	0
1	1	0	1	0
1	1	1	0	0
1	1	1	1	0

4006 18-Stage static shift register

The 4006 18-stage static shift register is comprised of four separate shift register sections of four stages, and two sections of five stages. Each section has an independent data input. Outputs are available at the fourth stage and the fifth stage of each section. A common clock signal is used for all stages. Data is shifted to the next stage on the negative-going (HIGH-to-LOW) transition of the clock. Through appropriate connections of inputs and outputs, multiple-register sections of 4, 5, 8, or 9 stages, or single-register sections of 10, 12, 13, 14, 16, 17, or 18 stages can be implemented using just one package.

Supply Voltage Range	3 to 15 V	
Noise Immunity	0.45 V_{DD} typ.	
Clock Input Capacitance	6 pF typ.	
Speed of Operation	10 MHz typ.	
	with V_{DD} = 10 V	

Truth Table

X = Don't care
Δ = Level change
NC = No change

D	CL^{Δ}	D+1
0		0
1		1
X		NC

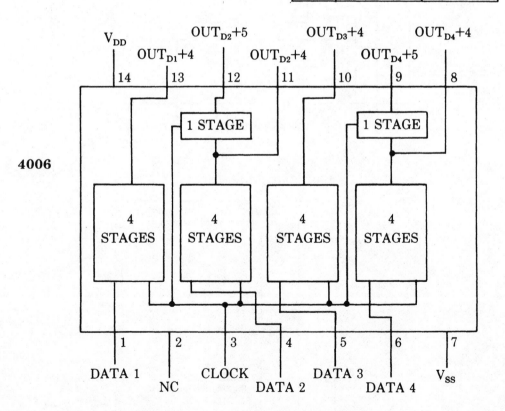

4007 Dual complementary pair plus inverter

The 4007 consists of three complementary pairs of N-channel and P-channel enhancement-mode MOS transistors suitable for series/shunt applications. All inputs are protected from static discharge by diode clamps to V_{DD} and V_{SS}.

For proper operation, the voltages at all pins must be constrained to be between $V_{SS} - 0.3$ V and $V_{DD} + 0.3$ V at all times.

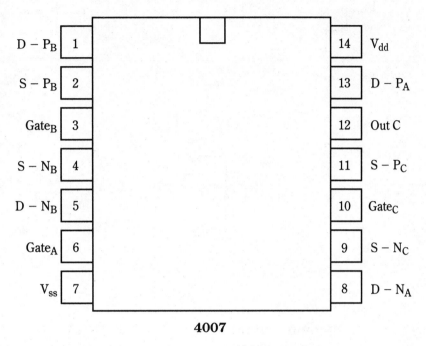

4007

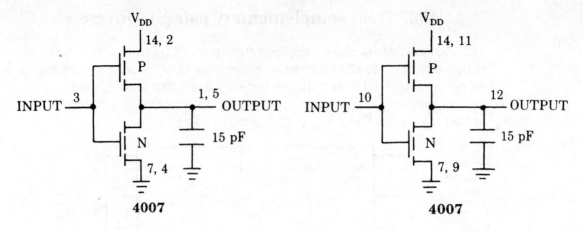

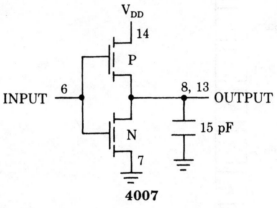

4007

Supply Voltage Range	**3 to 15 V**
Noise Immunity	**0.45 V$_{CC}$ typ.**

4008 4-Bit full adder

The 4008 consists of four full-adder stages with fast look ahead carry provision from stage to stage. Circuitry is included to provide a fast parallel carry out bit to permit high-speed operation in arithmetic sections using several 4008Bs. 4008B inputs include the four sets of bits to be added, A_1 to A_4 and B_1 to B_4, in addition to the carry in bit from a previous section. 4008B outputs include the four sum bits, S_1 and S_4, in addition to the high-speed parallel carry-out which can be utilized at a succeeding CD4008B section. All inputs are protected from damage as a result of static discharge by diode clamps to V_{DD} and GND.

Truth Table

Supply Voltage Range	3 to 15 V
Noise Immunity	0.45 V_{DD} typ.
TTL Compatibility	Fanout of 2 driving 74L
	or 1 driving 74LS
Quiescent Current	15 V
Maximum Input Leakage	1μA at 15 V

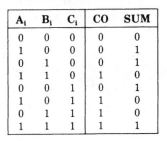

A_i	B_i	C_i	CO	SUM
0	0	0	0	0
1	0	0	0	1
0	1	0	0	1
1	1	0	1	0
0	0	1	0	1
1	0	1	1	0
0	1	1	1	0
1	1	1	1	1

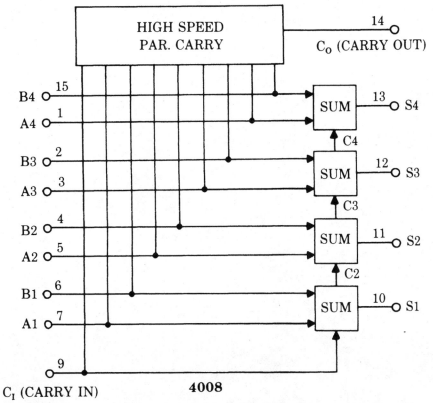

235

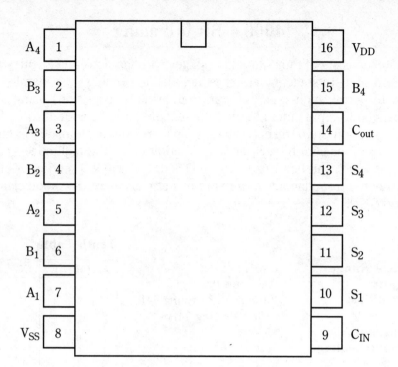

4009 Hex inverter/buffer

The 4009 contains six inverter buffer stages. Only the power-supply connections are common to all logic elements. A buffer is a sort of digital "signal amplifier" to prevent signal degradation in moderate to large logic systems. An inverter reverses the input signal state at its output. That is, a LOW input results in a HIGH output, and a HIGH input produces a LOW output.

Supply Voltage Range	3 to 15 V
Power	100 nW (typical)
Noise Immunity	0.45 V_{DD} (typical)
Current Sinking	
Capability	8 mA (min) at $V_O = 0.5$ V and $V_{DD} = 10$ V

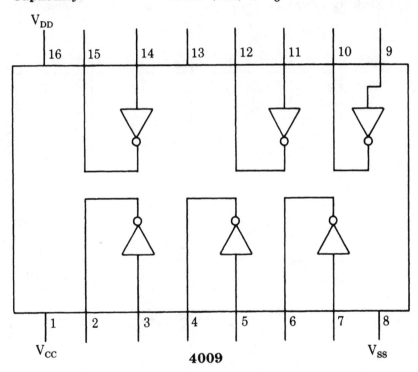

237

4010 Hex buffer (noninverting)

The 4010 is similar to the 4009, except it contains six noninverting buffer stages. Only the power supply connections are common to all logic elements. A buffer is a sort of digital "signal amplifier" to prevent signal degradation in moderate to large logic systems. The output of each buffer will match its input state. That is, a LOW input will result in a LOW output, and a HIGH input will generate a HIGH output.

Supply Voltage Range   3 to 15 V
Power   10 nW (typical)
Noise Immunity   0.45 V_{DD} (typical)
Current Sinking Capability   8 mA (min) at Vo = 0.5 V and V_{DD} = 10 V

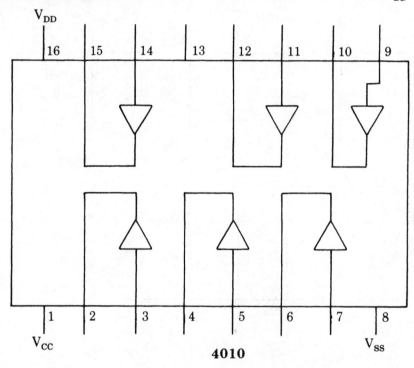

4010

4011 Quad two-input NAND gate

The 4011 contains four of NAND gates, each with two inputs. Only the power supply connections are common to both gates.

As with any NAND gate, the output is LOW if, and only if, all inputs are HIGH. If one or more inputs goes LOW, the output will go HIGH.

Supply Voltage Range	3 to 15 V
Power	10 nW (typical)
Noise Immunity	0.45 V_{DD} (typical)

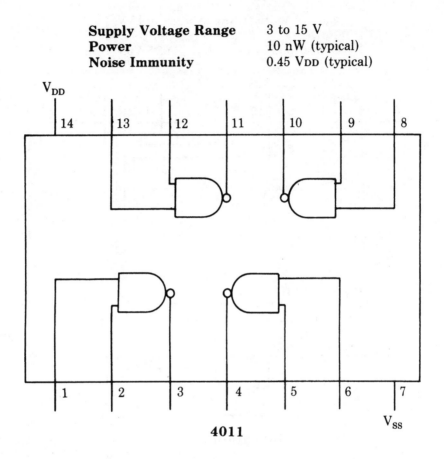

4011

4011 Truth Table

INPUTS		OUTPUT
0	0	1
0	1	1
1	0	1
1	1	0

4012 Dual four-input NAND gate

The 4011 contains a pair of NAND gates, each with four inputs. Only the power supply connections are common to both gates.

As with any NAND gate, the output is LOW if, and only if, all inputs are HIGH. If one or more inputs goes LOW, the output will go HIGH.

Supply Voltage Range   3 to 15 V
Power   10 nW (typical)
Noise Immunity   0.45 V_{DD} (typical)

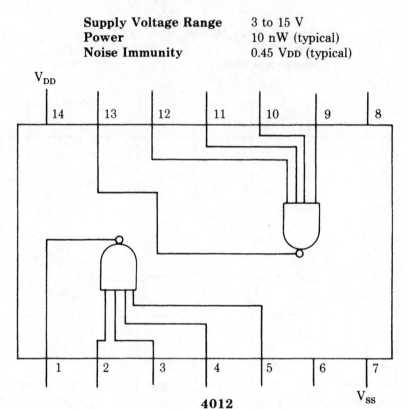

4012

V_{SS}

4012 Truth Table

INPUTS				OUTPUT
0	0	0	0	1
0	0	0	1	1
0	0	1	0	1
0	0	1	1	1
0	1	0	0	1
0	1	0	1	1
0	1	1	0	1
0	1	1	1	1
1	0	0	0	1
1	0	0	1	1
1	0	1	0	1
1	0	1	1	1
1	1	0	0	1
1	1	0	1	1
1	1	1	0	1
1	1	1	1	0

4013 BM dual-D flip-flop

The 4013 dual-D flip-flop is a monolithic complementary-MOS (CMOS) integrated circuit constructed with N-channel and P-channel enhancement transistors. Each flip-flop has independent data, set, reset, and clock inputs and Q and $\overline{Q}$ outputs. These devices can be used for shift register applications, and by connecting $\overline{Q}$ output to the data input, for counter and toggle applications. The logic level present at the D input is transferred to the Q output during the positive-going transition of the clock pulse. Setting or resetting is independent of the clock and is accomplished by a HIGH level on the set or reset line, respectively.

Supply Voltage Range   3 to 15 V
Noise Immunity   0.45 VDD typ.
TTL Compatibility   Fanout of 2 driving 74L or 1 driving 74LS

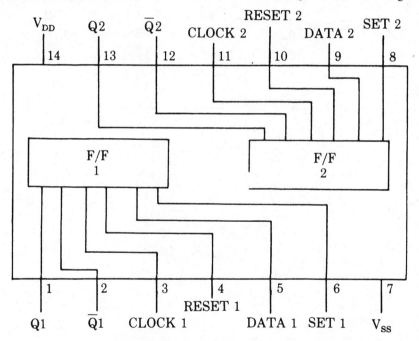

Truth Table

4013

CL†	D	R	S	Q	$\overline{Q}$
⟋	0	0	0	0	1
⟍	1	0	0	1	0
⟍	x	0	0	Q	$\overline{Q}$
x	x	1	0	0	1
x	x	0	1	1	0
x	x	1	1	1	1

No change
† = Level change
x = Don't care case

241

4014 Eight-stage static shift register

The 4014 is an eight-stage parallel input/serial output shift register. A parallel/serial control input enables individual jam inputs to each of the eight stages. Q outputs are available from the sixth, seventh, and eighth stages (though not from the earlier stages).

When the parallel/serial control input is LOW, data is serially shifted into the register synchronously with the positive-going (LOW-to-HIGH) transition of the clock. When the parallel/serial control input is HIGH, data is jammed into each stage of the register synchronously with the positive-going (LOW-to-HIGH) transition of the clock.

Supply Voltage Range	3 to 15 V
Noise Immunity	0.45 Vcc to typ.
Speed of Operation	5 MHz typ.

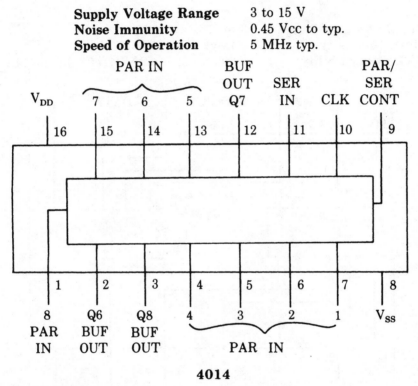

4014

Truth Table

CL^Δ	SERIAL INPUT	PARALLEL/ SERIAL CONTROL	Pl 1	Pl n	Q1 (INTERNAL)	Q_n	
$\nearrow$	X	1	0	0	0	0	
$\nearrow$	X	1	1	0	1	0	
$\nearrow$	X	1	0	1	0	1	
$\nearrow$	X	1	1	1	1	1	
$\nearrow$	0	0	X	X	0	Q_n 1	
$\nearrow$	1	0	X	X	1	Q_n 1	
$\searrow$	X	X	X	X	Q1	Q_n	No Change

Δ = Level change
x = Don't care case

242

4015 Dual 4-bit static register

The 4015 consists of two identical, independent, 4-stage serial-input/parallel-output registers. Each register has independent clock and reset inputs, as well as a single serial-data input. Q outputs are available from each of the four stages on both registers. All register stages are D-type, master-slave flip-flops. The logic level present at the data input is transferred into the first register stage and shifted over one stage at each positive-going clock transition. Resetting of all stages is accomplished by a HIGH level on the reset line. Register expansion to eight stages using one 4015 package, or to more than eight stages using additional 4015 is possible. All inputs are protected from static discharge by diode clamps to V_{DD} and V_{SS}.

Supply Voltage Range 3 to 15 V
Noise Immunity 0.45 Vcc typ.
Speed of Operation 9 MHz (typ.) clock rate at $V_{DD} - V_{SS} = 10$ V

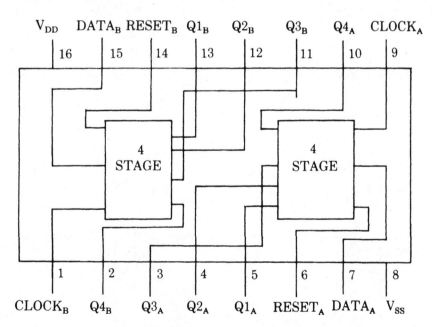

Truth Table

4015

CL$^\Delta$	D	R	Q1	Qn
⟋	0	0	0	$Q_n 1$
⟋	1	0	1	$Q_n 1$
⟍	X	0	Q1	Q_n
X	X	1	0	0

(No change)
$^\Delta$ Level change.
X Don't care case.

243

4016 Quad bilateral switch

The 4016 is a quad bilateral switch with an extremely HIGH off-resistance and LOW on-resistance. The switch will pass analog or digital signals in either direction, and is extremely useful in digital switching. In effect, it is the equivalent of an ordinary mechanical switch, except it is operated by logic signals on the appropriate control input pin, rather than physical motion or pressure.

Supply Voltage Range	3 to 15 V
Noise Immunity	0.45 Vcc typ.
Digital and Analog Levels	$\pm$ 7.5 V_{peak}
On Resistance	300Ω typ.
	$V_{DD} - V_{DD} = 15$ V
Switch Characteristics	Δ $R_{ON} = 40\Omega$ typ.
On/Off Output	65 dB typ.
Voltage Ratio	@ fis = 10 kHz R_L = 10 kΩ
Linearity	.5% distortion typ.
	@ fis = 1 kHz
Leakage	$V_{is} = 5$ V_{p-p}
	$V_{DD} - V_{SS} = 10$ V
	R_L = 10 kΩ

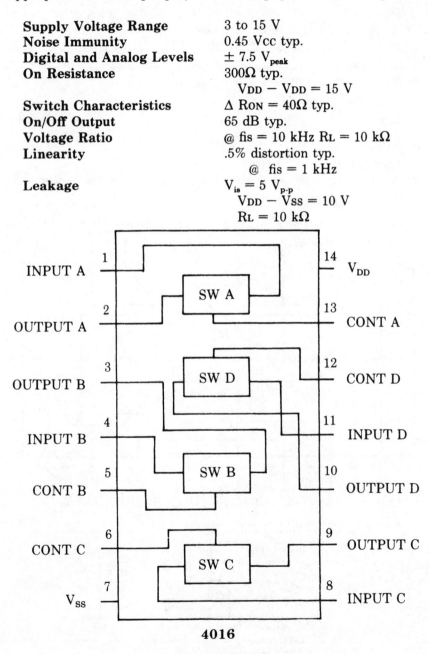

4016

4017 Decade counter/divider with 10 decoded outputs

The 4017 is a 5-stage divide-by-10 Johnson counter with 10 decoded outputs and a carry-out bit. The 4022 is a 4-stage divide-by-8 Johnson counter with eight decoded outputs and a carry-out bit.

These counters are cleared to their zero count by a logical 1 on their reset line. These counters are advanced on the positive edge of the clock signal when the clock-enable signal is in the logic 0 state.

The configuration of the 4017 and 4022 permits medium-speed operation and ensures a hazard-free counting sequence. The 10/8 decoded outputs are normally in the logic 0 state and go to the logic 1 state only at their respective time slot. Each decoded output remains high for one full clock cycle. The carry-out signal completes a full cycle for every 10/8 clock input cycles and is used as a ripple-carry signal to any succeeding stages.

Supply Voltage Range	3 to 15 V
Noise Immunity	0.45 V_{DD} typ.
TTL Compatibility	Fanout of 2 driving 74L or 1 driving 74LS
Speed of Operation	5.0 MHz typ. with 10 V V_{DD}
Power	10μ W typ.

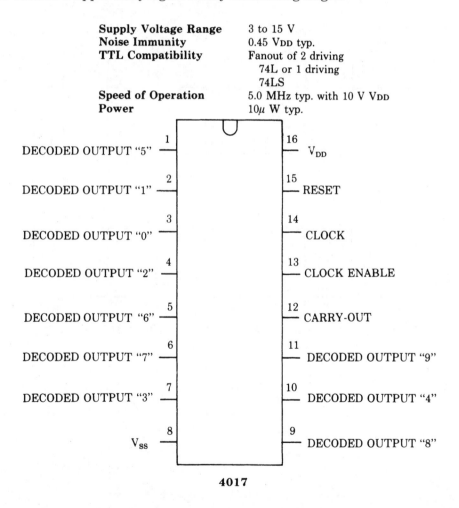

4017

4018 Presettable divide-by-N counter

The 4018B consists of five Johnson counter stages. A buffered $\overline{Q}$ output from each stage, clock, reset, data, preset enable, and five individual jam inputs are provided. The counter is advanced one count at the positive clock signal transition. A HIGH reset signal clears the counters to an all-zero condition. A HIGH preset-enable signal allows information on the jam inputs to preset the counter. Antilock gating is provided to assure the proper counting sequence.

Supply Voltage Range	3 to 15 V
Noise Immunity	0.45 VDD typ.
TTL Compatibility	Fanout of 2 driving 74L
	or 1 driving 74LS

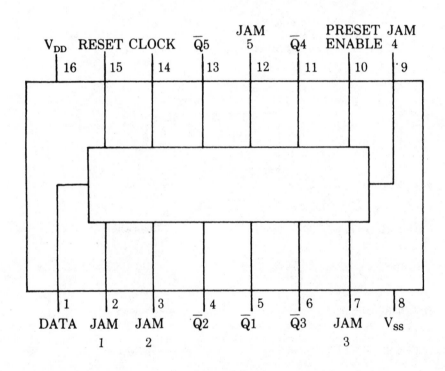

4018

4019 Quad and/or select gate

The 4019 contains four AND/OR select gate configurations. Each of these is made up of two 2-input AND gates driving a single 2-input OR gate. The Truth Table below outlines the operation of any one of these four compound gates. In effect, data selection is accomplished by the common control bits K_a and K_b.

All inputs are protected against static discharge damage.

Supply Voltage Range	3 to 15 V
Noise Immunity	0.45 V_{DD} typ.
TTL Compatibility	Driving 74L or 1 driving 74LS

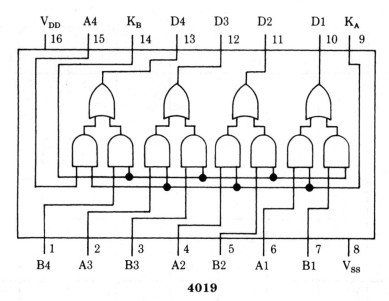

4019

4019 Truth Table

INPUTS				OUTPUT
K_a	A	K_b	B	D
0	0	0	0	0
0	0	0	1	0
0	0	1	0	0
0	0	1	1	1
0	1	0	0	0
0	1	0	1	0
0	1	1	0	0
0	1	1	1	1
1	0	0	0	0
1	0	0	1	0
1	0	1	0	0
1	0	1	1	1
1	1	0	0	1
1	1	0	1	1
1	1	1	0	1
1	1	1	1	1

4020 14-Stage ripple-carry binary counter

The 4020 (and the 4060, presented later in this section) is a 14-stage ripple-carry binary counter. The counter is advanced one count on the negative-going (HIGH-to-LOW) transition of each clock pulse. The counters are reset to the zero (all LOW) state by a HIGH signal at the reset input, regardless of the clock.

   The 4040 is similar to the 4020, except it has just 12 stages, instead of 14.

Supply Voltage Range	1 to 15 V
Noise Immunity	0.45 V_{DD} typ.
TTL Compatibility	Fanout of 2 driving 74L or 1 driving 74LS
Speed of Operation	8 MHz typ. at V_{DD} = 10 V

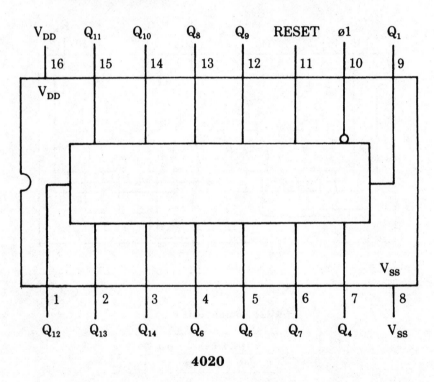

4020

4021 8-Stage static-shift register

The 4021 is an 8-stage parallel input/serial output shift register. A parallel/serial control input enables individual jam inputs to each of eight stages. Q outputs are available from the sixth, seventh, and eighth stages.

When the parallel/serial control input is in the logic 0 state, data is serially shifted into the register synchronously with the positive transition of the clock. When the parallel/serial control is in the logic 1 state, data is jammed into each stage of the register asynchronously with the clock.

Supply Voltage Range   3 to 15 V
Noise Immunity   0.45 Vcc typ.
Speed of Operation   5 MHz typ.

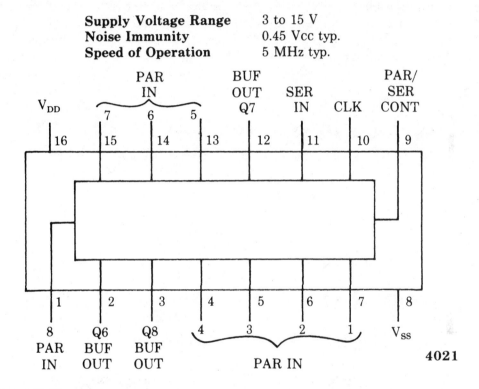

Truth Table

CL$^\Delta$	SERIAL INPUT	PARALLEL/ SERIAL CONTROL	Pl 1	Pl n	Q1 (INTERNAL)	Q_n
X	X	1	Q	0	0	0
X	X	1	0	1	0	1
X	X	1	1	0	1	0
X	X	1	1	1	1	1
↗	0	0	X	X	0	Q_n 1
↗	1	0	X	X	1	Q_n 1
↘	X	0	X	X	Q1	Q_n

No Change

$^\Delta$ Level change

X Don't care case

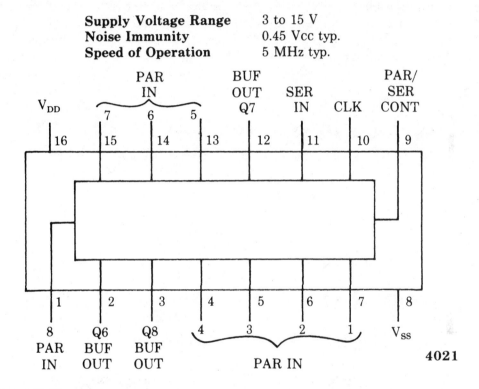

4022 Divide-by 8 counter/divider
with 8 decoded outputs

The 4017 is a 5-stage divide-by-10-Johnson counter with 10 decoded outputs and a carry-out bit. The 4022 is a 4-stage divide-by-8 Johnson counter with eight decoded outputs and a carry-out bit.

These counters are cleared to their zero count by a logic 1 on their reset line. These counters are advanced on the positive edge of the clock signal when the clock-enable signal is in the logic 0 state.

The configuration of the 4017 and 4022 permits medium-speed operation and ensures a hazard-free counting sequence. The 10/8 decoded outputs are normally in the logic 0 state and go to the logic 1 state only at their respective time slot. Each decoded output remains high for one full clock cycle. The carryout signal completes a full cycle for every 10/8 clock input cycles and is used as a ripple-carry signal to any succeeding stages.

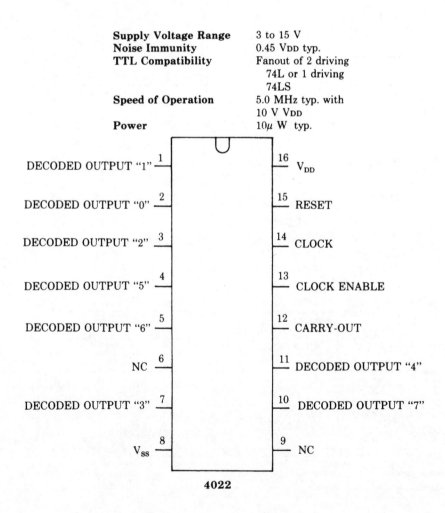

Supply Voltage Range	3 to 15 V
Noise Immunity	0.45 V_{DD} typ.
TTL Compatibility	Fanout of 2 driving 74L or 1 driving 74LS
Speed of Operation	5.0 MHz typ. with 10 V V_{DD}
Power	10μ W typ.

DECODED OUTPUT "1" — 1

DECODED OUTPUT "0" — 2

DECODED OUTPUT "2" — 3

DECODED OUTPUT "5" — 4

DECODED OUTPUT "6" — 5

NC — 6

DECODED OUTPUT "3" — 7

V_{ss} — 8

16 — V_{DD}

15 — RESET

14 — CLOCK

13 — CLOCK ENABLE

12 — CARRY-OUT

11 — DECODED OUTPUT "4"

10 — DECODED OUTPUT "7"

9 — NC

4022

4023 Triple 3-input NAND gate

The 4023 contains three independent NAND gates, each with three inputs. The output of a NAND (or "Not AND") gate is LOW if, and only if, all inputs are HIGH. If one or more inputs goes LOW, the output will go HIGH.

Only the power supply connections are common to all three gates on this chip. All inputs are protected against static discharge and latching conditions.

Supply Voltage Range   3 to 15 V
Power   10 nW (typical)
Noise Immunity   0.45 V_{DD} (typical)

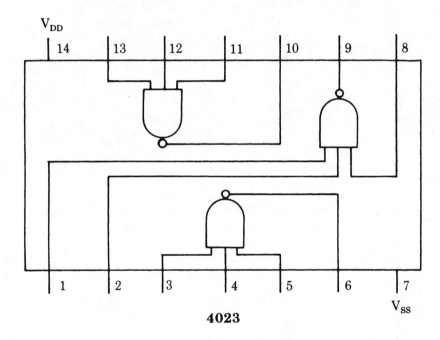

4023

4023 Truth Table

INPUTS			OUTPUT
a	b	c	
0	0	0	1
0	0	1	1
0	1	0	1
0	1	1	1
1	0	0	1
1	0	1	1
1	1	0	1
1	1	1	0

4024 Seven-stage ripple-carry binary counter

The 4024 is a ripple-carry binary counter with seven stages. Buffered outputs are externally available from stages 1 through 7. The counter is reset to 0 (all LOWs) by a HIGH signal on the reset input. The counter is advanced one count on the negative-going (HIGH-to-LOW) transition of each clock pulse.

Supply Voltage Range	3 to 15 V
Noise Immunity	0.45 V_{DD} typ.
TTL Compatibility	Fanout of 2 driving 74L
	or 1 driving 74LS
Speed	12 MHz (typ.)
	input pulse rate
	$V_{DD} - V_{SS} = 10$ V

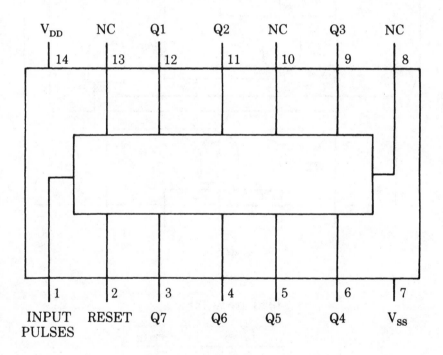

4024

4025 Triple 3-input NOR gate

The 4025 contains three independent NOR gates, each with three inputs. The output of a NOR (or "Not OR") gate is HIGH if, and only if, all inputs are LOW. If one or more inputs goes HIGH, the output will go LOW.

Only the power supply connections are common to all three gates on this chip. All inputs are protected against static discharge and latching conditions.

Supply Voltage Range	3 to 15 V
Noise Immunity	0.45 V_{DD} typ.
TTL Compatibility	Fanout of 2 driving 74L or 1 driving 74LS
Maximum Input Leakage	1 μA at 15 V

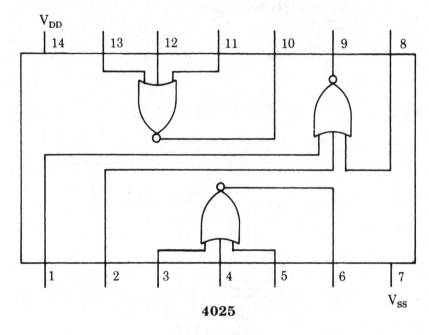

4025

4025 Truth Table

INPUTS			OUTPUT
a	b	c	
0	0	0	1
0	0	1	0
0	1	0	0
0	1	1	0
1	0	0	0
1	0	1	0
1	1	0	0
1	1	1	0

4027 Dual JK master/slave flip-flop

These dual JK flip-flops are monolithic complementary-MOS (CMOS) integrated circuits constructed with N-channel and P-channel enhancement-mode transistors. Each flip-flop has independent J, K, set, reset and clock inputs, and buffered Q and $\overline{Q}$ outputs. These flip-flops are edge-sensitive to the clock inputs and change state on the positive-going transition of the clock pulses. Set or reset is independent of the clock and is accomplished by a high level on the respective input. All inputs are protected against damage as a result of static discharge by diode clamps to V_{DD} and V_{SS}.

Supply Voltage Range	3 to 15 V
Noise Immunity	0.45 V_{DD} typ.
TTL Compatibility	Fanout of 2 driving 74L or 1 driving 74LS
Power	50 nW typ.
Speed of Operation	12 MHz typ. with 10 V supply

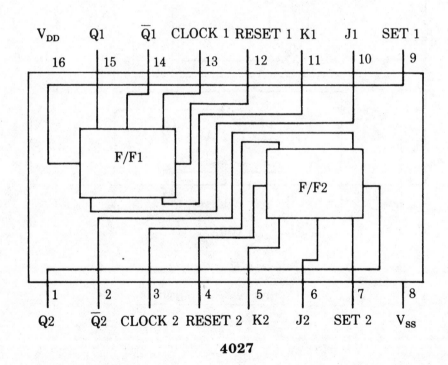

4027

4028 BCD-to-decimal decoder

The 4028 is a BCD (Binary Coded Decimal) to decimal or binary-to-octal decoder. It consists of four inputs, decoding logic gates, and 10 output buffers. A BCD code applied to the four inputs (A, B, C, and D) results in a HIGH level at the selected one-of-10 decimal decoded outputs (0–9). Similarly, a three-bit binary code applied to inputs A, B, and C is decoded in octal at outputs 0 through 7. A HIGH signal at the D input inhibits octal decoding and causes outputs 0 through 7 to go LOW.

All inputs are protected against static discharge damage by diode clamps to V_{DD} and V_{SS}.

Supply Voltage Range	3 to 15 V
Noise Immunity	0.45 V_{DD} typ.
TTL Compatibility	Fanout of 2 driving 74L
Power	or 1 driving 74LS

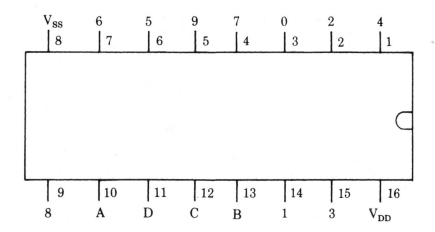

4028

Truth Table

D	C	B	A	0	1	2	3	4	5	6	7	8	9
0	0	0	0	1	0	0	0	0	0	0	0	0	0
0	0	0	1	0	1	0	0	0	0	0	0	0	0
0	0	1	0	0	0	1	0	0	0	0	0	0	0
0	0	1	1	0	0	0	1	0	0	0	0	0	0
0	1	0	0	0	0	0	0	1	0	0	0	0	0
0	1	0	1	0	0	0	0	0	1	0	0	0	0
0	1	1	0	0	0	0	0	0	0	1	0	0	0
0	1	1	1	0	0	0	0	0	0	0	1	0	0
1	0	0	0	0	0	0	0	0	0	0	0	1	0
1	0	0	1	0	0	0	0	0	0	0	0	0	1

4029 Presettable binary/decade up/down counter

The 4029 is a presettable up/down counter that counts in either binary or decade mode depending on the voltage level applied at binary/decade input. When binary/decade is at logic 1, the counter counts in binary; otherwise; it counts in decade. Similarly, the counter counts up when the up/down input is at logic 1 and vice versa.

At logic 1 preset enable signal allows information at the jam inputs to preset the counter to any state asynchronously with the clock. The counter is advanced one count at the positive-going edge of the clock if the carry in and preset enable inputs are at logic 0. Advancement is inhibited when either or both of these two inputs are at logic 1. The carry-out signal is normally at the logic 1 state and goes to the logic 0 state when the counter reaches its maximum count in the Up mode or the minimum count in the Down mode, provided the carry input is at logic 0 state. All inputs are protected against static discharge by diode clamps to both V_{DD} and V_{SS}.

Supply Voltage Range	3 to 15 V
Noise Immunity	0.45 V_{DD} typ.
TTL Compatibility	Fanout of 2 driving 74L or 1 driving 74LS

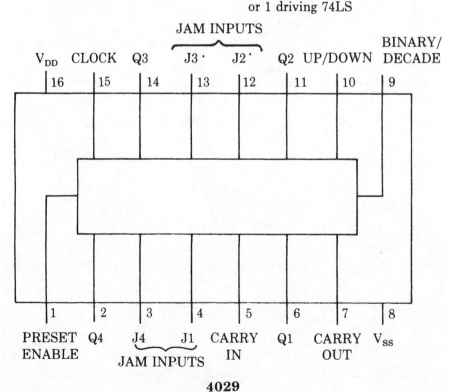

4029

4030 Quad Exclusive-OR gate

The 4030 contains four two-input Exclusive-OR (X-OR) gates, of two inputs each. The output of each gate goes HIGH if, and only if, one of the inputs is HIGH, but not both. If both inputs are LOW, or if both inputs are HIGH, the output will go LOW. Because the output is HIGH only when the two inputs are in opposite states, an X-OR gate can serve as a simple one-bit digital comparator or difference detector.

Only the power supply connections are common to all four gates on this chip. All inputs are protected against static discharge damage with diode clamps to V_{DD} and V_{SS}.

Supply Voltage Range	3 to 15 V
Power	100 nW (typ.)
Speed of Operation	40 ns (typ.)
Noise Immunity	0.45 Vcc (typ.)

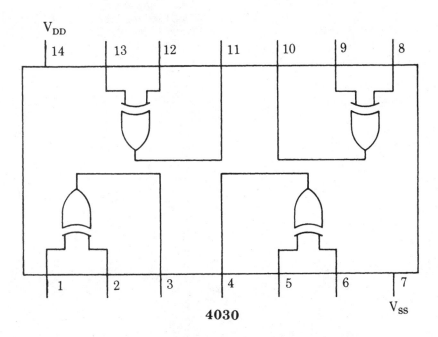

4030

4030 Truth Table

INPUTS		OUTPUT
0	0	0
0	1	1
1	0	1
1	1	0

Truth Table

A	B	J
0	0	0
1	0	1
0	1	1
1	1	0

Where: "1" = High Level
"0" = Low Level

257

4031 64-Stage static-shift register

The 4031 is an integrated, complementary-MOS (CMOS), 64-stage, fully static-shift register. Two data inputs, data in and recirculate in, and a mode-control input are provided. Data at the data input (when mode control is LOW) or data at the recirculate input (when mode control is HIGH), which meets the setup and hold time requirements, are entered into the first stage of the register and are shifted one stage at each positive transition of the clock.

Data output is available in both true and complement forms from the 64th stage. Both the data-out (Q) and data-out ($\overline{Q}$) outputs are fully buffered.

The clock input of the 4031BM/4031BC is fully buffered and presents only a standard input load capacitance. However, a delayed clock output (CL_D) allows reduced clock drive-fanout and transition time requirements when cascading packages.

Supply Voltage Range	3 to 15 V
Noise Immunity	0.45 V_{DD} typ.
TTL Compatibility	Fanout of 2 driving 74L
	or 1 driving 74LS
Range of Operation	Dc to 8 MHz
	(typical @ V_{DD} = 10 V)
Clock Input	5 pF (typ.)
	Input Capacitance
High Current Sinking Capability, Q Output	1.6 mA
	@ V_{DD} = 5 V and 25°C

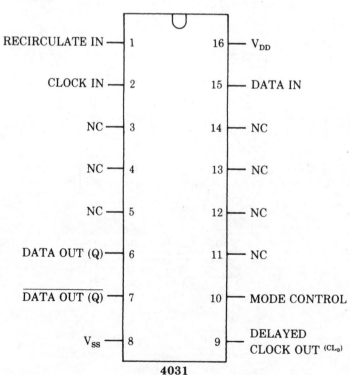

4031

MODE CONTROL (data selection)

MODE CONTROL	DATA IN	RECIRCULATE IN	DATA INTO FIRST STAGE
0	0	X	0
0	1	X	1
1	X	0	0
1	X	1	1

EACH STAGE

D_n	CL	Q_n
0	╱	0
1	╱	1
X	╲	NC

X = irrelevant
NC = no change
╱ = Low to High level transition
╲ = High to Low level transition

4032 Triple serial adder

The 4032 contains three serial adders, with common clock and carry reset inputs. The carry is added on the positive-going (LOW-to-HIGH) clock transition. The outputs are buffered.

The 4032 is intended for applications as digital correlator, serial arithmetic units, datalink computers, and servo-control systems. This device is similar to the 4032, except the triggering polarity of the clock pulse is reversed.

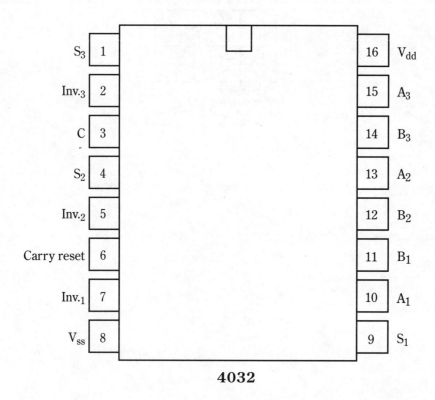

Pin	Label		Pin	Label
1	S_3		16	V_{dd}
2	Inv.$_3$		15	A_3
3	C		14	B_3
4	S_2		13	A_2
5	Inv.$_2$		12	B_2
6	Carry reset		11	B_1
7	Inv.$_1$		10	A_1
8	V_{ss}		9	S_1

4032

4034 8-Stage 3-state bidirectional parallel/ serial input/output bus register

The 4034 is an 8-bit CMOS static-shift register with two parallel bidirectional data ports (A and B) which, when combined with serial-shifting operations, can be used to bidirectionally transfer parallel data between two buses, convert serial data to parallel form and direct them to either of two buses, store (recirculate) parallel data, or accept parallel data from either of two buses and convert them to serial form. These operations are controlled by five control inputs:

- **A enable (AE)** A data port is enabled only when AE is at logic 1. This action allows the use of a common bus for multiple packages.
- **A-bus-to-B-bus/B-bus-to-A-bus (A/B)** This input controls the direction of data flow. When at logic 1, data flows from port A to B (A is input and B is output). When at logic 0, the data-flow direction is reversed.
- **Asynchronous/synchronous (A/S)** When A/S is at logic 0, data transfer occurs at positive transition of the clock. When A/S is at logic 1, data transfer is independent of the clock for parallel operation. In the serial mode, A/S input is internally disabled such that the operation is always synchronous. Asynchronous serial operation is not possible.
- **Parallel/serial (P/S)** A logic 1 P/S input allows data transfer into the registers via A or B port (synchronous if A/S = logic 0 and asynchronous if A/S = logic 1). A logic 0 P/S allows serial data to transfer into the register synchronously with the positive transition of the clock, independent of the A/S input.
- **Clock** Single phase, enabled only in synchronous mode. Either P/S = logic 1 and A/S = logic 0 or P/S = logic 0.

All register stages are D-type master-slave flip-flops with separate master and slave clock inputs, generated internally to allow synchronous or asynchronous data transfer from master to slave. All inputs are protected against damage as a result of static discharge by diode clamps to V_{DD} and V_{SS}.

Supply Voltage Range	3 to 18 V	
Noise Immunity	0.45 V_{DD} typ.	
TTL Compatibility	Fanout of 2 driving 74L	
	or 1 driving 74LS	

```
      V_DD    A8     A7     A6     A5     A4     A3     A2     A1   CLOCK   A/S    P/S

       24     23     22     21     20     19     18     17     16     15     14     13
    ┌─────┬──────┬──────┬──────┬──────┬──────┬──────┬──────┬──────┬──────┬──────┬──────┐
    │                                                                                   │
    │                                                                                   │
    │                                                                                   │
    │                                                                                   │
    └─────┬──────┬──────┬──────┬──────┬──────┬──────┬──────┬──────┬──────┬──────┬──────┘
       1      2      3      4      5      6      7      8      9     10     11     12

      B8     B7     B6     B5     B4     B3     B2     B1     AE  SERIAL   A/B    V_SS
                                                                 DATA
```

4034

4038 Triple serial adder

The 4038 contains three serial adders, with common clock and carry reset inputs. The carry is added on the negative-going (HIGH-to-LOW) clock transition. The outputs are buffered.

The 4038 is intended for applications as digital correlators, serial arithmetic units, datalink computers, and servo-control systems. This device is similar to the 4032, except the triggering polarity of the clock pulse is reversed.

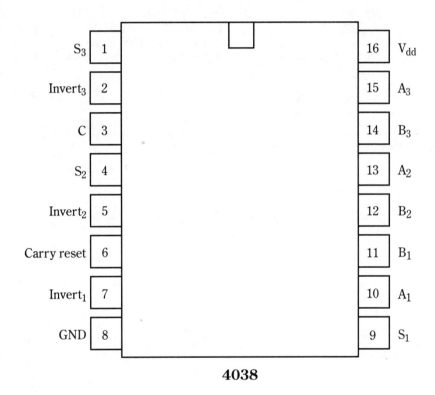

4038

4040 14-Stage ripple-carry binary counter

The 4020 and 4060 are 14-stage ripple-carry binary counters, and the 4040BM/4040BC is a 12-stage ripple-carry binary counter. The counters are advanced one count on the negative transition of each clock pulse. The counters are reset to the zero state by a logic 1 at the reset input that is independent of clock.

Supply Voltage Range	1 to 15 V
Noise Immunity	0.45 V_{DD} typ.
TTL Compatibility	Fanout of 2 driving 74L or 1 driving 74LS
Speed of Operation	8 MHz typ. at $V_{DD} = 10$ V

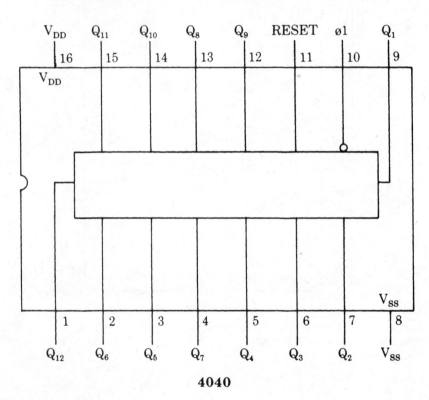

4040

4041 Quad true/complement buffer

The 4041 is a quad true/complement buffer consisting of N-channel and P-channel enhancement-mode transistors that have low-channel resistance and high-current (sourcing and sinking) capability. The 4041 is intended for use as a buffer, line driver, or CMOS-to-TTL driver. All inputs are protected from static discharge by diode clamps to V_{DD} and V_{SS}.

Supply Voltage Range	3 to 15 V
Noise Immunity	40% V_{DD} typ.
True Output:	
High Current Source and Sink Capability	8 mA (typ.) @ V_O = 9.5 V, V_{DD} = 10 V
	3.2 mA (typ.) @ V_O = 0.4 V, V_{DD} = 5 V (two TTL loads)
Complement Output:	Medium current source and sink capability
	3.6 mA (typ.) @ V_O = 9.5 V, V_{DD} = 10 V
	1.6 mA (typ.) @ V_O = 0.4 V, V_{DD} = 5 V

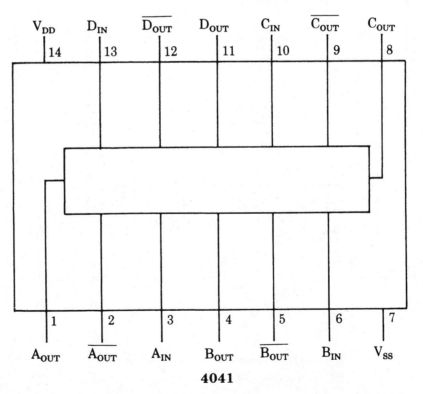

4041

4042 Quad clocked D latch

The 4042 contains four clocked D-type latches. Complementary outputs Q and $\overline{Q}$ either latch or follow the data input, depending on the clock level that is programmed by the polarity input. If the polarity input is made LOW, the information at the data inputs is transferred to the appropriate Q and $\overline{Q}$ outputs during the LOW clock level. Feeding a HIGH control signal into the polarity input reverses this, and the data is transferred from the D inputs to the appropriate Q and $\overline{Q}$ outputs during the HIGH portion of the clock pulse.

When a clock transition occurs (positive-going for polarity-LOW, or negative-going for polarity-HIGH), the information present at the input during the clock transition is latched and retained until an opposite clock transition occurs.

Supply Voltage Range	3 to 15 V
Noise Immunity	0.45 VDD typ.
TTL Compatibility	Fanout of 2 driving 74 L or 1 driving 74LS

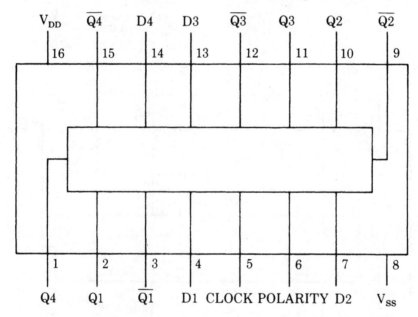

4042

Truth Table

CLOCK	POLARITY	Q
0	0	D
⌐	0	Latch
1	1	D
⌐_	1	Latch

4043 Quad 3-state NOR R/S latches

The 4043 is a quad cross-coupled tri-state CMOS NOR latch. It is similar to the CD4044, which is a quad cross-coupled tri-state CMOS NAND latch. Each latch on this chip has a separate Q output and individual set and reset inputs. It has a common tri-state enable input for all four latches. A HIGH on this enable input effectively disconnects the latch states from the Q outputs, resulting in an open-circuit condition or a high-impedance (Z) state on the Q outputs. A LOW signal must be fed to the enable input to permit normal operation. The tri-state feature allows common bussing of the outputs.

Supply Voltage Range   3 to 15 V
Power   100 nW typ.
Noise Immunity   0.45 V_{DD} typ.

Truth Table

S	R	E	Q
X	X	0	OC
0	0	1	NC
1	0	1	1
0	1	1	0
1	1	1	Δ

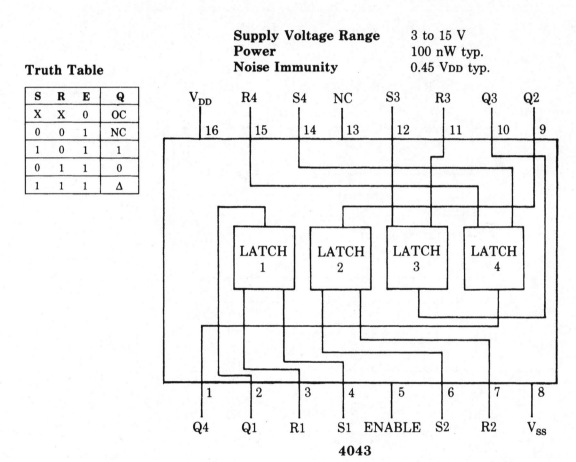

4043

4044 Quad 3-state NAND R/S latches

The 4044 is a quad cross-coupled tri-state CMOS NAND latch. It is similar to the CD4043, which is a quad cross-coupled tri-state CMOS NOR latch. Each latch on this chip has a separate Q output and individual set and reset inputs. It has a common tri-state enable input for all four latches. A HIGH on this enable input effectively disconnects the latch states from the Q outputs, resulting in an open-circuit condition or a high-impedance (Z) state on the Q outputs. A LOW signal must be fed to the enable input to permit normal operation. The tri-state feature allows common bussing of the outputs.

Supply Voltage Range	3 to 15 V
Power	100 nW typ.
Noise Immunity	0.45 V_{DD} typ.

Truth Table

S	R	E	Q
X	X	0	OC
1	1	1	NC
0	1	1	1
1	0	1	0
0	0	1	ΔΔ

OC — TRI-STATE
NC — No change
X — Don't care
Δ — Dominated by
       S=1 input
ΔΔ — Dominated by
        R=0 input

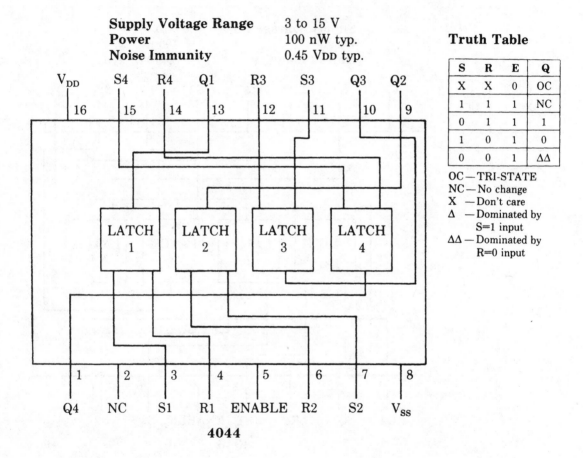

4044

268

4046 Micropower phase-locked loop

The 4046 micropower phase-locked loop (PLL) consists of a low-power, linear voltage-controlled oscillator (VCO), a source follower, a zener diode, and two phase comparators. The two phase comparators have a common signal input and a common comparator input. The signal input can be directly coupled for a large voltage signal, or capacitively coupled to the self-biasing amplifier at the signal input for a small voltage signal.

Phase comparator 1, an exclusive OR gate, provides a digital error signal (phase comparator 1 out) and maintains 90-degree shifts at the VCO center frequency. Between the signal input and the comparator input (both at 50-percent duty cycle), it can lock onto the signal input frequencies that are close to harmonics of the VCO center frequency.

Phase comparator 11 is an edge-controlled digital memory network. It provides a digital error signal (phase comparator 11 out) and lock-in signal (phase pulses) to indicate a locked condition and maintains a 0-degree phase shift between signal input and comparator input. The linear voltage-controlled oscillator (VCO) produces an output signal (VCO$_{out}$) whose frequency is determined by the voltage at the VCO$_{IN}$ input and the capacitor and resistors connected to pin C1$_A$, C1$_B$, R$_1$, and R$_2$.

The source-follower output of the VCO$_{IN}$ (demodulator out) is used with an external register of 10 kΩ or more. The inhibit input, when HIGH, disables the VCO and source follower to minimize standby power consumption. The zener diode is provided for power-supply regulation, if necessary.

Supply Voltage Range	3 to 18 V
Dynamic Power Consumption	70μ W (typ.) at fo = 10 kHz, V_{DD} = 5 V
VCO Frequency	1.3 MHz (typ.) at V_{DD} = 10 V
Frequency Drift	0.06%/°C at V_{DD} = 10 V
VCO Linearity	1% (typ.)

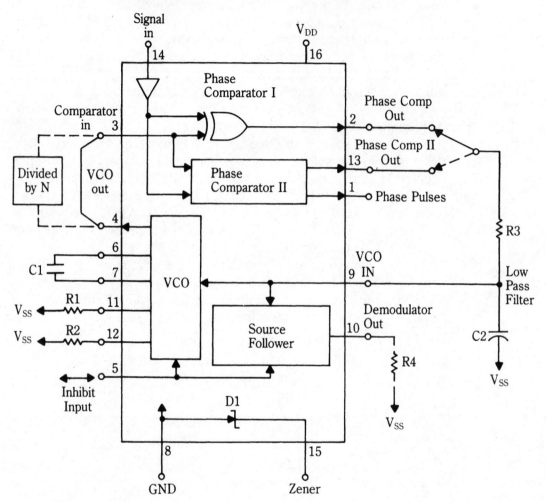

4047 Low power monostable/astable multivibrator

The 4047 is capable of operating in either the monostable or astable mode. It requires an external capacitor between pins 1 and 3 and an external resistor between pins 2 and 3 to determine the output pulse width in the monostable mode, and the output frequency in the astable mode.

Astable operation is enabled by a HIGH level on the astable input or LOW level on the astable input. The output frequency (at 50-percent duty cycle) at the Q and $\overline{Q}$ outputs is determined by the timing components. A frequency twice that of Q is available at the oscillator output. A 50-percent duty cycle is not guaranteed.

Monostable operation is obtained when the device is triggered by LOW-to-HIGH transition at + trigger input or HIGH-to-LOW transition at – trigger input. The device can be retriggered by applying a simultaneous LOW-to-HIGH transition to both the + trigger and retrigger inputs. A HIGH level on reset input resets the outputs Q to LOW and $\overline{Q}$ to HIGH.

Supply Voltage Range	3 to 15 V
Noise Immunity	0.45 V_{DD} typ.
TTL Compatibility	Fanout of 2 driving 74L or driving 74LS

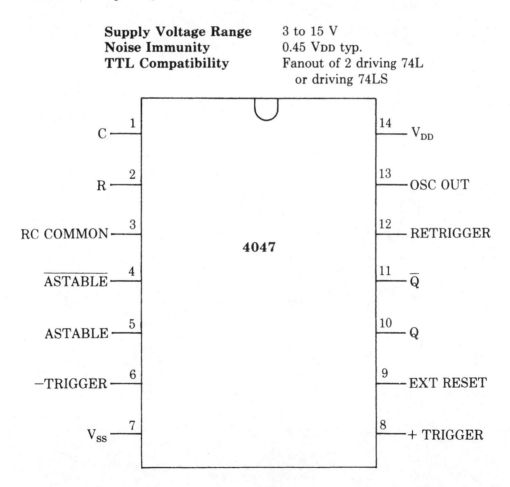

4048 Tri-state expandable eight-function eight-input gate

The 4048 is a programmable eight-input gate. The user can determine the logic function. Three binary control lines (K_a, K_b, and K_c) determine the eight different logic functions available with this gate. These functions are OR, NOR, AND, NAND, OR/AND, OR/NAND, AND/OR, and AND/NOR.

A fourth control input (K_d) is the tri-state function control. When K_d is HIGH, the output is enabled. When K_d is LOW, the output is forced into the third high-impedance (Z) state. This feature allows the user to connect the device to a common bus line.

The expanded input permits the user to increase the number of gate inputs. For example, two eight-input 4048s can be cascaded into a 16-input multi-function gate. When the expand input is not used, it should be tied to V_{ss}. All inputs on this chip are buffered and protected against electrostatic effects.

Supply Voltage Range   3 to 15 V
Noise Immunity   0.45 V_{DD} typ.
TTL Compatibility   Drives 1 standard
        TTL load at $V_{CC} = 5$ V, over full
        temperature range.

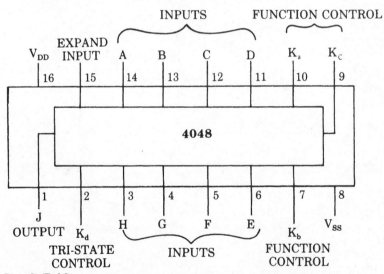

Truth Table

OUTPUT FUNCTION	BOOLEAN EXPRESSION	CONTROL INPUTS				UNUSED INPUTS
		K_a	K_b	K_c	K_d	
NOR	$J = A + B + C + D + E + F + G + H$	0	0	0	1	V_{ss}
OR	$J = A + B + C + D + E + F + G + H$	0	0	1	1	V_{ss}
OR/AND	$J = (A + B + C + D) \cdot (E + F + G + H)$	0	1	0	1	V_{ss}
OR/NAND	$J = (A + B + C + D) \cdot (E + F + G + H)$	0	1	0	1	V_{ss}
AND	$J = A \cdot B \cdot C \cdot D \cdot E \cdot F \cdot G \cdot H$	1	0	0	1	V_{DD}
NAND	$J = A \cdot B \cdot C \cdot D \cdot E \cdot F \cdot G \cdot H$	1	0	1	1	V_{DD}
AND/NOR	$J = (A \cdot B \cdot C \cdot D) + (E \cdot F \cdot G \cdot H)$	1	1	0	1	V_{DD}
AND/NOR	$J = (A \cdot B \cdot C \cdot D) + (E \cdot F \cdot G \cdot H)$	1	1	1	1	V_{DD}
H_1-Z		X	X	X	0	X

Positive logic 0 = low level, 1 = high level, X = irrelevant, EXPAND input tied to V_{ss}.

4049 Hex inverting buffer

The 4049 contains six independent inverting buffers. Only the power supply connections are common to all six inverters. The output of each stage is the opposite state as its input. That is, a LOW input results in a HIGH output, and a HIGH input produces a LOW output.

These devices feature logic-level conversion using only one supply voltage (V_{DD}). Unlike most digital devices, the input-signal HIGH level can exceed the V_{DD} supply voltage when these units are used for logic-level conversions. These devices can be used as regular inverting buffers, CMOS-to-TTL converters, or as CMOS current drivers. If V_{DD} is +5 volts, each inverter can directly drive two TTL loads over the full operating temperature range. The 4049 is similar to the 4050, except for the signal inversion performed by this chip.

Supply Voltage Range	3 to 15 V
TTL Compatibility	Direct drive to 2 TTL loads at 5 V over full temperature range.

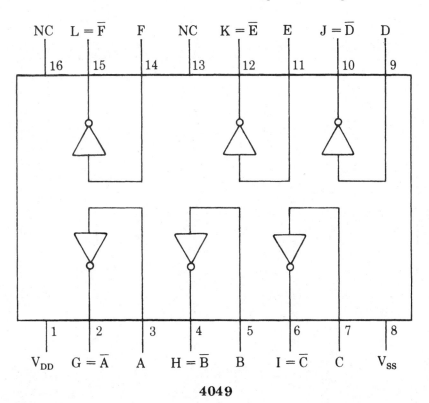

4049

273

4050 Hex inverting buffer

The 4050 contains six independent noninverting buffers. Only the power supply connections are common to all six inverters. The output of each stage matches the opposite state as its input. That is, a LOW input results in a LOW output, and a HIGH input produces a HIGH output.

These devices feature logic-level conversion using only one supply voltage (V_{DD}). Unlike most digital devices, the input-signal HIGH level can exceed the V_{DD} supply voltage when these units are used for logic-level conversions. These devices can be used as regular noninverting buffers, CMOS to TTL converters, or as CMOS current drivers. If V_{DD} is +5 volts, each inverter can directly drive two TTL loads over the full operating temperature range. The 4050 is similar to the 4049, except no signal inversion is performed by this chip.

Supply Voltage Range	3 to 15 V
TTL Compatibility	Direct drive to 2 TTL loads at 5 V over full temperature range.

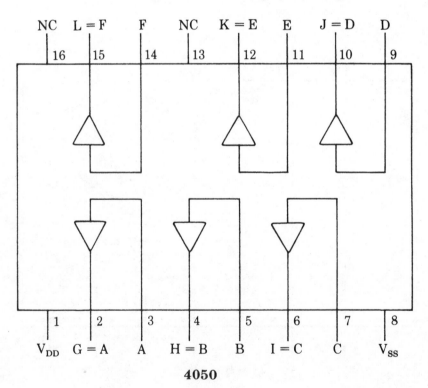

4050

4051 Single 8-channel analog multiplexer/demultiplexer

These analog multiplexers/demultiplexers are digitally controlled switches with low impedance and very low OFF leakage currents. Control of analog signals up to 15 V_{p-p} can be achieved by digital signal amplitudes of 3 to 15 V. For example, if $V_{DD} = 5$ V, $V_{SS} = 0$ V, and $V_{EE} = -5$ V, analog signals from –5 to + 5 V can be controlled by digital inputs of 0 to 5 V. The multiplexer circuits dissipate extremely low quiescent power over the full V_{DD} to V_{SS} and V_{DD} to V_{EE} supply-voltage ranges, independent of the logic state of the control signals. When a logic 1 is present at the inhibit input terminal, all channels are off.

- 4051 is a single 8-channel multiplexer having three binary control inputs, A, B, and C, and an inhibit input. The three binary signals select one of eight channels to be turned on and connect the input to the output.
- 4052 is a differential 4-channel multiplexer having two binary control inputs, A and B, and an inhibit input. The two binary input signals select one of four pairs of channels to be turned on, and connect the differential analog inputs to the differential outputs.
- 4053 is a triple 2-channel multiplexer having three separate digital control inputs, A, B, and C, and an inhibit input. Each control input selects one of a pair of channels that are connected in a single-pole double-throw configuration.

Range of digital and analog signal levels: Digital 3 to 15 V, analog to 15 V p-p.
On resistance: 80 Ω (typ.) over entire 15 Vp-p signal-input range for $V_{DD} - V_{EE} = 15$ V.
Off resistance: Channel leakage of ±10 pA (typ.) at $V_{DD} - V_{EE} = 10$ V.
Logic level conversion for digital addressing signals of 3 to 15 V ($V_{DD} - V_{SS} = 3$ to 15 V) to switch analog signals to 15 Vp-p ($V_{DD} - V_{EE} = 15$ V).
Logic level conversion for digital addressing signals of 3 – 15 V ($V_{DD} - V_{SS} = 3$ to 15 V) to switch analog signals to 15 Vp-p ($V_{DD} - V_{EE} = 15$ V).
Matched switch characteristics: $\Delta R_{ON} = 5$ Ω (Typ.) for $V_{DD} - V_{EE} = 15$ V.
Quiescent power dissipation under all digital control input and supply conditions:
1 μ W (typ.) at $V_{DD} - V_{SS} = V_{DD} - V_{EE} = 10$ V.
Binary address decoding on chip.

CD4051BM/CD4051BC

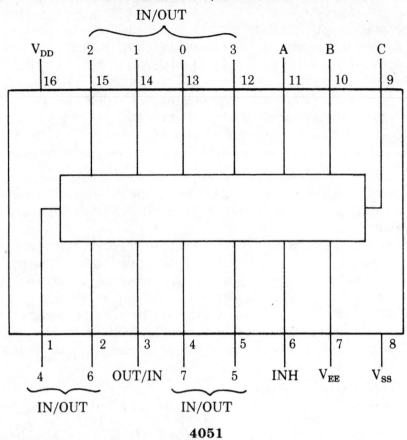

4051

Truth Table 4051, 4052, 4053

INPUT STATES				"ON" CHANNELS		
INHIBIT	C	B	A	CD4051B	CD4052B	CD4053B
0	0	0	0	0	0X, 0Y	cx, bx, ax
0	0	0	1	1	1X, 1Y	cx, bx, ay
0	0	1	0	2	2X, 2Y	cx, by, ax
0	0	1	1	3	3X, 3Y	cx, by, ay
0	1	0	0	4		cy, bx, ax
0	1	0	1	5		cy, bx, ay
0	1	1	0	6		cy, by, ax
0	1	1	1	7		cy, by, ay
1	*	*	*	NONE	NONE	NONE

*Don't Care Condition

4052 Dual 4-channel multiplexer/demultiplexer

These analog multiplexers/demultiplexers are digitally controlled analog switches having low ON impedance and very low OFF leakage currents. Control of analog signals up to 15 V_{p-p} can be achieved by digital signal amplitudes of 3 to 15 V. For example, if V_{DD} = 5 V, V_{SS} = 0 V and V_{EE} = –5 V, analog signals from –5 to + 5 V can be controlled by digital inputs of 0 to 5 V. The multiplexer circuits dissipate extremely low quiescent power over the full V_{DD} to V_{SS} and V_{DD} to V_{EE} supply voltage ranges, independent of the logic state of the control signals. When a logic 1 is present at the inhibit input terminal, all channels are off.

- 4052 is a differential 4-channel multiplexer having two binary control inputs, A and B, and an inhibit input. The two binary input signals select one of four pairs of channels to be turned on, and connect the differential analog inputs to the differential outputs.

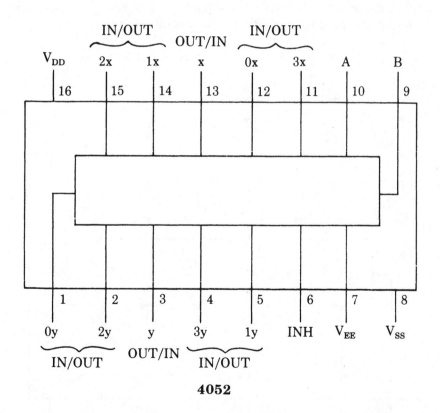

4052

4053 Triple two-channel analog multiplexer/demultiplexer

The 4053 is a triple two-channel multiplexer/demultiplexer with three separate digital control inputs (A, B, and C) and an inhibit input. Each control input selects one of a pair of channels that are connected in a single-pole, double-throw (SPDT) configuration.

The 4053 analog multiplexers/demultiplexers are digitally controlled analog switches, with low ON-impedance and very low OFF-leakage currents. Three two-channel units are included on this chip.

Control of analog signals up to 15 volts (peak-to-peak) can be achieved by digital signal amplitudes of 3 to 15 V. For example, if $V_{DD} = 5$ V, $V_{SS} = 0$ V, and $V_{EE} = 5$ V, analog signals from –5 to +5 V can be controlled by digital inputs of 0 to 5 V.

The multiplexer circuits dissipate extremely low quiescent power over the full V_{DD}-to-V_{SS} and V_{DD}-to-V_{EE} supply voltage ranges, independent of the logic state of the control signals. When the inhibit input terminal is made HIGH, all channels are turned off. The 4052 is similar, except it contains two analog multiplexer/demultiplexers of four-channels each.

CD4053BM/CD4053BC

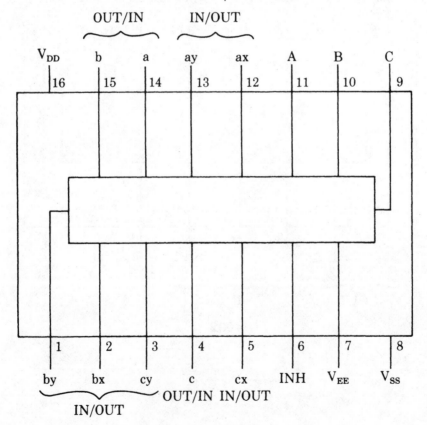

4060 14-Stage ripple carry binary counter

The 4060 (and the 4020) is a 14-stage ripple-carry binary counter. The counter is advanced one count on the negative-going (HIGH-to-LOW) transition of each clock pulse. The counters are reset to the zero (all LOW) state by a HIGH signal at the reset input, regardless of the clock. The 4040 is similar to the 4060, except it has just 12 stages, instead of 14.

Supply Voltage Range	1 to 15 V
Noise Immunity	0.45 V_{DD} typ.
TTL Compatibility	Fanout of 2 driving 74L or 1 driving 74LS
Speed of Operation	8 MHz typ. at $V_{DD} = 10$ V

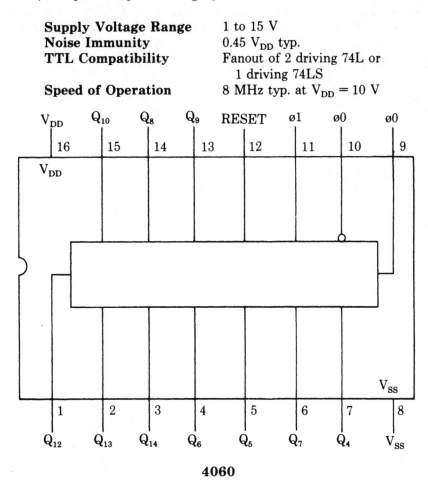

4060

4066 Quad bilateral switch

The 4066 is a quad bilateral switch with an extremely high OFF-resistance and low ON-resistance. The switch will pass analog or digital signals in either direction, and it is extremely useful in digital switching. In effect, it is the equivalent of an ordinary mechanical switch, except it is operated by logic signals on the appropriate control input pin, rather than physical motion or pressure.

The 4066 is pin-for-pin compatible with the 4016 (discussed earlier in this section), but it has a much lower on-resistance, and its on-resistance is relatively constant over the input-signal range. Essentially, the 4066 is simply an improved version of the 4016.

Supply Voltage Range	3 to 15 V
Noise Immunity	0.45 V_{DD} typ.
Range of Digital and Analog Switching	± 7.5 V_{PEAK}
On Resistance for 15 V Operation	80 Ω typ.
Matched on Resistance Over 15 V Signal Input	Δ RON = 5 Ω typ.
On/off Output Voltage Ratio	65 dB typ.
Linearity	0.4% distortion typ.
Off Switch Leakage	0.1 nA typ.
Control Input Impedance	10^{12} Ω typ.
Crosstalk Between Switches	-50 dB typ.
	@ fis = 0.9 MHz, R_L = 1 KΩ
Frequency Response, Switch On	40 MHz typ.

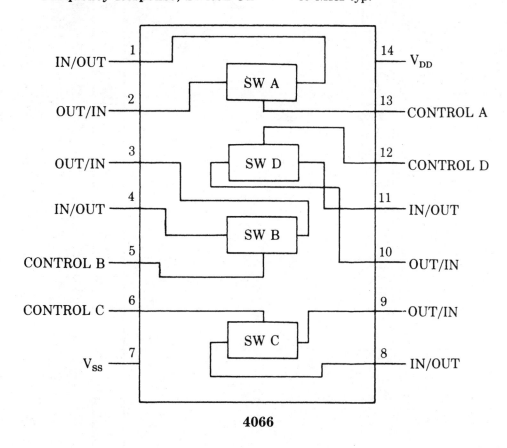

4066

4067 16-Channel analog multiplexer/demultiplexer

The 4067 consists of 16 digitally controlled analog switches, which can be used in either analog or digital applications. These switches have very low ON-resistance and low OFF-leakage current.

The 4067 has four binary control inputs (A, B, C, and D), which select one-of-16 channels by turning on the appropriate switch. Only one switch is on (closed) at any given time. In effect, this device permits the selection of any of 16 separate input signals. There is also an inhibit input, which turns off all 16 switches so that there is no output signal.

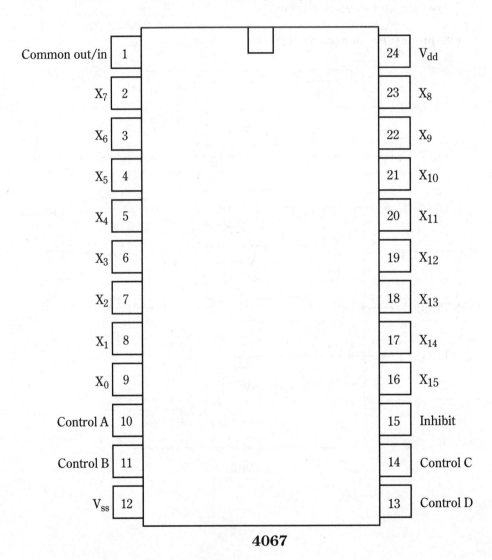

Common out/in	1	24	V_{dd}
X_7	2	23	X_8
X_6	3	22	X_9
X_5	4	21	X_{10}
X_4	5	20	X_{11}
X_3	6	19	X_{12}
X_2	7	18	X_{13}
X_1	8	17	X_{14}
X_0	9	16	X_{15}
Control A	10	15	Inhibit
Control B	11	14	Control C
V_{ss}	12	13	Control D

4067

4067 Truth Table

CONTROL INPUTS					SELECTED CHANNEL
A	B	C	D	Inh	
x	x	x	x	1	NONE
0	0	0	0	0	0
0	0	0	1	0	1
0	0	1	0	0	2
0	0	1	1	0	3
0	1	0	0	0	4
0	1	0	1	0	5
0	1	1	0	0	6
0	1	1	1	0	7
1	0	0	0	0	8
1	0	0	1	0	9
1	0	1	0	0	10
1	0	1	1	0	11
1	1	0	0	0	12
1	1	0	1	0	13
1	1	1	0	0	14
1	1	1	1	0	15

x = Don't care

4069 Hex inverter

The 4069 contains six inverters that are designed for very high noise immunity and low power dissipation. Each of the six inverters is a single-stage circuit to minimize propagation delays.

The output of each inverter has the opposite logic state as its input. That is, a LOW input produces a HIGH output, and a HIGH input results in a LOW output.

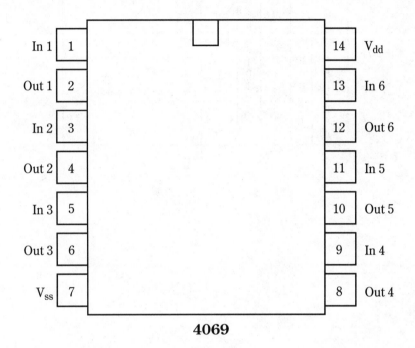

4069

4070 Quad 2-input Exclusive-OR gate

The 4070 contains four Exclusive-OR (X-OR) gates, each with two inputs. The output of each gate is HIGH if, and only if, one of the inputs is HIGH, but not both. In other words, as the Truth Table shows, the output is HIGH if the two inputs are at opposite logic states, but LOW if they are the same. An X-OR gate can be used as a one-bit digital comparator.

Truth Table

INPUTS		OUTPUTS
A	**B**	**Y**
L	L	L
L	H	H
H	L	H
H	H	L

Supply Voltage Range 3 to 15 V
Noise Immunity 0.45 V_{DD} typ.
TTL Compatibility Fanout of 2 driving 74L
   or 1 driving 74LS

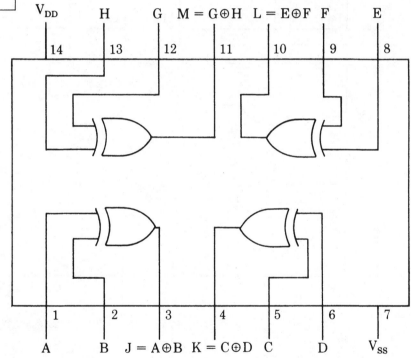

4070

4070 Truth Table

INPUTS		OUTPUT
A	**B**	
0	0	0
0	1	1
1	0	1
1	1	0

4071 Quad 2-input OR buffered B series gate

The 4071 contains four OR gates with two inputs each. The output of each gate is HIGH if either or both of the inputs is HIGH. The output is LOW if, and only if, all inputs are LOW, as shown in the Truth Table. The 4071's gates also have buffered outputs that improve transfer characteristics by providing very high gain.

TTL Compatibility	Fanout of 2 driving 74L or 1 driving 74LS
Parametric Ratings	5 — 10 — 15 V
Maximum Input Leakage	1 μA at 15 V

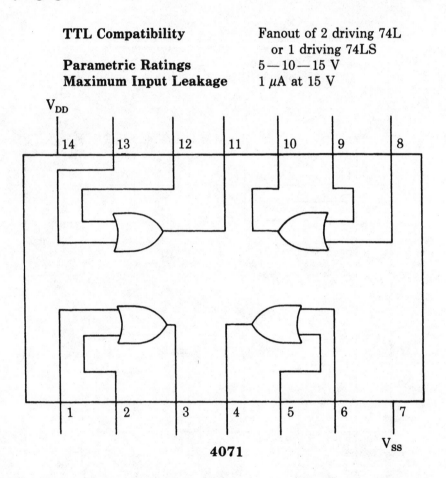

4071

4071 Truth Table

INPUTS		OUTPUT
A	B	
0	0	0
0	1	1
1	0	1
1	1	1

4072 Dual 4-input gate

The 4072 contains two OR gates with four inputs each. The output of each gate is HIGH if any one or more of the inputs are HIGH. The output is LOW if, and only if, all inputs are LOW.

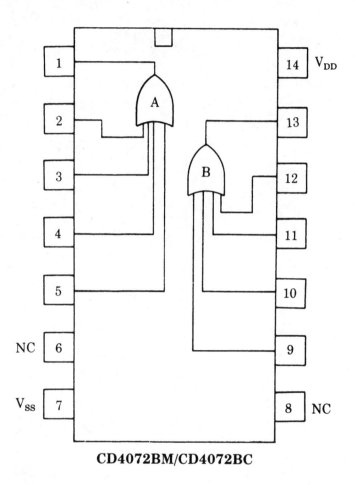

CD4072BM/CD4072BC

Truth Table

INPUTS				OUTPUT
A	B	C	D	
0	0	0	0	0
0	0	0	1	1
0	0	1	0	1
0	0	1	1	1
0	1	0	0	1
0	1	0	1	1
0	1	1	0	1
0	1	1	1	1
1	0	0	0	1
1	0	0	1	1
1	0	1	0	1
1	0	1	1	1
1	1	0	0	1
1	1	0	1	1
1	1	1	1	1

4073 Double-buffered 3-input NAND gate

The 4073 contains three OR gates with three inputs each. The output of each gate is HIGH if any one or more of the inputs are HIGH. The output is LOW if, and only if, all inputs are LOW, as shown in the Truth Table. The 4073's gates also have buffered outputs that improve transfer characteristics by providing very high gain.

Supply Voltage Range   3 to 15 V
Noise Immunity $0.45\ V_{DD}$ typ.
TTL Compatibility   2 driving 74L
                         or 1 driving 74LS

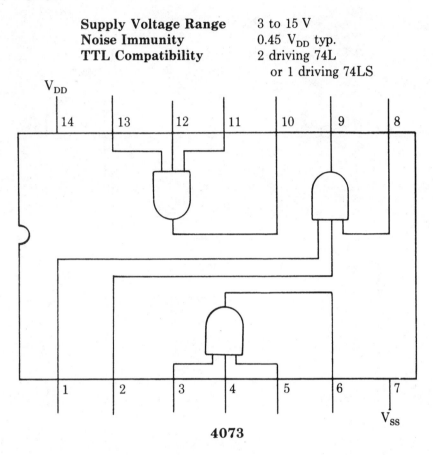

4073

4073 Truth Table

INPUTS			OUTPUT
A	B	C	
0	0	0	0
0	0	1	1
0	1	0	1
0	1	1	1
1	0	0	1
1	0	1	1
1	1	0	1
1	1	1	1

4075 Double-buffered triple 3-input NOR gate

The 4075 contains three NOR gates with three inputs each. The output of each gate is LOW if any one or more of the inputs are HIGH. The output is HIGH if, and only if, all inputs are LOW, as shown in the Truth Table. The 4075's gates also have buffered outputs that improve transfer characteristics by providing very high gain.

Supply Voltage Range	3 to 15 V
Noise Immunity	0.45 V_{DD} typ.
TTL Compatibility	2 driving 74L
	or 1 driving 74LS

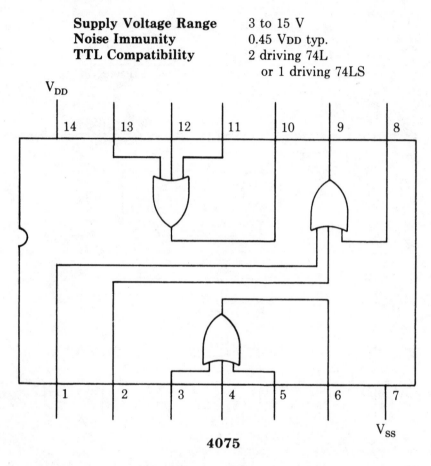

4075

4075 Truth Table

INPUTS			OUTPUT
A	B	C	
0	0	0	1
0	0	1	0
0	1	0	0
0	1	1	0
1	0	0	0
1	0	1	0
1	1	0	0
1	1	1	0

4076 Four-bit D-type register with tri-state outputs

The 4076 is a four-bit register, made up of four D-type flip-flops, which operate synchronously from a common clock. A pair of OR-gated output disable inputs can be used to force the outputs into the third high impedance (Z or OFF) state. The tri-state feature is very useful in bus organized systems.

The 4076 also has two OR-gated data-disable inputs which cause the Q outputs to be fed back to the D inputs of the flip-flops. When this is done, the flip-flops are inhibited from changing state while the clocking process remains undisturbed.

An asynchronous master reset input is also available on the 4076. This input force clears all four flip-flop stages simultaneously, regardless of any and all signals on the clock, data, or disable inputs.

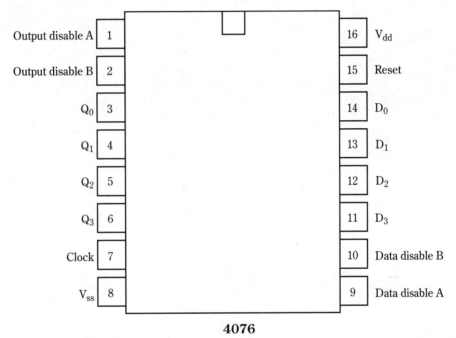

4076

4076 Function Table

	INPUTS				OUTPUT
RESET	**CLOCK**	**DATA DISABLE**		**DATA**	
		A	**B**	**D**	**Q**
1	x	x	x	x	0
0	0	x	x	x	Qn
0	T	1	x	x	Qn
0	T	x	1	x	Qn
0	T	0	0	0	0
0	T	0	0	1	1

x   = Don't care
T   = LOW-to-HIGH- transition
Qn = Previous output state

4077 Quad 2-input Exclusive-NOR gate

The 4077 contains four Exclusive-NOR gates, each with two inputs. An Exclusive-NOR gate is the opposite of an Exclusive-OR gate. The output of each gate is LOW if, and only if, one of the inputs is HIGH, but not both. In other words, as the Truth Table shows, the output is LOW if the two inputs are at opposite logic states, but HIGH if they are the same. An Exclusive-NOR gate can be used as a one-bit digital comparator.

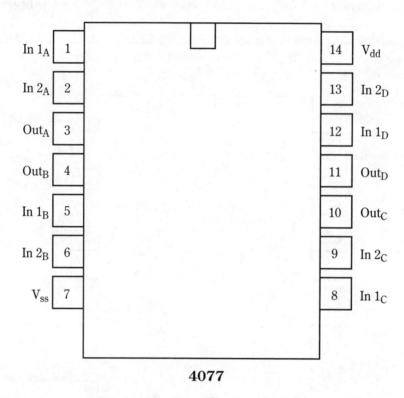

4077

4077 Truth Table

INPUTS		OUTPUT
A	B	
0	0	1
0	1	0
1	0	0
1	1	1

4081 Quad 2-input AND buffered B-series gate

The 4081 contains four AND gates with two inputs each. The output of each gate is HIGH if, and only if, both of the inputs are HIGH. The output is LOW if either input (or both) is LOW. The 4081's gates also have buffered outputs that improve transfer characteristics by providing very high gain.

TTL Compatibility   Fanout of 2 driving 74L
                              or 1 driving 74LS
Parametric Ratings   5 – 10 – 15 V
Maximum Input Leakage   1 μA at 15 V

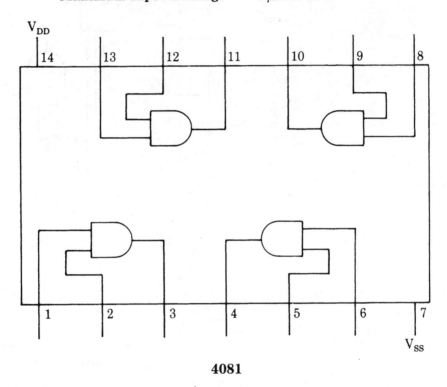

4081

4081 Truth Table

INPUTS		OUTPUT
A	B	
0	0	0
0	1	0
1	0	0
1	1	1

4082 Dual 4-input AND gate

The 4082 contains two AND gates with four inputs each. The output of each gate is HIGH if, and only if, all of the inputs are HIGH. The output is LOW if any one input (or more) is LOW.

Truth Table

INPUTS				OUTPUT
A	B	C	D	
0	0	0	0	0
0	0	0	1	0
0	0	1	0	0
0	0	1	1	0
0	1	0	0	0
0	1	0	1	0
0	1	1	0	0
0	1	1	1	0
1	0	0	0	0
1	0	0	1	0
1	0	1	0	0
1	0	1	1	0
1	1	0	0	0
1	1	0	1	0
1	1	1	1	1

4082

4093 Quad 2-input NAND Schmitt trigger

The 4093 contains four Schmitt trigger circuits, each with two inputs, which are combined in a NAND configuration. Each of these gates switches at different points for positive-going, and negative-going signals. The difference between the positive switching voltage (V_t+) and the negative switching voltage (V_t-) is the hysteresis voltage (V_h). All outputs on this chip have equal source and sink currents, and conform to standard B-series output drive specifications.

Supply Voltage Range	3 to 15 V
Noise Immunity	Greater than 50%
Hysteresis Voltage (any input)	$T_A = 25°C$
Typical	$V_{DD} = 5$ V $V_H = 1.5$ V
	$V_{DD} = 10$ V $V_H = 2.2$ V
	$V_{DD} = 15$ V $V_H = 2.7$ V
Guaranteed	$V_H = 0.1\ V_{DD}$

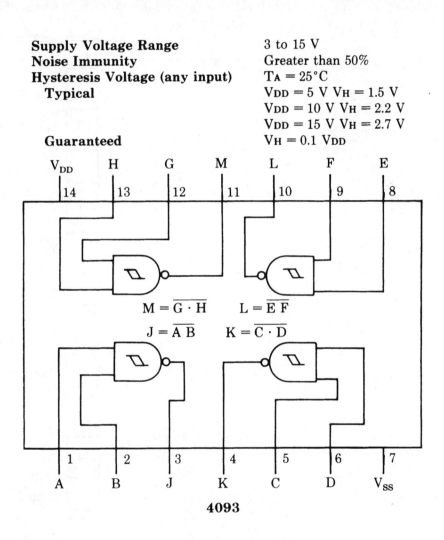

$$M = \overline{G \cdot H}$$
$$L = \overline{E\ F}$$
$$J = \overline{A\ B}$$
$$K = \overline{C \cdot D}$$

4093

4097 8-Channel analog multiplexer/demultiplexer

The 4097 consists of eight digitally controlled analog switches, which can be used in either analog or digital applications. These switches have very low ON resistance and low OFF leakage current.

The 4067 has three binary control inputs (A, B, and C), which select one-of-eight channels by turning on the appropriate switch. Only one switch is ON (closed) at any given time. In effect, this device permits the selection of any of eight separate input signals. It also has an inhibit input, which turns off all eight switches so that there is no output signal.

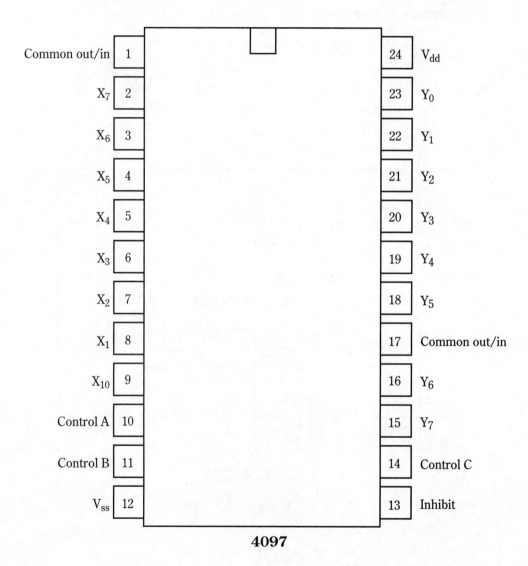

4097

4097 Truth Table

CONTROL INPUTS				SELECTED CHANNEL
A	B	C	Inh	
x	x	x	1	None
0	0	0	0	0
0	0	1	0	1
0	1	0	0	2
0	1	1	0	3
1	0	0	0	4
1	0	1	0	5
1	1	0	0	6
1	1	1	0	7

x = Don't care

4106B Hex Schmitt trigger

Six inverting Schmitt trigger circuits are included within the 4106B. Any inverter, of course, reverses the logic level from its input to its output. For example, a LOW input results in a HIGH output, and a HIGH input produces a LOW output.

Each of these inverters switches at different points for positive-going, and negative-going signals. The difference between the positive switching voltage (V_t+) and the negative switching voltage (V_t-) is the hysteresis voltage (V_h). The hysteresis voltage is increased over the older 4584B.

Because each inverter has a built-in Schmitt trigger circuit, it can be used to replace devices, such as the 4069 in applications where greater noise immunity is required or where slowly changing input waveforms might be encountered. A Schmitt trigger, in effect, "squares up" a slowly changing signal.

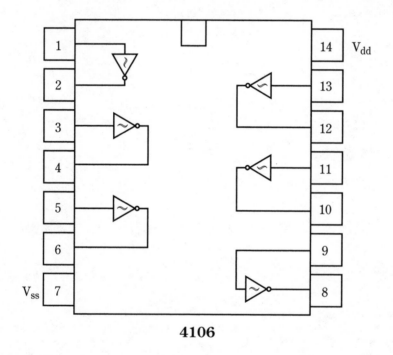

4106

4174B Hex D-type flip-flop

The 4174B consists of six D-type flip-flop stages, operated from a common clock, and an active LOW master reset. Data on the D inputs meeting the set-up time requirements is transferred to the Q outputs when the clock has a LOW-to-HIGH transition. All inputs and outputs on the 4171B are buffered.

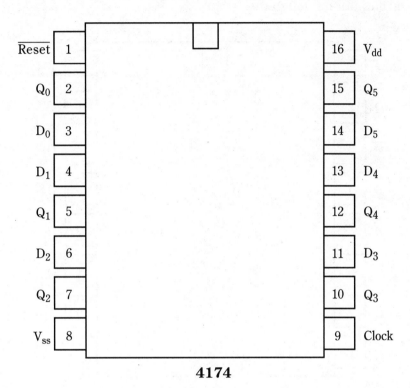

4174

4171B Truth Table

INPUTS			OUTPUT
CLOCK	**RESET**	**DATA**	
x	0	x	0
LH	1	0	0
LH	1	1	1
HL	1	x	Q

x = Don't care
Q = previous state (no change)
LH = LOW-to-HIGH clock transistion
HL = HIGH-to-LOW clock transistion

4175B Quad D-type flip-flop

The 4175B contains four D-type flip-flop stages with both true (Q) and complementary ($\overline{Q}$) outputs. All flip-flop stages on this chip share a common clock and an asynchronous active LOW master reset. The flip-flops are triggered by the LOW-to-HIGH transition of the clock pulse. All of the 4175B's inputs and outputs are buffered.

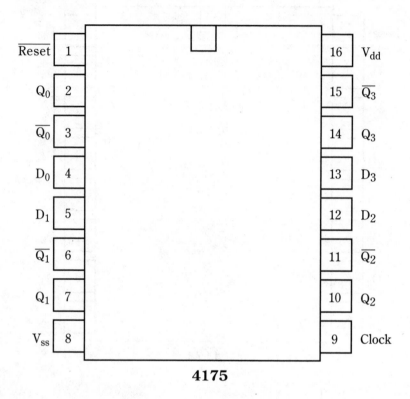

4175

4175B Truth Table

INPUTS			OUTPUTS	
CLOCK	RESET	DATA	Q	$\overline{Q}$
x	0	x	0	1
LH	1	0	0	1
LH	1	1	1	0
HL	1	x	Q	Q

x = Don't care
Q = previous state (no change)
LH = LOW-to-HIGH clock transistion
HL = HIGH-to-LOW clock transistion

4194B 4-Bit bidirectional universal shift register

The MC4194B is a four-bit static shift register, which can operate in several shift modes, namely, parallel load, serial shift left, serial shift right, and hold. The mode is selected via two control inputs (S_0 and S_1), as outlined in the Truth Table. Both the serial and parallel shift functions are triggered on LOW-to-HIGH clock transitions. The asynchronous Master Reset input forces all outputs LOW whenever it is fed a LOW logic level, regardless of any other input conditions.

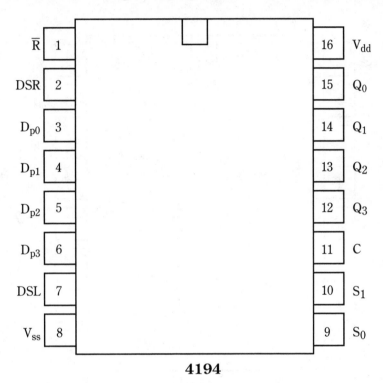

4194

4194 Truth Table

INPUTS						OUTPUTS				MODE
R	S_0	S_1	DSR	DSL	DPO-3	Q_0	Q_1	Q_2	Q_3	
0	x	x	x	x	x	0	0	0	0	Reset
1	0	0	x	x	x	q0	q1	q2	q3	Hold
1	0	1	0	x	x	0	q0	q1	q2	Shift Right
1	0	1	1	x	x	1	q0	q1	q2	" "
1	1	0	x	0	x	q1	q2	q3	0	Shift Left
1	1	0	x	1	x	q1	q2	q3	1	" "
1	1	1	x	x	0	0	0	0	0	Parallel
1	1	1	x	x	1	1	1	1	1	"

x = Don't care
q_n = Previous state of Q output of stage n

4415 Quad precision timer/driver

The 4415 consists of four very precise timer/driver stages. Each stage's output pulse has a width determined by the frequency of the input clock signal. Once the output buffer is set (turned on) by the appropriate input sequence, it will remain on for 100 clock pulses, then it will be reset (turned off). The inputs to this device can be set up to respond to either LOW-to-HIGH or HIGH-to-LOW transitions to suit the specific intended application.

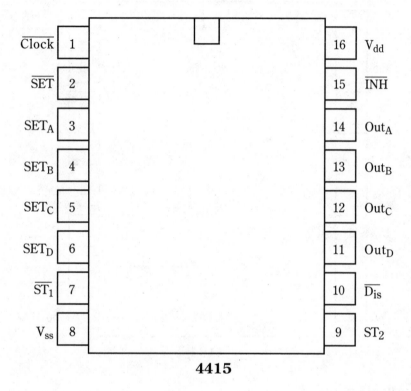

4415

4504B Hex level shifter for
TTL-to-CMOS or CMOS-to-CMOS

The 4504B can adapt TTL signals by shifting their levels so that they can reliably drive standard CMOS gates, for any CMOS supply voltage from 5 to 15 volts. A control input also permits the device to adjust levels from CMOS to CMOS at different logic levels. Either up or down level shifting can be done simply by selecting the appropriate voltages for the V_{CC} and V_{DD} supply voltages. The input signal levels are determined by the V_{CC} voltage, and the V_{dd} voltage selects the output voltage levels. Six level shifter stages are included in the 4504B.

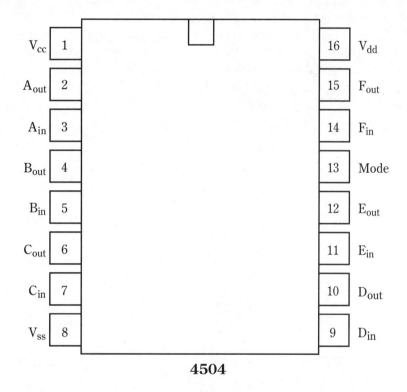

V_{cc} 1	16 V_{dd}
A_{out} 2	15 F_{out}
A_{in} 3	14 F_{in}
B_{out} 4	13 Mode
B_{in} 5	12 E_{out}
C_{out} 6	11 E_{in}
C_{in} 7	10 D_{out}
V_{ss} 8	9 D_{in}

4504

4510 BCD up/down counter

The 4510 is an up/down counter, which counts in the BCD (Binary Coded Decimal) format. The count goes up when the up/down input is HIGH, and down, if this input is LOW. A HIGH preset-enable signal allows data at the parallel inputs to preset the counter to any state asynchronously with the clock. The counter is advanced one count on each LOW-to-HIGH transition of the $\overline{\text{clock}}$ input signal, if the $\overline{\text{carry-in}}$, $\overline{\text{present-enable}}$, and $\overline{\text{reset}}$ inputs are all LOW. Advancement of the count is inhibited if any of these control inputs is brought HIGH.

The $\overline{\text{carry-out}}$ signal is normally HIGH, but it goes LOW when the counter reaches its maximum count in the Up mode, or its minimum count in the Down mode, provided that the $\overline{\text{carry}}$ input is LOW.

All counter stages are cleared (reset to 0) synchronously by applying a HIGH signal to the $\overline{\text{reset}}$ input. All inputs on this chip are protected against static discharge by internal diode clamps to both V_{DD} and V_{SS}. The 4510 is similar to the 4516, except that the 4510 counts in BCD and the 4516 counts in binary.

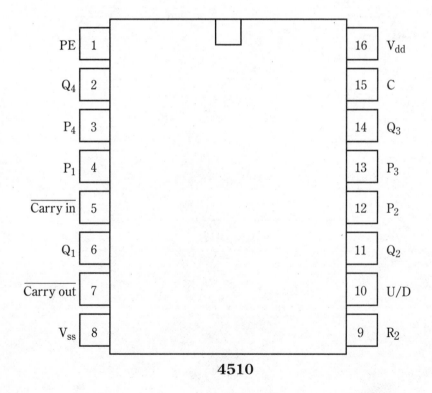

4510

4516 Binary up/down counter

The 4516 is an up/down counter which counts in the binary format. The count goes up when the up/down input is HIGH, and down, if this input is LOW. A HIGH preset-enable signal allows data at the parallel inputs to preset the counter to any state asynchronously with the clock. The counter is advanced one count on each LOW-to-HIGH transition of the clock input signal, if the $\overline{\text{carry-in}}$, $\overline{\text{present-enable}}$, and $\overline{\text{reset}}$ inputs are all LOW. Advancement of the count is inhibited if any of these control inputs is brought HIGH.

The $\overline{\text{carry-out}}$ signal is normally HIGH, but it goes LOW when the counter reaches its maximum count in the Up mode, or its minimum count in the Down mode, provided that the carry input is LOW.

All counter stages are cleared (reset to 0) synchronously by applying a HIGH signal to the reset input. All inputs on this chip are protected against static discharge by internal diode clamps to both V_{DD} and V_{SS}. The 4516 is similar to the 4510, except that the 4510 counts in BCD and the 4516 counts in binary.

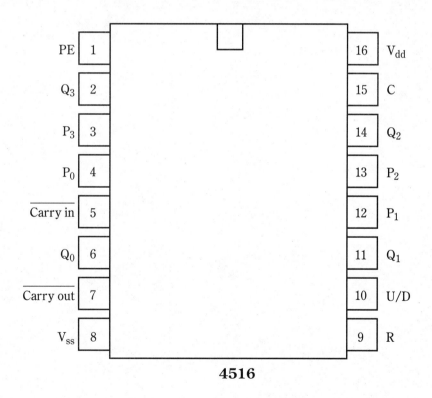

4516

4527 BCD rate multiplier

The 4089 is a 4-bit binary rate multiplier that provides an output pulse rate that is the input clock pulse rate multiplied by $\frac{1}{16}$ times the binary input number. For example, if 5 is the binary input number, there will be 5 output pulses for every 16 clock pulses.

The 4527 is a 4-bit BCD rate multiplier that provides an output pulse rate that is in the input clock pulse rate multiplied by $\frac{1}{10}$ times the BCD input number. For example, if 5 is the BCD input number, there will be 5 output pulses for every 10 clock pulses.

These devices can be used to perform arithmetic operations. These operations include multiplication and division, A/D and D/A conversion and frequency division.

Supply Voltage Range   3 to 15 V
Noise Immunity   0.45 V_{DD} typ.
TTL Compatibility   Fanout of 2 driving 74L
        or 1 driving 74LS

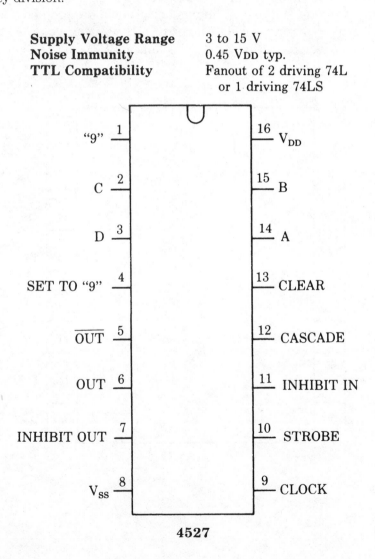

4527

3
SECTION

Special-purpose digital

Some special-purpose digital devices don't fit into either the 74xx (see Section 1) or 4xxx (see Section 2) numbering schemes. These devices, (mostly CMOS), which are presented in this section, are generally more sophisticated than those in the preceding sections.

DS1012 Digital delay line

The DS1012 can be used to create simple delays, or multiple delays combined with eight logic functions (AND, NAND, OR, NOR, XOR, NXOR, HALF-XOR, and HALF-XNOR). This device, which is compatible with either TTL or CMOS circuitry has two inputs and four buffered delay outputs. The basic operation of this chip is indicated in the functional diagram. Output 1 is a straight delay of input 1 and output 2 is a straight delay of input 2. Either of these outputs can be configured to provide complementary (inverted) delayed outputs, if this suits the intended application.

The unusual part of the DS1012 is in outputs 3 and 4. These outputs combine both inputs 1 and 2 with a selected logic relationship.

Because this IC has just eight pins, it is very easy to use in practical circuits. In addition to the two power supply pins, there are just the two signal input pins (1 and 7), and the four buffered delay outputs (3, 6, 2, and 5).

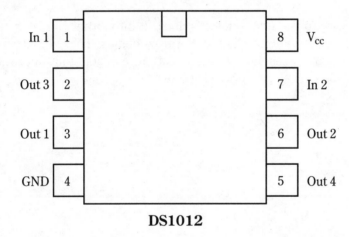

DS1012

DS1386/DS1486 Watchdog time-keeper

Dallas Semiconductor's Watchdog time-keeper family of devices provide both an accurate real-time clock and interrupt capabilities at specific, user-programmable times. It also provides an upgradeable nonvolatile RAM path.

The DS1386-08 module has 8K of nonvolatile RAM, the DS1386-32 has 32K, and the DS1486 has 128K. Except for the memory size, these modules have the same features and functions. For convenience, only the DS1486 is directly referred to from here on.

The DS1486 is designed around a very accurate low-power oscillator. An internal lithium battery can provide up to 10 years of continuous operation in the absence of external system power, so this device is well suited for portable or remote applications, which need to keep operating without direct supervision—even in the event of general power failure.

The accuracy of the time-keeper is within +/- 1 minute per month. The real-time clock registers are mapped directly into the device's on-board RAM. The external system's CPU or microcontroller can easily access any of the time-keeping registers, just as if there were comparable addresses in ordinary system memory.

Two types of interrupt outputs are provided by the Watchdog time-keepers. The first is a time-of-day alarm. The device can be programmed to generate an interrupt at any desired time of day for a specific day of the week. The alarm registers feature special mask bits, so this interrupt can also be generated once a minute, once an hour, or once a day, to suit the intended application.

The other type of interrupt generated by the Watchdog time-keeper is the "Watchdog Alarm," which provides a periodic interrupt at a user-programmed interval ranging from every 0.01 second to every 100 seconds, in 0.01 second increments. This Watchdog Alarm can be used as a periodic interrupt or as a microprocessor monitor to ensure that the microprocessor does not go out of control for any reason. In this type of application, the microprocessor is programmed to periodically "check in" with the Watchdog Time-keeper by reading or writing to any of its alarm registers. If the microprocessor does not check in within the allotted time, an "alarm" interrupt will be generated. If the microprocessor does check in when it is supposed to, the DS1486 will reset itself, and no interrupt will occur until the next "check in" time is missed.

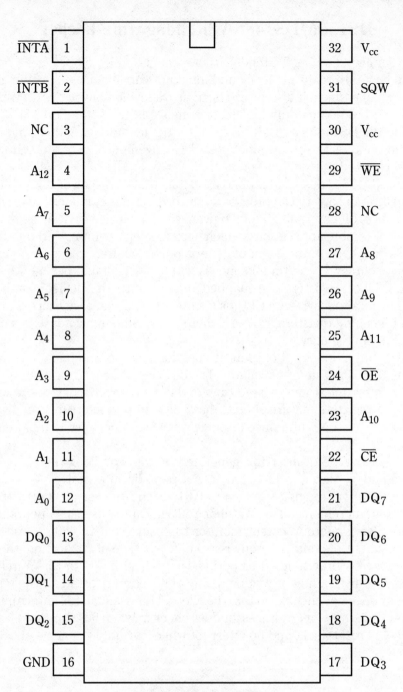

DS1386-08 8KX8

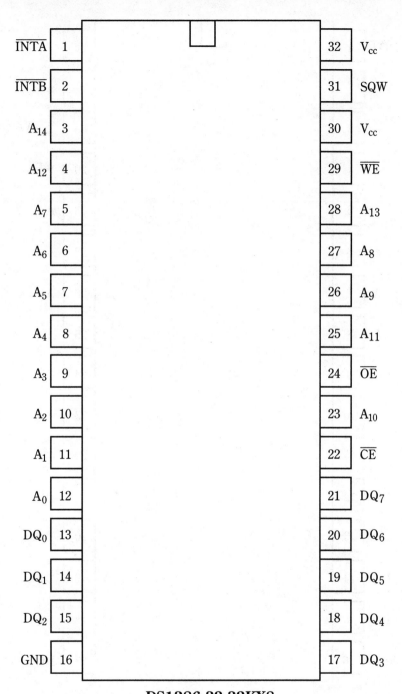

DS1386-32 32KX8

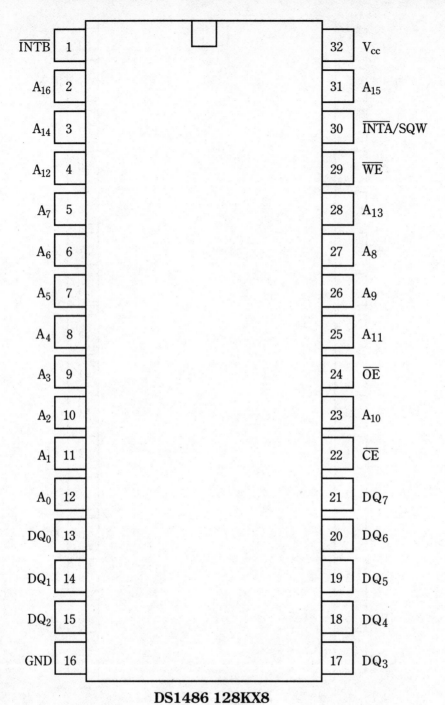

DS1486 128KX8

MC1408, MC1508 Eight-bit multiplying D/A converter

The MC1408/MC1508 is a digital-to-analog converter that accepts an eight-bit digital word at its inputs, and converts it into a proportional linear output current. The unusual feature of this device is that it also has an additional analog input. The voltage applied to this input serves as a constant multiplier. The output current of the MC1408/MC1508 is the linear product of the eight-bit digital input and the analog input voltage.

This chip is very fast and accurate in its operation. Its noninverting digital inputs are both TTL and CMOS compatible.

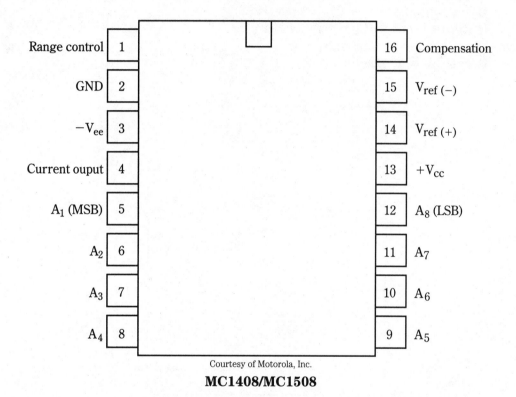

Courtesy of Motorola, Inc.

MC1408/MC1508

Standard Supply Voltages

V_{cc}	+4.5 to +5.5 V
V_{ee}	−5.0 to −15 V
Output voltage swing	+0.4 to −5.0 V
Settling time	300 ns typ.
Multiplying input slew rate	4.0 mA/μS

MC1488 Quad MDTL line driver (EIA-232D)

The 1488 contains four line-driver stages that are designed to interface data terminal equipment and data communications equipment in conformance to the specifications of EIA Standard EIA-232.

The supply voltage range for this device is flexible and the slew rate can be simply controlled with an external capacitor. The 1488 is fully compatible with TTL and DTL devices. The outputs are current limited (typically ±10 mA).

	Minimum	Typical	Maximum
Supply Voltage			±15 V
Input Signal Voltage	−15 V		+7.0 V
Output Signal Voltage	−15 V		+15 V
Input Current (Logic LOW)		1.0 mA	1.6 mA
Input Current (Logic HIGH)			10 μA

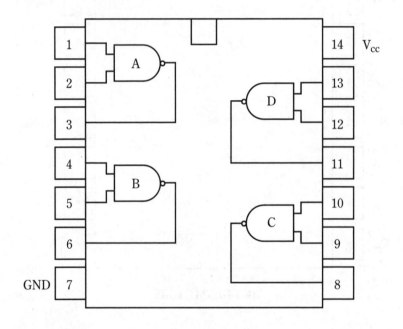

314

MC1489 Quad MDTL line receiver (EIA-232D)

The 1489 contains four line-receiver stages that are designed to interface data terminal equipment and data communications equipment, in conformance to the specifications of EIA Standard EIA-232.

   The supply voltage range for this device is flexible and the slew rate can be simply controlled with an external capacitor. The 1489 is fully compatible with TTL and DTL devices. The outputs are current limited (typically ±10 mA).

	Minimum	Typical	Maximum
Supply Voltage			±15 V
Input Signal Voltage	−15 V		+7.0 V
Output Signal Voltage	−15 V		+15 V
Input Current (Logic LOW)		1.0 mA	1.6 mA
Input Current (Logic HIGH)			10 µA

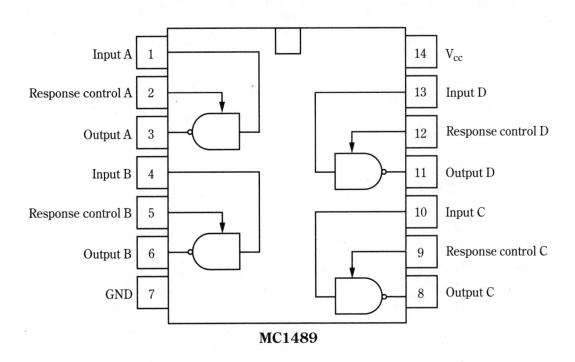

MC1489

315

DS1602 System timer

The DS1602 system timer is actually more than just a simple system clock generator, although it can certainly be used in such applications. This chip is a sophisticated and versatile time-keeping and tracking device. It contains a 32-bit counter that can run continuously from power-up. The counter is incremented one bit per second, so the maximum available count is 4.29×10 seconds = about 136 years, which should be more than enough for almost any practical application. In fact, for most real-world applications, this would clearly be overkill. In such applications, some of the unused counter bits can be put to other use, rather than being simply wasted. For example, these bits can be used to store power-up cycling information.

The DS1602 can be read from and written to directly by a microprocessor or microcontroller. Only simple software is required for basic functions. More specialized software can be used to track the time of day, day of week, number of days, months, or years. The power-up time and number of power-up cycles can also be tracked via appropriate software. The DS1603 is a module version of the DS1602.

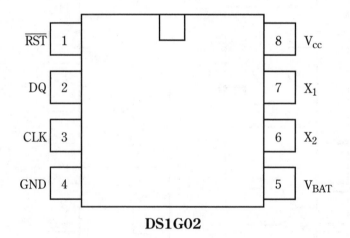

DS1G02

DS1620 Digital thermometer/thermostat

The DS1620 is a digital thermometer/thermostat, providing nine-bit temperature readings as its output. It features three alarm outputs, making the DS1620 appropriate for thermostat and simple thermometer applications.

The user can program desired switching temperature values into on-chip nonvolatile memory. The DS1620 can be pre-programmed before it is inserted into a system, thereby simplifying the actual circuitry. This device is designed for use within computer-controlled systems or in stand-alone dedicated applications without a CPU. Temperature settings and temperature readings for this device are all sent to and from the chip over a simple three-wire interface.

The DS1620 has three alarm outputs, which respond by comparing the sensed temperature to the pre-programmed switching temperature levels. Th (pin 7) goes HIGH is the sensed temperature exceeds the pre-programmed switching temperature value. Tl (pin 6) works in exactly the opposite way. This output pin goes LOW if the sensed temperature drops below the pre-programmed switching temperature value. Separate switching temperatures can be set for the Th and Tl outputs. The Tc (pin 5) goes HIGH if the sensed temperature exceeds the Th high temperature, and stays latched HIGH until the sensed temperature drops below the Tl reference temperature point.

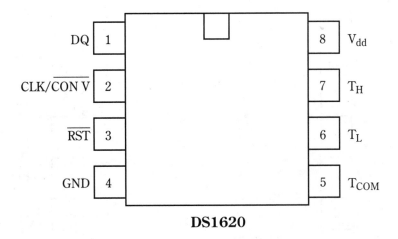

DS1620

MC3448A Quad tri-state
bus transceiver with termination networks

The 3448 is a bidirectional bus transceiver that is designed to interface TTL or MOS circuitry, and the IEEE standard Instrumentation Bus (488-1978, GPIB). Internal bus termination is provided on-chip.

This device contains four driver/receiver pairs—each of which is a complete interface between the bus and an instrument. The signal direction is determined by the logic signal fed to the Send/Receive Input pin. When the driver is enabled, the receiver is disabled, and its output is forced to the third high-impedance state. Conversely, enabling the receiver disables the driver, and switches its output to the third high-impedance state.

Optionally, the driver outputs can be operated in an open collector or active pull-up configuration. Hysteresis (typically about 600 mV) is added to the receiver inputs to improve the noise margin.

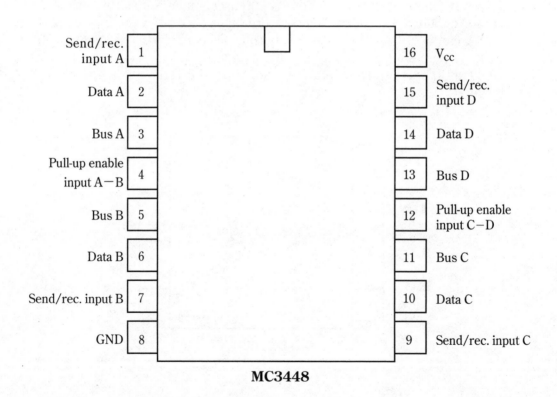

MC3448

MC3453 Quad line driver with common inhibit input

The 3453 contains four line drivers (SN75110 type). All four driver stages are served by a common inhibit input (pin 6). When this pin is HIGH, a constant output current is switched between each pair of outputs, as controlled by the logic level of that particular channel's input. In other words, the drivers are enabled, and active. However, when the common inhibit input is brought LOW, the output transistors in each driver channel are biased to cut-off, making the outputs nonconductive, and disabling all channels. This inhibit condition is particularly useful in larger systems to minimize unnecessary loading when many drivers share the same line. Other than the common inhibit and the power supply connections, the four driver stages in the 3453 are functionally independent.

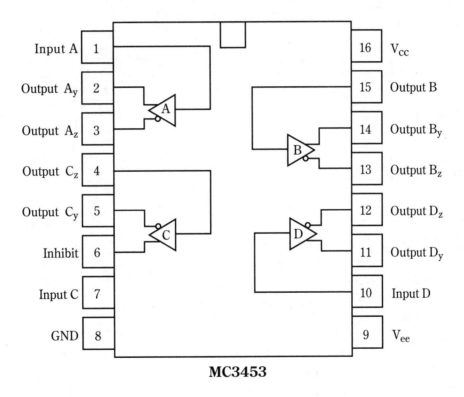

MC3453

MC3487 Quad line driver with tri-state outputs

The 3487 consists of four independent driver chains of the EIA-422 standard type. Tri-state outputs are provided on all four driver stages. The output is turned off (high-impedance state) when that driver's control input is brought LOW. Making a stage's control input HIGH enables that driver.

All inputs on this device are PNP buffered to give HIGH input impedance and minimize input loading effects. During power-up or power-down, internal circuitry automatically drives all outputs into the high-impedance (OFF) state.

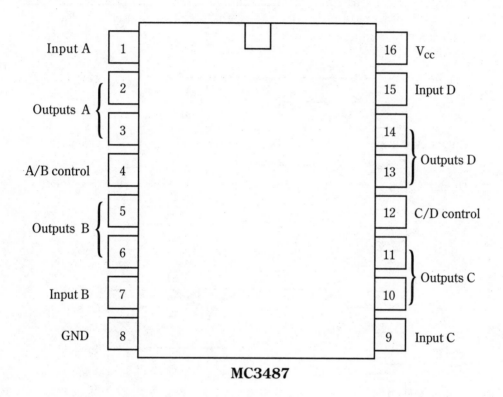

MC3487

MM5368 CMOS oscillator divider

The MM5368 accepts a square-wave input signal from an oscillator (or other signal source) and divides the signal frequency to three much lower output frequencies.

The input signal at pin 6 should be a floating 16 kHz (16,000 Hz) square wave. The same signal is buffered by the MM5368 and can be tapped off at pin 5. A 10-kHz (10,000 Hz) signal is available at pin 3, and pin 4 offers a 1-kHz (1,000 Hz) signal. The signal frequency at pin 1 can be either 50 Hz or 60 Hz, depending on the logic state at the select input (pin 7).

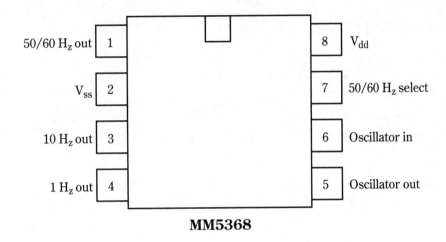

MM5368

Supply voltage range	3 to 15 V
Voltage at any pin	-0.3 V to V_{dd} to $+0.3$ V
Maximum input frequency	
$V_{dd} = 3$ V	64 kHz
$V_{dd} = 15$ V	500 kHz

MM5450 LED display driver

The 5450 is a 40-pin display driver, designed to drive common anode-type (separate cathode) LED digit displays. The brightness of the LED display can be adjusted by varying the resistance between pin 19 and V_{DD}.

Although MOS circuitry is used in this chip, it is also TTL compatible. The 5450 can be operated off a fairly wide range of supply voltages, from 4.75 up to 11 volts. This chip has 34 outputs, each with the capacity of sinking up to 15 mA.

The 5450 is very similar to the 5451, except that this device has a LOW activated DATA ENABLE input (pin 23) and one less output.

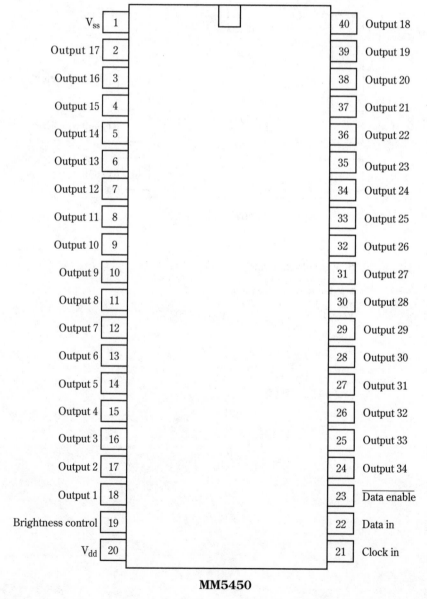

V_{ss}	1		40	Output 18
Output 17	2		39	Output 19
Output 16	3		38	Output 20
Output 15	4		37	Output 21
Output 14	5		36	Output 22
Output 13	6		35	Output 23
Output 12	7		34	Output 24
Output 11	8		33	Output 25
Output 10	9		32	Output 26
Output 9	10		31	Output 27
Output 8	11		30	Output 28
Output 7	12		29	Output 29
Output 6	13		28	Output 30
Output 5	14		27	Output 31
Output 4	15		26	Output 32
Output 3	16		25	Output 33
Output 2	17		24	Output 34
Output 1	18		23	$\overline{\text{Data enable}}$
Brightness control	19		22	Data in
V_{dd}	20		21	Clock in

MM5450

MM5451 LED display driver

The 5451 is a 40-pin display driver, designed to drive common anode-type (separate cathode) LED digit displays. The brightness of the LED display can be adjusted by varying the resistance between pin 19 and V_{DD}.

Although MOS circuitry is used in this chip, it is also TTL compatible. The 5451 can be operated off a fairly wide range of supply voltages, from 4.75 up to 11 volts. This chip has 35 outputs, each with the capacity of sinking up to 15 mA. The 5451 is very similar to the 5450, except that this device does not have a DATA ENABLE input, which is replaced with an extra output line.

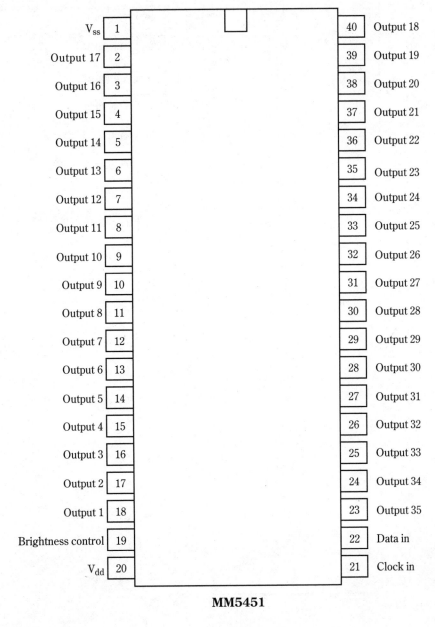

V_{ss}	1		40	Output 18
Output 17	2		39	Output 19
Output 16	3		38	Output 20
Output 15	4		37	Output 21
Output 14	5		36	Output 22
Output 13	6		35	Output 23
Output 12	7		34	Output 24
Output 11	8		33	Output 25
Output 10	9		32	Output 26
Output 9	10		31	Output 27
Output 8	11		30	Output 28
Output 7	12		29	Output 29
Output 6	13		28	Output 30
Output 5	14		27	Output 31
Output 4	15		26	Output 32
Output 3	16		25	Output 33
Output 2	17		24	Output 34
Output 1	18		23	Output 35
Brightness control	19		22	Data in
V_{dd}	20		21	Clock in

MM5451

MM5483 Liquid crystal display driver

The 5483 is a MOS device designed to drive up to 31 segments of a LCD (Liquid Crystal Display). If more display segments are needed, multiple devices can be cascaded. A 5483 can drive a 4½ digit 7-segment display with only a minimum of interface between the display and the drive (data) source.

This chip features serial data input and output, and is TTL compatible. Display data can be latched within the 5483 itself, and held until another load pulse is received.

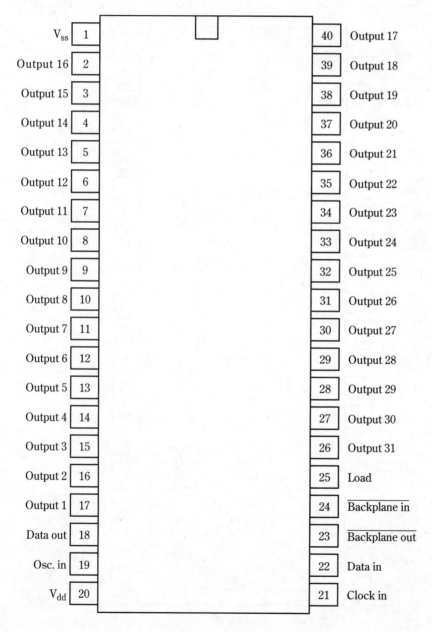

V_{ss}	1	40	Output 17
Output 16	2	39	Output 18
Output 15	3	38	Output 19
Output 14	4	37	Output 20
Output 13	5	36	Output 21
Output 12	6	35	Output 22
Output 11	7	34	Output 23
Output 10	8	33	Output 24
Output 9	9	32	Output 25
Output 8	10	31	Output 26
Output 7	11	30	Output 27
Output 6	12	29	Output 28
Output 5	13	28	Output 29
Output 4	14	27	Output 30
Output 3	15	26	Output 31
Output 2	16	25	Load
Output 1	17	24	$\overline{\text{Backplane in}}$
Data out	18	23	$\overline{\text{Backplane out}}$
Osc. in	19	22	Data in
V_{dd}	20	21	Clock in

MM5484 16-Segment LED display driver

The 5484 display driver can control up to 16 LED segments in common anode (separate cathode) displays. Each of the 16 outputs can sink up to 15 mA. Although MOS circuitry is used in this device, it is capable of TTL compatibility. The 16 segments available can drive up to 2½ digits of standard 7-segment LED displays.

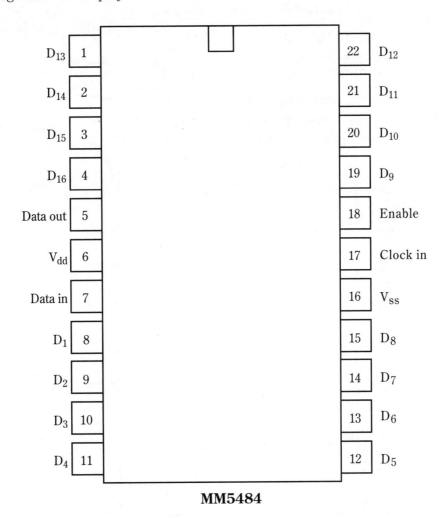

MM5484

DS75494 Hex digit driver

The 7594 is a hex (six) digit driver intended to interface common-cathode LED digit displays with MOS circuitry. The outputs supply a low voltage at relatively high operating currents. A common enable input shuts down all six outputs when brought HIGH. The outputs are enables, when this enable input is LOW.

Absolute Maximum Ratings

Supply Voltage	10 Volts
Input Voltage	10 Volts
Output Voltage	10 Volts

Maximum Power Dissipation

Cavity Package	1433 mW
Molded Package	1362 mW

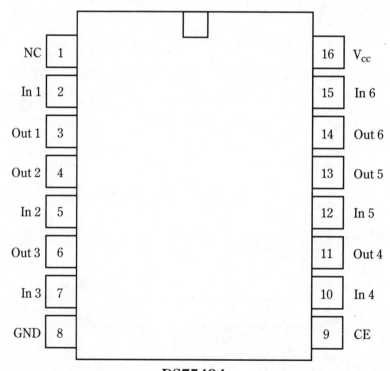

DS75494

Truth Table

ENABLE	Vin	V_{out}
0	0	1
0	1	0
1	x	1

x = Don't care

14410 Tone encoder

The 14410 2-of-8 tone encoder is constructed with complementary-MOS (CMOS) enhancement-mode devices. It is designed to accept digital inputs in a 2-of-8 code format and to digitally synthesize the high-band and low-band sine waves specified by telephone tone dialing systems. The inputs are normally originated from a 4-by-4 matrix keypad, which generates four row and four column input signals in a 2-of-8 code format (1 row and 1 column are simultaneously connected to V_{SS}). The master clocking for the 14410 is achieved from a crystal-controlled oscillator that is included on the chip. Internal clocks, which operate the logic, are enabled only by one or more row and column signals being activated simultaneously. The two sine-wave outputs have NPN bipolar structures on the same substrate, which allows for LOW output impedance and large source currents. This device can be used in telephone tone dialers, radio and mobile telephones, process control, point-of-sale terminals and credit-card verification terminals.

Noise Immunity	45% of V_{dd} Typical
Supply Voltage Range	4.4 to 6.0 Vdc
Frequency Accuracy	±0.2%

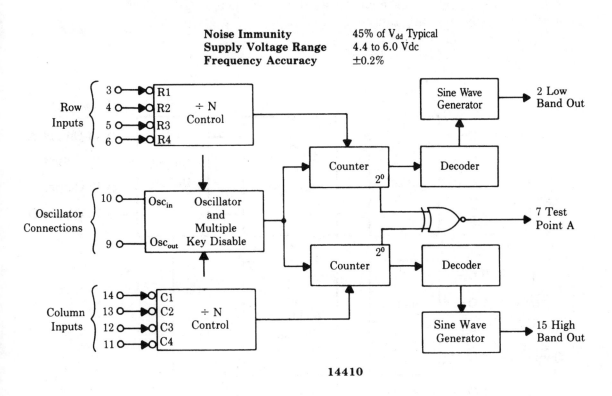

14410

14415 Timer/driver

The 14415 quad timer/driver is constructed with complementary-MOS (CMOS) enhancement-mode devices. The output pulse width of each digital timer is a function of the input clock frequency. Once the proper input sequence is detected, the output buffer is set (turned ON), and after 100 clock pulses are counted, the output buffer is reset (turned OFF).

The 14415 was designed specifically for application in high-speed line printers to provide the critical timing of the hammer drivers, but can be used in many applications requiring precision pulse widths. It features:

- Four precision digital time delays
- Schmitt-trigger clock conditioning
- NPN bipolar output drivers
- Tuning disable capability using inhibit output
- Positive- or negative-edge strobing on the inputs
- Synchronous polynomical counters used for delay counting
- Power-supply operating range
  = 3.0 to 18 V_{dc} (14415EFL)
  = 3.0 to 16 V_{dc} (14415FL/FP)
  = 3.0 to 6.0 V_{dc} (14415EVL/VL/VP)

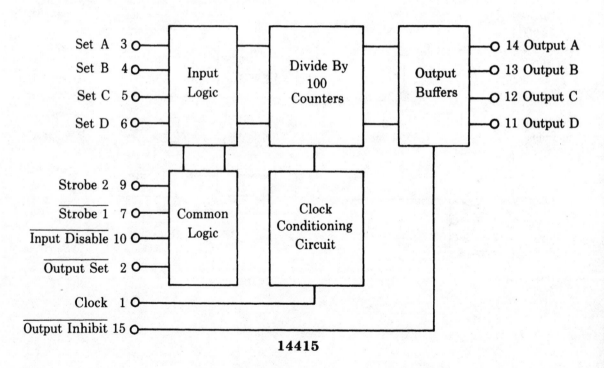

14415

14490 Contact bounce eliminator

The 14490 is constructed with complementary-MOS (CMOS) enhancement-mode devices, and is used for the elimination of extraneous level changes that result when interfacing with mechanical contacts. The digital contact bounce-eliminator circuit takes an input signal from a bouncing contact and generates a clean digital signal four clock periods after the input has stabilized. The bounce-eliminator circuit will remove bounce on both the make and the break of a contact closure.

The clock for operation of the 14490 is derived from an internal RC oscillator which requires only an external capacitor to adjust for the desired operating frequency (bounce delay). The clock can also be driven from an external clock source or the oscillator of another 14490.

- Noise immunity = 45% of V_{DD} Typical
- Supply voltage range
  - = 3.0 to 18 V_{DC} (14490EFL)
  - = 3.0 to 16 V_{DC} (14490FL/FP)
  - = 3.0 to 6.0 V_{DC} (14490EVL/VL/VP)

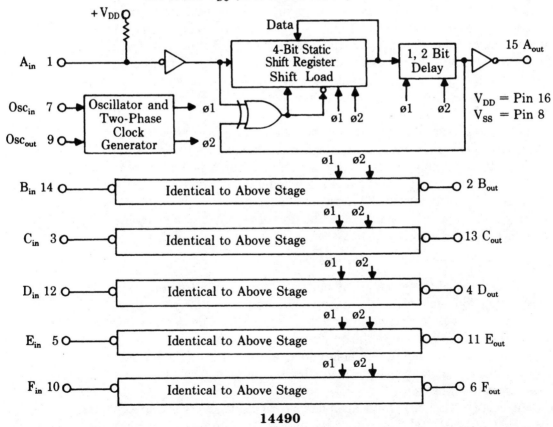

14490

329

4
SECTION

Operational amplifiers

One of the most common elements in modern linear electronic circuits is the *operational amplifier (op amp)*. At one time, this was a fairly esoteric and uncommon type of circuit, used only in highly sophisticated analog computer circuits. It was expensive and difficult to design and build op amps using discrete components. The advent of the integrated circuit changed all that. It is no more difficult to fabricate an operational amplifier than any other type of circuit on a chip, and once initial research costs were covered, it became quite inexpensive. Today, the op amp is probably the most widely used single type of linear IC around.

Very simply, an operational amplifier is an analog amplifier with two inputs. The signal at one input is inverted, and the signal to the second input is noninverted. Roughly speaking, the inverted input signal is "subtracted" from the noninverted input signal, before being multiplied by the amplifier's gain and fed to the output. If we call the inverting input B, and the noninverting input A, and the gain G, the function of an operational amplifier can be expressed algebraically:

$$Output = (A - B)G$$

Assuming that there is no feedback path from output to input, the theoretical open loop gain of an operational amplifier is infinite. True infinite gain is not possible with a real-world circuit, of course, but the open loop gain is still extremely high. Because the output voltage can't exceed the supply voltage, a very, very small differential input signal can saturate the output, so in most cases, we can reasonably assume that the open loop gain is functionally equivalent to infinity.

In most (but not all) practical op-amp applications, a lower gain would be desired, if not essential. This can be accomplished by connecting a resistance element in a feedback path between the output and the inverting input, creating a closed negative feedback loop. The closed loop gain is determined by the feedback resistance and the basic circuit configuration. Using an ordinary fixed resistor, the gain will be linear, and frequency independent. However, other, nonlinear components can be used in the feedback path to create special responses. For example, a capacitor in the feedback loop will result in a gain that varies with signal frequency (useful in filter and oscillator circuits). A transistor in the feedback path will give a logarithmic gain response curve, rather than the straight-line linear response of a simple resistor. Countless circuits have been designed around operational amplifiers.

The de facto standard for op-amp ICs seems to be the 741. Most single op-amp ICs are pin-for-pin compatible with this popular device, and their specifications are usually compared to it. The 741 is far from state-of-the-art. In fact, it is notably inferior to most modern op amps. But it is still good enough for the vast majority of practical applications. It is inexpensive, and widely available. In high precision applications, newer, more improved operational amplifiers would be called for.

Operational amplifiers

input bias current The average of the two input currents.

input offset current The absolute value of the difference between the two input currents for which the output will be driven higher than or lower than specified voltages.

input offset voltage The absolute value of the voltage between the input terminals required to make the output voltage greater than or less than specified voltages.

input voltage range The range of voltage on the input terminals (common-mode) over which the offset specifications apply.

logic threshold voltage The voltage at the output of the comparator at which the loading logic circuitry changes it digital state.

negative output level The negative dc output voltage with the comparator saturated by a differential input equal to or greater than a specified voltage.

output leakage current The current into the output terminal with the output voltage within a given range and the input drive equal to or greater than a given value.

output resistance The resistance seen looking into the output terminal with the dc output level at the logic threshold voltage.

output sink current The maximum negative current that can be delivered by the comparator.

positive output level The high output voltage level with a given load and the input drive equal to or greater than a specified value.

power consumption The power required to operate the comparator with no output load. The power will vary with signal level, but is specified as a maximum for the entire range of input signal conditions.

response time The interval between the application of an input step function and the time when the output crosses the logic threshold voltage. The input step drives the comparator from some initial, saturated input voltage to an input level just barely in excess of that required to bring the output from saturation to the logic threshold voltage. This excess is referred to as the voltage overdrive.

saturation voltage The low output voltage level with the input drive equal to or greater than a specified value.

strobe current The current out of the strobe terminal when it is at the zero logic level.

strobed output level The dc output voltage, independent of input conditions, with the voltage on the strobe terminal equal to or less than the specified LOW state.

strobe on voltage The maximum voltage on either strobe terminal required to force the output to the specified HIGH state independent of the input voltage.

strobe off voltage   The minimum voltage on the strobe terminal that will guarantee that it does not interfere with the operation of the comparator.

strobe release time   The time required for the output to rise to the logic threshold voltage after the strobe terminal has been driven from 0 to the logic 1 level.

supply current   The current required from the positive or negative supply to operate the comparator with no output load. The power will vary with input voltage, but is specified as a maximum for the entire range of input voltage conditions.

voltage gain   The ratio of the change in output voltage to the change in voltage between the input terminals producing it.

LM11 Operational amplifier

The LM11 is one of the oldest operational amplifier ICs still available. It's specifications don't quite match those of some more modern devices, but it can still be used in a great many practical applications, with excellent results.

In most applications, the LM11 will require external frequency compensation. A special pin for this purpose is included on the chip. Internal output protection permits the output to be short-circuited indefinitely, without damage to the IC.

Absolute maximum ratings

Total supply voltage	40 V	(± 20 V)
Input Current	± 10 mA	
Power Dissipation	500 mW	
Common-Mode Rejection		
LM11	130 dB	typ.
LM11C	130 dB	typ.
LM11CL	110 dB	typ.
Supply-Voltage Rejection		
LM11	118 dB	typ.
LM11C	118 dB	typ.
LM11CL	100 dB	typ.
Supply Current	0.3 mA	typ.

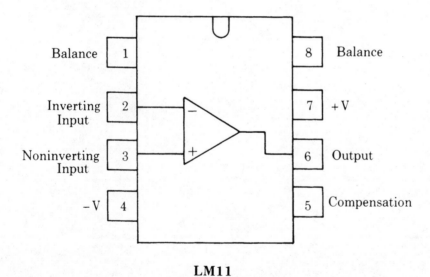

LM11

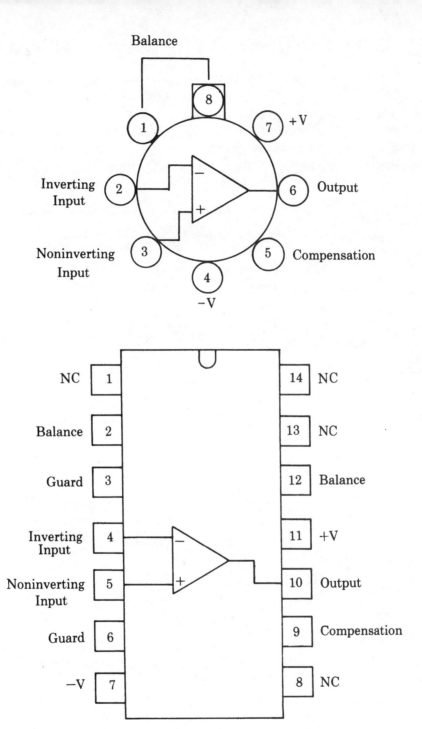

TL062 Dual low-power JFET input operational amplifier

The TL062 contains two JFET input type operational amplifiers designed for LOW-power applications. The supply current drawn by this device is typically just 200 µA per amplifier.

    The operational amplifiers in the TL062 offer very good specifications, including HIGH slew rate, gain bandwidth product, input impedance, and output voltage swing, as well as LOW input bias, and input offset currents.

	Minimum	Typical	Maximum
Supply Voltage			±36 V
Differential Input Voltage			±30 V
Offset Voltage		3.0 mV	
Input Offset Current		0.5 pA	
Input Bias Current		3.0 pA	
Input Resistance		10	
Slew Rate		6.0 V/µS	
Output Voltage Swing			±14 V

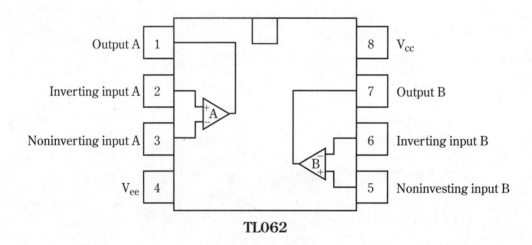

TL062

LM124/LM224/LM324, LM124A/ LM224A/LM324A, LM2902 Low-power quad operational amplifiers

The LM124 series consists of four independent, HIGH-gain, internally frequency-compensated operational amplifiers that were designed specifically to operate from a single power supply over a wide range of voltages. Operation from split power supplies is also possible and the LOW power-supply current drain is independent of the magnitude of the power-supply voltage.

Application areas include transducer amplifiers, dc-gain blocks and all the conventional op-amp circuits, which now can be more easily implemented in single power-supply systems. For example, the LM124 series can be directly operated from the standard +5 V_{dc} power-supply voltage that is used in digital systems. The voltage will easily provide the required interface electronics without requiring additional +15 V_{dc} power supplies.

In the linear mode, the input common-mode voltage range includes ground and the output voltage can also swing to ground, even though operated from only a single power-supply voltage. The unity-gain cross frequency is temperature-compensated. The input bias current is also temperature-compensated.

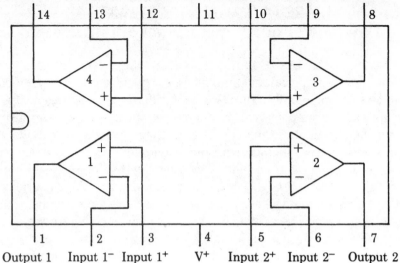

Output 4 Input 4⁻ Input 4⁺ GND Input 3⁺ Input 3⁻ Output 3

Output 1 Input 1⁻ Input 1⁺ V⁺ Input 2⁺ Input 2⁻ Output 2

Top View

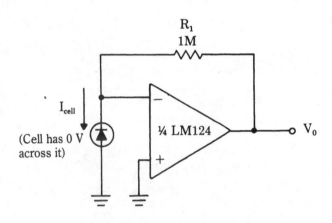

R_1
1M

I_{cell}

(Cell has 0 V across it)

¼ LM124

V_0

LM124

Dc Voltage Gain	100 dB
Bandwidth (Unity Gain)	1 MHz
Power Supply Range:	
Single Supply	3 to 30 Vdc
or Dual Supplies	± 1.5 to ± 15 Vdc
Supply Current Drain	800 μA – Essentially independent of supply voltage (1 mW/op amp at + 5 Vdc)
Input Biasing Current	45 nAdc
Input Offset Voltage	2 mVdc
Offset Current	5 nAdc
Output Voltage Swing	0 to V⁺ – 1.5 Vdc

LM307 Internally compensated monolithic operational amplifier

The 307 is an improvement over the popular 741 operational amplifier in that it can be used with lower input currents. The improved input characteristics of this chip allow for greater accuracy in such applications as sample-and-hold circuits and long-interval integrators. The 307 is internally compensated.

	Minimum	Typical	Maximum
Supply Voltage			±18 V
Differential Input Voltage			±30 V
Offset Voltage		2.0 mV	7.5 mV
Input Offset Current		3.0 nA	50 nA
Input Bias Current		70 nA	250 nA
Input Resistance	0.5 M	2.0 M	

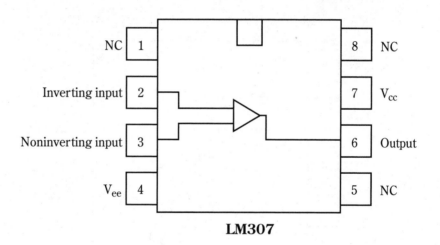

LM307

LM308A Super gain precision operational amplifier

A special "Super-Beta" processing technology is used in the manufacture of the 308 to create superior specifications, including LOW noise, HIGH input impedance, LOW input offset, and LOW temperature drift. This operational amplifier can also operate over a wide range of power supply voltages.

For high-speed applications, the slew rate can be improved by feedforward compensation techniques. This does not degrade the other performance characteristics of the operational amplifier.

	Minimum	Typical	Maximum
Supply Voltage			±18 V
Input Voltage			±15 V
Offset Voltage		0.3 mV	0.5 mV
Input Offset Current		0.2 nA	1.0 nA
Input Bias Current		1.5 nA	7.0 nA
Input Resistance	10 M	40 M	

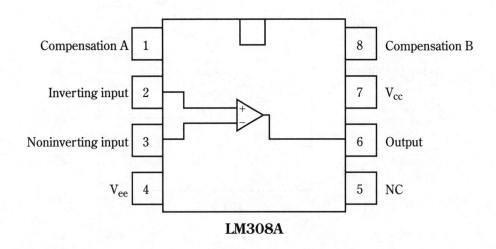

LM308A

353 Wide-bandwidth dual-JFET operational amplifier

The 353 contains two high-quality operational-amplifier stages in a single 8-pin package. These op amps are easy to use because they have no special-purpose pins. All eight pins are taken up with the supply-voltage connections, and the inputs and outputs for each of the two operational-amplifier stages. Except for the power-supply connections, the two op-amp stages are completely independent of each other.

The 353 features full internal-frequency and phase compensation.

The 353 operational amplifiers use well-matched high-voltage JFET input devices, so they exhibit an extremely HIGH input impedance, and very LOW input-bias and offset currents. Other advantages of the 353 are minimal offset-voltage drift and very LOW noise. These features make op amps very well suited to HIGH-fidelity audio and data-transmission applications.

Thanks to internal output-protection circuitry, the output of either op amp in the 353 can be safely short-circuited for an indefinite period of time.

Supply Voltage	36 V ($\pm$18 V) max.
Power Dissipation	500 mW typ.
Input Voltage	$\pm$ 15 V max.
Differential-Input Voltage	$\pm$ 30 V max.
Wide-Gain Bandwidth	4 MHz
Slew Rate	13 V/μS

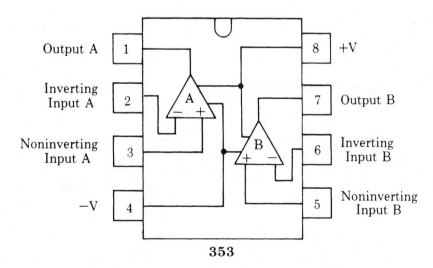

353

LM709/LM709A/LM709C Operational amplifier

The LM709 series is a monolithic operational amplifier intended for general-purpose applications. Operation is completely specified over the range of voltages commonly used for these devices. The design, in addition to providing HIGH gain, minimizes both offset voltage and bias currents. Further, the class-B output stage gives a large output capability with minimum power drain.

External components are used to frequency-compensate the amplifier. Although the specified unity-gain compensation network will make the amplifier unconditionally stable in all feedback configurations, compensation can be tailored to optimize high-frequency performance for any gain setting.

Since the amplifier is built on a single silicon chip, it provides LOW offset and temperature drift at a minimum cost. It also ensures negligible drift as a result of temperature gradients in the vicinity of the amplifier.

The LM709C is the industrial version of the LM709. It is identical to the LM709/LM709A, except that it is specified for operation from 0°C to +70°C.

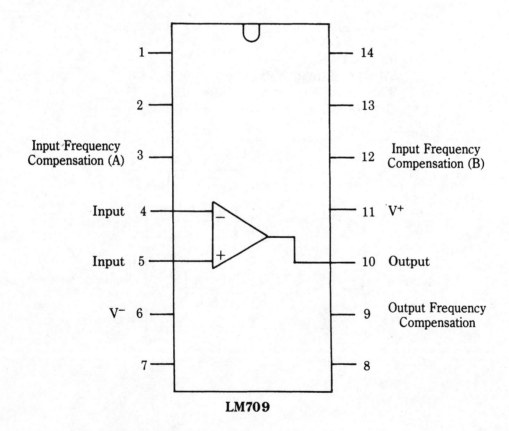

LM709

BA718 Dual operational amplifier

The 718 contains a pair of independent operational amplifiers in a single 9-pin SIP (single inline package) housing. These op amps are very easy to use because the only connections made to the chip are the supply voltages and the inverting and noninverting inputs and output for each op amp.

The 718's op amp feature internal phase compensation. The op amp outputs are short-circuit protected.

This device can be operated from either a dual-polarity or a single-polarity power supply. Even if a single-ended (positive) supply voltage is used with the 718, a negative voltage can be included in the common-mode input voltage range.

Supply Voltage	3 to 18 V
	$\pm$ 1.5 to $\pm$ 9 V
Current Consumption	1.5 mA (typ.)
Power Dissipation	450 mW (max.)
Common-Mode Rejection Ratio	90 dB (typ.)
Differential Input Voltage	18 V (max.)

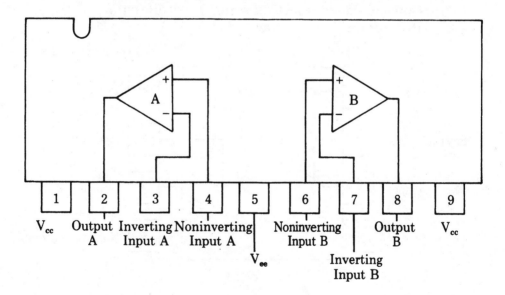

BA718

BA728 Dual operational amplifier

The 728 contains a pair of independent operational amplifiers in a single 8-pin DIP (Dual-Inline Package) housing. Other than the packaging, the 728 is identical to the 718.

The operational amplifiers within the 728 are very easy to use, because the only connections made to the chip are the supply voltages, the inverting and noninverting inputs, and the output for each op amp.

The 728's operational amplifiers feature internal phase compensation. The op-amp outputs are short-circuit protected. This device can be operated from either a dual-polarity (floating ground) or a single-polarity power supply. Even if a single-ended (positive) supply voltage is used with the 728, a negative voltage can still be included in the common-mode input voltage range.

Supply voltage	3 to 18 V
	±1.5 to ±9 V
Current consumption	1.5 mA (typ.)
Power dissipation	450 mW (max.)
Common-mode rejection ratio	90 dB (typ.)
Differential input voltage	18 V (max.)

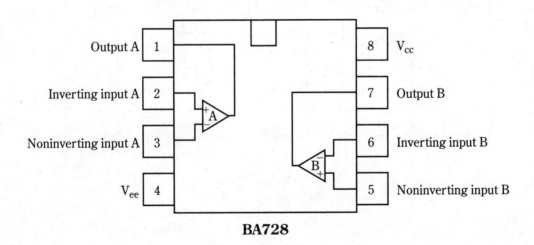

BA728

LM741/LM741A/LM741C
/LM741E Operational amplifier

The LM741 series are general-purpose operational amplifiers that feature improved performance over industry standards, like the LM709. They are direct, plug-in replacements for the 709C, LM201, MC1439, and 748 in most applications.

The amplifiers offer many features that make their application nearly foolproof: overload protection on the input and output, no latch-up when the common-mode range is exceeded, as well as freedom from oscillations. The LM741C/LM741E is identical to the LM741/LM741A, except that the LM741C/LM741E has its performance guaranteed over a 0°C to +70°C temperature range, instead of –55°C to +125°C.

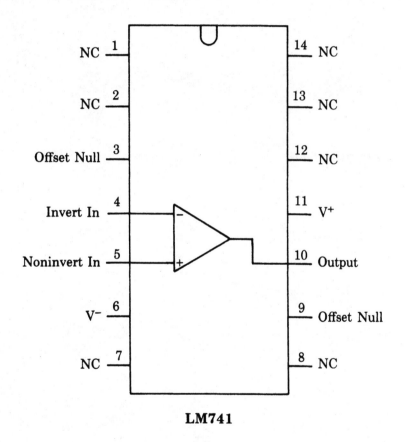

LM741

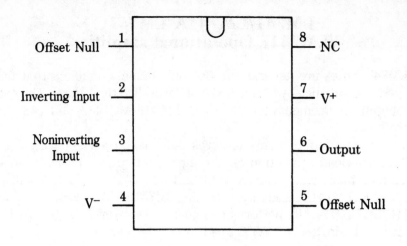

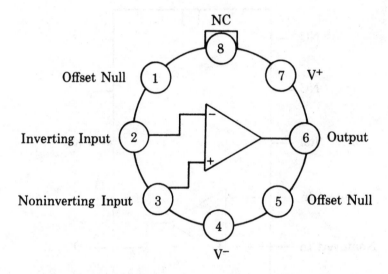

Note: Pin 4 connected to case.

LM747/LM747A/LM747C/ LM747E Dual operational amplifier

The LM747 series are general-purpose dual operational amplifiers. The two amplifiers share a common-bias network and power-supply leads. Otherwise, their operation is completely independent. Features include:

- No frequency compensation required
- Short-circuit protection
- Wide common-mode and differential voltage ranges
- Low power consumption
- No latch-up
- Balanced offset null

The LM747C/LM747E is identical to the LM747/LM747A except that the LM747C/LM747E has its specifications guaranteed over the temperature range from 0°C to +70°C instead of –55°C to +125°C.

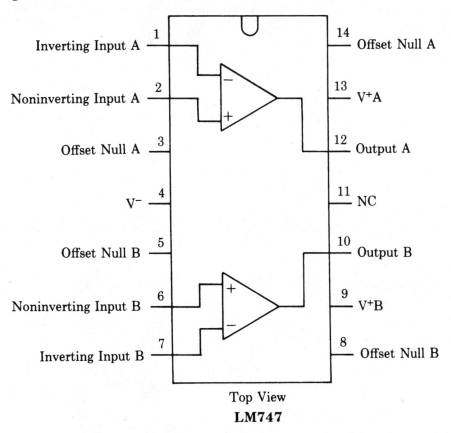

Top View
LM747

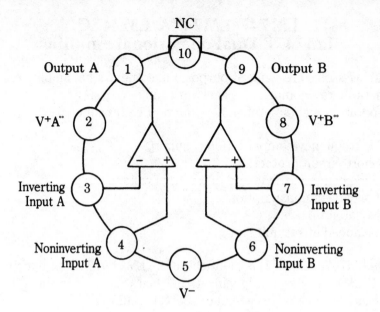

NC

Output A   1   10   9   Output B

$V^+A^{\circ\circ}$ 2

8   $V^+B^{\circ\circ}$

Inverting
Input A   3

7   Inverting
Input B

Noninverting   4
Input A

6   Noninverting
Input B

5

V^-

Top View

LM748/LM748C Operational amplifier

The LM748/LM748C is a general-purpose operational amplifier built on a single silicon chip. The resulting close match and tight thermal coupling gives LOW offsets and temperature drift as well as fast recovery from thermal transients. In addition, the device features:

- Frequency compensation with a single 30-pF capacitor
- Operation from ±5 to ±20 V
- Low current drain—1.8 mA at ±20 V
- Continuous short-circuit protection
- Operation as a comparator with differential inputs as high as ±30 V
- No latch-up when common mode range is exceeded.
- Same pin configuration as the LM101.

The unity-gain compensation specified makes the circuit stable for all feedback configurations, even with capacitive loads. However, it is possible to optimize compensation for best high-frequency performance at any gain. As a comparator, the output can be clamped at any desired level to make it compatible with logic circuits.

The LM748 is specified for operation over the –55°C to +125°C military temperature range. The LM748C is specified for operation over the 0°C to + 70°C temperature range.

Inverting Amplifier with Balancing Circuit

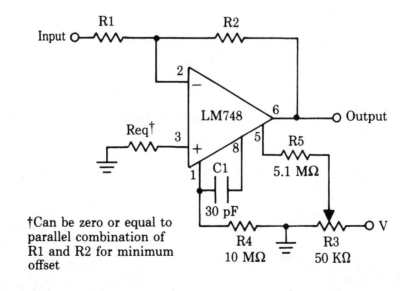

†Can be zero or equal to parallel combination of R1 and R2 for minimum offset

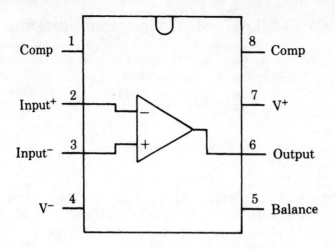

Top View

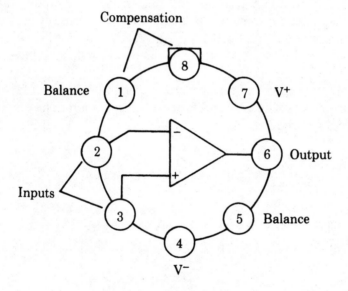

Note: Pin 4 connected to case

LM748

LM833 Dual audio operational amplifier

The 833 contains a pair of op amps designed specifically for high-quality performance with audio-frequency signals. This chip is internally compensated for all closed loop gains, and is pin-for-pin compatible with industry standard dual operational amplifiers.

	Minimum	Typical	Maximum
Supply Voltage			±36 V
Differential Input Voltage			±30 V
Offset Voltage		0.3 mV	5.0 mV
Input Offset Current		10 nA	200 nA
Input Bias Current		300 nA	1000 nA
Power Dissipation			500 mW
Power Bandwidth	120 kHz		
High Gain-Bandwidth Product	15 MHz		
(Minimum rating)	10 MHz		
Distortion	0.002%		
Slew Rate	7 V/μS		
(Minimum rating)	5 V/μS		
Dynamic Range	>140 dB		

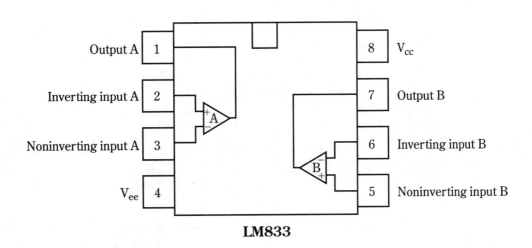

LM833

LM1458/LM1558 Dual operational amplifier

The LM1458 and LM1558 are general-purpose dual operational amplifiers. The two amplifiers share a common-bias network and power-supply leads. Otherwise, their operation is completely independent. Features include:

- No frequency compensation required
- Short-circuit protection
- Wide common-mode and differential voltage ranges
- Low-power consumption
- 8-lead TO-5 and 8-lead mini-DP
- No latch-up when the input common-mode range is exceeded

The LM1458 is identical to the LM1558, except that the LM1458 has its specifications guaranteed over the temperature range from 0°C to 70°C instead of the −55°C to +125°C range of the LM1558.

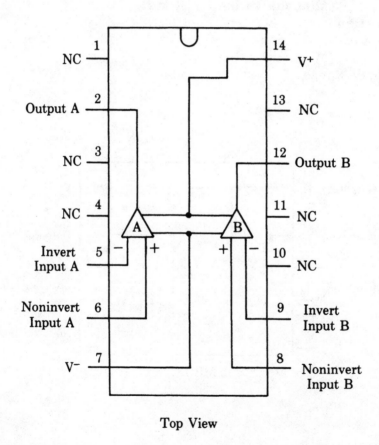

Top View

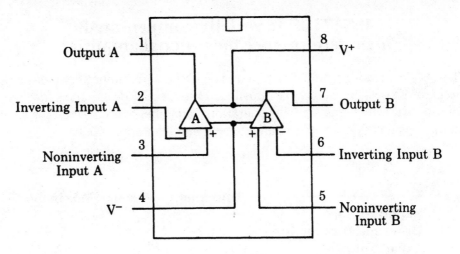

Top View

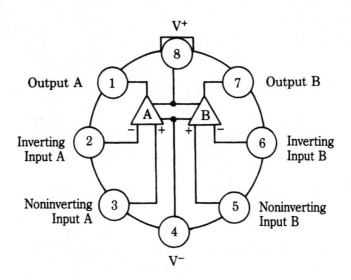

Top View

LM1558

MC1741C Internally compensated, high-performance operational amplifier

The 1741 operational amplifier requires no external frequency compensation. It features LOW power consumption, no latch-up, and wide common-mode and differential voltage ranges. This operational amplifier was designed for use in summing amplifier, integrator, and general amplifier applications.

	Minimum	Typical	Maximum
Supply Voltage			±18 V
Differential Input Voltage			±30 V
Offset Voltage		2.0 mV	6.0 mV
Input Offset Current		20 nA	200 nA
Input Bias Current		80 nA	500 nA
Input Resistance	0.3 M	2.0 M	

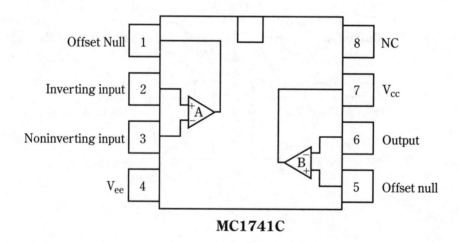

MC1741C

MC1748C High-performance operational amplifier

The 1748 is an uncompensated version of the 1741. Only a single 30-pF capacitor is required to externally compensate this device for unity gain. This chip features LOW power consumption, no latch-up, and wide common-mode and differential voltage ranges. This operational amplifier was designed for use in summing amplifier, integrator, and general amplifier applications.

	Minimum	Typical	Maximum
Supply Voltage			±18 V
Differential Input Voltage			±30 V
Offset Voltage		2.0 mV	6.0 mV
Input Offset Current		20 nA	200 nA
Input Bias Current		80 nA	500 nA
Input Resistance	0.3 M	2.0 M	

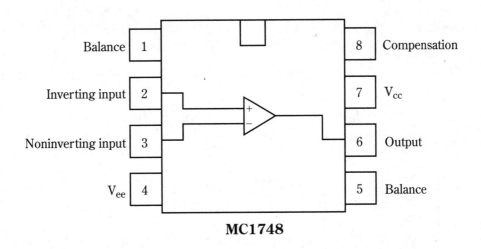

MC1748

LM1900/LM2900/LM3900, LM3301, LM3401 Quad amplifier

The LM1900 series consists of four independent, dual input, internally compensated amplifiers that were designed specifically to operate from a single power supply voltage and to provide a large output voltage swing. These amplifiers make use of a current mirror to achieve the noninverting input function. They are used as ac amplifiers; RC active filters; low-frequency triangle-square- and pulse-wave generation circuits; tachometers and low-speed high-voltage digital logic gates. Features include:

- Wide single supply voltage range or dual supplies 4 to 36 V_{dc}
  ± 2 to ± 18 V_{dc}
- Supply current drain independent of supply voltage
- Low input-biasing current   30 nA
- High open-loop gain   70 dB
- Wide bandwidth   2.5 MHz (unity gain)
- Large output voltage swing   (V + −1) V_{p-p}
- Internally frequency compensated for unity gain
- Output short-circuit protection

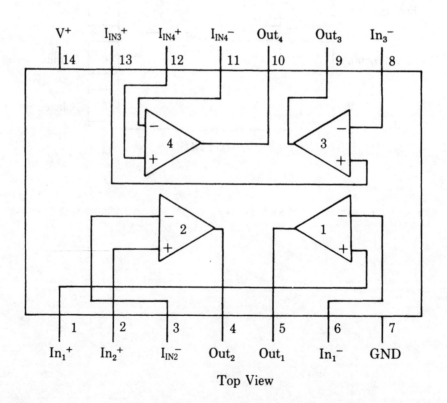

Top View

356

BA4560 Dual high slew-rate operational amplifier

The 4560 contains a pair of high-quality operational amplifiers in a simple 8-pin DIP housing. These two op amps are very easy to use because there are no special-purpose pins. All eight pins are taken up with the supply-voltage connections, and the inputs (inverting and noninverting) and outputs of the two operational amplifier stages. Except for the power-supply connections, the two op amps on this chip are completely independent of each other. The operational amplifiers within the 4560 offer excellent frequency response, slew rate, and high output current capability.

Supply Voltage	36 V ($\pm$ 18 V) max.
Power Dissipation	500 mW max.
Differential Input Voltage	$\pm$ 30 V max.
Slew Rate	4 V/μS
Gain-Bandwidth Product	10 MHz
Common-Mode Rejection Ratio	90 dB typ.

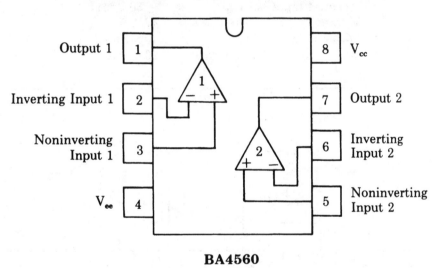

BA4560

BA4561 Dual high slew-rate operational amplifier

The 4561 contains a pair of high-quality operational amplifiers in a simple 9-pin SIP (Single-Inline Pin) housing. All nine pins are arranged in one continuous row. The packaging is the only difference between the 4561 and the 4560. Electrically, they are identical. The V_{cc} connection is provided twice on the 4561, accounting for the extra pin.

These two op amps are very easy to use because there are no special-purpose pins. All eight pins are taken up with the supply-voltage connections, and the inputs (inverting and noninverting) and outputs of the two operational amplifier stages. Except for the power-supply connections, the two op amps on this chip are completely independent of each other. The operational amplifiers within the 4561 offer excellent frequency response, slew rate, and HIGH output current capability.

Supply Voltage	36 V ($\pm$ 18 V) max.
Power Dissipation	500 mW max.
Differential Input Voltage	$\pm$ 30 V max.
Slew Rate	4 V/μS
Gain-Bandwidth Product	10 MHz
Common-Mode Rejection Ratio	90 dB typ.

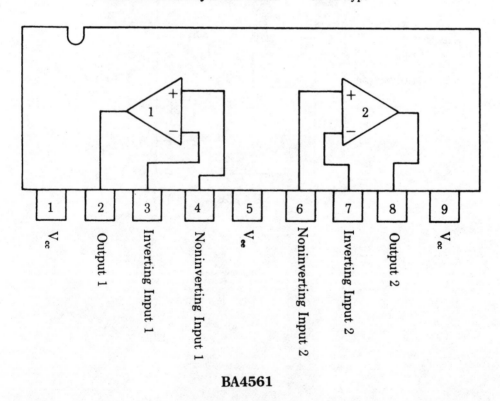

BA4561

358

BA6110 Voltage-controlled operational amplifier

The 6110 is contained within a simple 9-pin SIP (single-inline package) housing. All nine pins are arranged in one continuous row.

The forward transconductance (gm) of the 6110 op amp is programmable over a wide range with high linearity. In effect, the op amp's gain can be controlled via an external signal; thus, this is a voltage-controlled op amp. Actually, the control signal is accepted by the chip in the form of a current, but by simply adding a resistor in series with the control input (pin 4), a more convenient control voltage can be used.

The open-loop gain of the 6110 is set by the applied control current and external gain resistor (R1).

The 6110 also features an on-chip low-impedance output buffer. This internal buffer can be used to simplify external city and to minimize the overall parts count of the complete circuit.

The op amp used in the 6110 is a high-quality device, featuring low noise, low distortion, and low offset.

Supply Voltage	34 V ($\pm$17 V) max.
Power Dissipation	500 mW max.
Quiescent Current	0.9 mA min.
	3.0 mA typ.
	6.0 mA max.
Forward Transconductance	4800 $\mu\Omega$ min.
	8000 $\mu\Omega$ typ.
	12000 $\mu\Omega$ max.

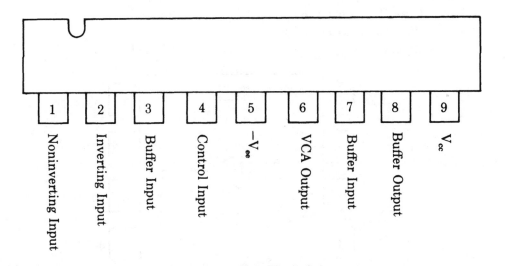

BA6110

359

LM6161/LM6261/LM6361 High-speed operational amplifier

The 6161 (and its companion devices, the 6261, and 6361) is a very high-speed operational amplifier, designed for use at high frequencies, making it well suited to RF and video applications.

The 6161 meets military temperature specifications. The 6361 meets the less stringent commercial temperature specifications. The 6261 is designed to meet an intermediate level of industrial temperature standards. Otherwise, these three devices are essentially identical, although there are some operational variations on the maximum ratings for a few specifications.

This operational amplifier typically draws just a mere 5 mA of supply current, yet it can provide unity-gain stability with signal frequencies up to 50 MHz, and a fast slew rate of 300 V/µS. This chip accepts a wide range of power supply voltages, ranging from a low of +5 volts to an absolute maximum of 36 volts.

	Minimum	Typical	Maximum
Supply Voltage	+4.75V		+36 V
Differential Input Voltage			±8 V
Offset Voltage (6161)		5.0 mV	7 mV
Offset Voltage (6261)		5.0 mV	7 mV
Offset Voltage (6361)		5.0 mV	22 mV
Input Offset Current (6161)		150 nA	350 nA
Input Offset Current (6261)		150 nA	350 nA
Input Offset Current (6361)		150 nA	1500 nA
Input Bias Current (6161)		2 nA	3 nA
Input Bias Current (6261)		2 nA	3 nA
Input Bias Current (6361)		2 nA	5 nA
Input Resistance		325 kΩ	

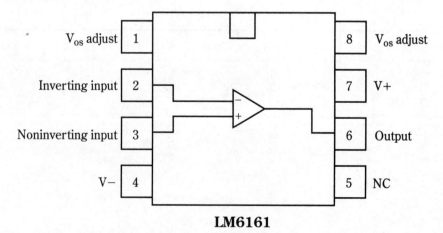

LM6161

MC3303/MC3403 Quad low-power operational amplifiers

The 3303 and 3403 are low-cost ICs, each containing four operational amplifiers with true differential inputs. The 3303 is designed to operate over a wider temperature range (military specifications) than the 3403 (commercial specifications). The 3403 is more than adequate for almost any consumer application. Other than the operating temperature range, these two devices are virtually identical, and have electrical characteristics similar to the 1741. The outputs are short circuit protected.

	Minimum	Typical	Maximum
Supply Voltage			
(Single sided)	3.0 V		36 V
(Split supply)	±1.5 V		±18 V
Differential Input Voltage			±36 V
Offset Voltage		2.0 mV	10 mV
Input Offset Current		30 nA	50 nA
Input Bias Current		−200 nA	−500 nA
Input Resistance	0.3 M	1.0 M	

MC3303
MC3403

MC3476 Low-cost programmable operational amplifier

The 376 is a very versatile, yet low-cost operational amplifier, with LOW power consumption and HIGH input impedance. The quiescent currents within this op amp can be externally programmed with a resistor of an appropriate value or a current source applied to the I_{set} input. This programmability permits the device's operating characteristics to be customized to suit the input current and power-consumption requirements of the intended application—even over wide variations in power-supply voltages. No external frequency compensation is required with the 3476.

	Minimum	Typical	Maximum
Supply Voltage	±6.0 V		±18 V
Differential Input Voltage			±30 V
Programming Current (I_{set})			200 μA
Input Offset Current		20 nA	25 nA
Input Bias Current		15 nA	50 nA
Input Resistance		5.0 M	

Typical R_{set} Values

Supply Voltage	$I_{set} = 1.5\ \mu A$	$I_{set} = 15\ \mu A$
±6.0 V	3.6 M	360 k
±10 V	6.2 M	620 k
±12 V	7.5 M	750 k
±15 V	10 M	1.0 M

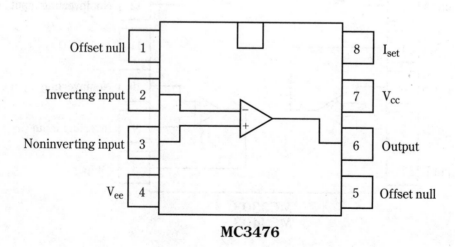

MC3476

MC4558 Dual wide-bandwidth operational amplifiers

The 4558 is a dual operational amplifier, similar in features to the 1458, but this device can offer up to three times the gain bandwidth of the industry standard. This IC is internally compensated and protected against short circuits. It offers LOW power consumption, as well as gain and phase matching between the two operational amplifiers on the chip. The 4558 is available in two versions, with slightly different specifications. The main difference is that the MC4558C has a guaranteed unity-gain bandwidth of 2.0 MHz, and the MC4558AC has a guaranteed unity-gain bandwidth of 2.5 MHz.

MC4558AC Specifications

	Minimum	Typical	Maximum
Supply Voltage			±22 V
Differential Input Voltage			±30 V
Offset Voltage		1.0 mV	5.0 mV
Input Offset Current		20 nA	200 nA
Input Bias Current		80 nA	500 nA
Input Resistance	0.3 M	2.0 M	
Unity Gain Bandwidth		2.5 MHz	

MC4558C Specifications

	Minimum	Typical	Maximum
Supply Voltage			±18 V
Differential Input Voltage			±30 V
Offset Voltage		2.0 mV	6.0 mV
Input Offset Current		20 nA	200 nA
Input Bias Current		80 nA	500 nA
Input Resistance	0.3 M	2.0 M	
Unity Gain Bandwidth		2.0 MHz	

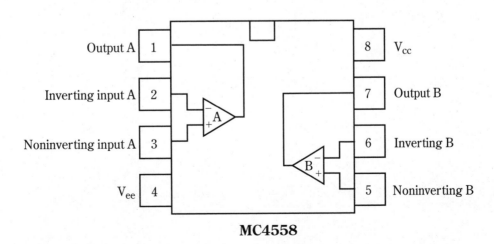

MC4558

MC4741C Differential-input operational amplifier

The 4741 is essentially four independent 1741s on a single chip. It is especially useful in applications where high packing density and/or close amplifier matching is called for. Each of the operational amplifiers on this chip is functionally equivalent to the 1741, with true differential inputs, internal frequency compensation, and short-circuit protection. These operational amplifiers also feature a class-AB output stage to minimize crossover distortion. Each op amp draws just 0.6 mA of current from the power supply.

	Minimum	Typical	Maximum
Supply Voltage			±18 V
Differential Input Voltage			±36 V
Offset Voltage		2.0 mV	6.0 mV
Input Offset Current		20 nA	200 nA
Input Bias Current		80 nA	500 nA
Input Resistance	0.3 M	2.0 M	

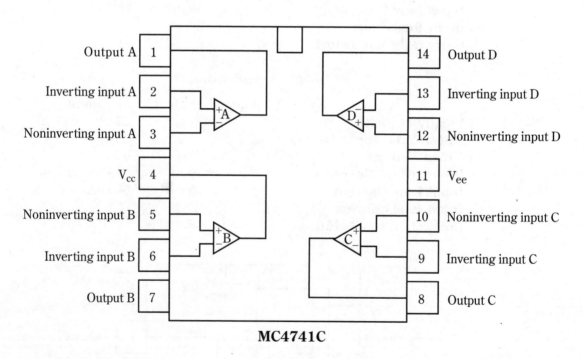

MC4741C

MC33102 Dual Sleep-Mode operational amplifier

The unique feature of the 33102 dual operational-amplifier IC is the special Sleep-Mode technology. The "Sleep-Mode" is a micropower state intended to minimize power consumption when the operational amplifier is not required to output much signal. The amplifier is active and alert to input signals in this mode. When the input signal is sufficient to cause the operational amplifier to source or sink 160 μA (typical) to the load, the device will automatically switch itself into the "Awake-Mode" for high-performance operation. This transition from "Sleep-Mode" to "Awake-Mode" typically requires just 4 μS, so it can be considered virtually instantaneous in most practical applications.

	Minimum	Typical	Maximum
Supply Voltage			+36 V
Differential Input Voltage	Between V_{ee} and V_{cc}		

Typical Specifications

	Sleep-Mode	Awake-Mode
Input Offset Voltage	0.15 mV	0.15 mV
Current Drain	45 μA	750 μA
Gain Bandwidth (at 20 kHz)	0.33 MHz	4.6 MHz
Slew Rate	0.16 V/μS	1.7 V/μS
Output Current Capability	0.15 mA	50 mA

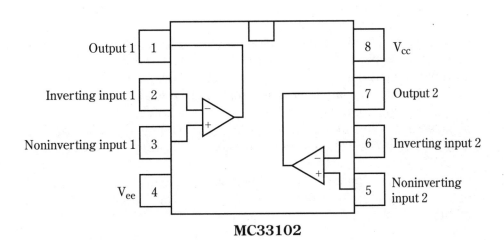

MC33102

5
SECTION

Audio amplifiers

Perhaps the most commonly used type of active electronic circuit is the amplifier. Usually, an amplifier accepts an input signal, and produces a higher amplitude replica of the input signal at the output. The amount of increase is called the *gain*. To be absolutely literal, the term *amplifier* implies positive gain: that is, the output signal is larger (higher amplitude) than the input signal. In practical electronics, this is not always the case. Some amplifiers actually have negative gain so that the output signal is lower in amplitude than the input signal. This type of "reverse amplification" is also known as *attenuation*. Some amplifier circuits have unity gain, which means the output signal has the same amplitude as the input signal. This might sound like a useless circuit, but it can be very useful in interfacing equipment or circuits. A unity-gain amplifier limits potential loading problems and/or can provide impedance matching. A unity-gain amplifier is also sometimes referred to as a "buffer."

In this section, we will look at ICs designed primarily for the amplification of signals within the audio-frequency (AF) range (nominally about 20 Hz to 20 kHz). Section 7 moves on to radio-frequency (RF) amplifiers, and related devices.

Incidentally, the operational amplifiers discussed in Section 5 are special variations on the basic amplifier. Op amps are sometimes used in audio or RF amplifier applications.

LM380 Audio power amplifier

The LM380 is a power audio amplifier for consumer application. In order to hold system cost to a minimum, gain is internally fixed at 34 dB. A unique input stage allows inputs to be ground referenced. The output is automatically self-entering to one-half of the supply voltage.

The output is short-circuit proof with internal thermal limiting. The package outline is standard dual-in-line. A copper lead frame is used with the center three pins on either side, comprising a heat sink. This makes the device easy to use in standard PC layout.

Uses include simple phonograph amplifiers, intercoms, line drivers, teaching-machine outputs, alarms, ultrasonic drivers, TV sound systems, AM/FM radio, small servo drivers, power converters, etc.

A selected part for more power on higher supply voltages is available as the LM384. Features include:

- Wide supply-voltage range
- Low quiescent power drain
- Voltage gain fixed at 50
- High peak current capability
- Input referenced to GND
- High input impedance
- Low distortion
- Quiescent output voltage is at one-half of the supply voltage
- Standard dual-in-line package

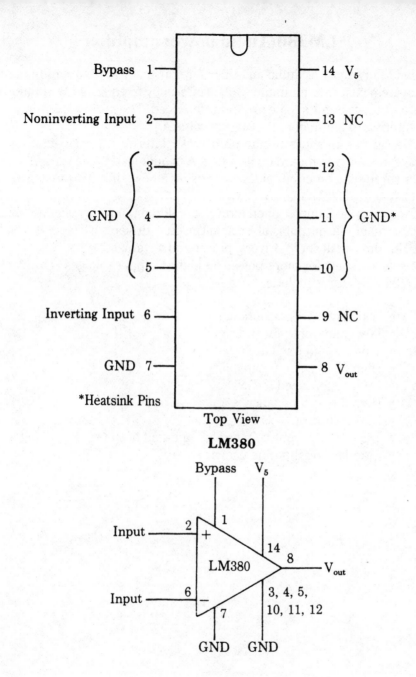

Bypass 1

Noninverting Input 2

GND { 3
       4
       5

Inverting Input 6

GND 7

*Heatsink Pins

14 V_5

13 NC

12 }
11 } GND*
10 }

9 NC

8 V_{out}

Top View

LM380

Bypass V_5

Input ——— 2 + 1

                14
LM380   8 V_{out}

Input ——— 6 −

              7   3, 4, 5,
                  10, 11, 12

GND   GND

LM384 5-Watt audio power amplifier

The 384 is an inexpensive medium-power audio amplifier intended for use in general consumer applications. The gain is internally fixed at 34 dB. The voltage gain is fixed at 50 dB.

This chip can be operated from a fairly wide range of supply voltages, from 12 V to 26 V (28 V, absolute maximum). It has a high peak-current capability, over 1 ampere, but low quiescent power drain. The internal circuitry is short-circuit proof, and thermally limited. The amplifier has high input impedance and low distortion.

A unique input stage permits the inputs to be referenced to ground. The output signal automatically centers itself to one half the supply voltage.

384 Specifications

	Minimum	Typical	Maximum
V+ Supply Voltage	12 V		28 V
Av Gain	40 V/V	50 V/V	60 V/V
Output Power	5 W	5.5 W	
Power Dissipation			1.67 W
Peak Current			1.3 A
THD (Total Harmonic Distortion)		0.25%	1.0%
Quiescent Output Voltage		11 V	
Quiescent Supply Current		8.5 mA	25 mA

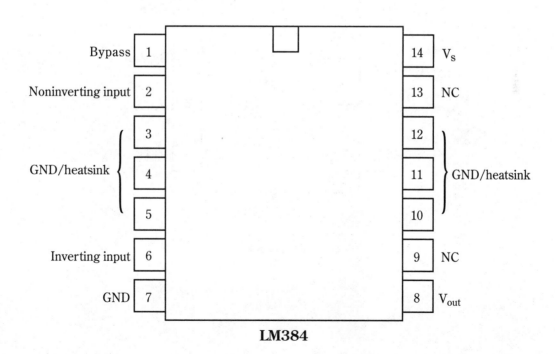

LM384

LM386 Low-voltage audio power amplifier

The '386 is closely related to the '380. The big advantage of the '386 is that it is designed for very low power requirements, making this chip an excellent choice for battery-powered applications.

The '386 can be reliably driven from a power source as low as 4 volts. In the quiescent (no input signal) state, the power consumption is a mere 18 mW, assuming a 6-volt power supply is being used.

In many practical circuits using the '386, overall parts count tends to be quite low. The amplifier's gain is internally set to 20. If more gain is required, the circuit can easily be modified by adding an external resistor and capacitor between pins 1 and 8. The gain can be increased by up to 200.

The '386's inputs are referenced to ground. Internal circuitry within the device automatically bias the output signal to one half the supply voltage.

The amplifier is designed to drive low-impedance loads, such as 8-Ω or 16-Ω speakers, directly.

Supply Voltage	+15 V Max.
Power Dissipation	660 mW Max.
Quiescent Current Consumption	3 mA typ.

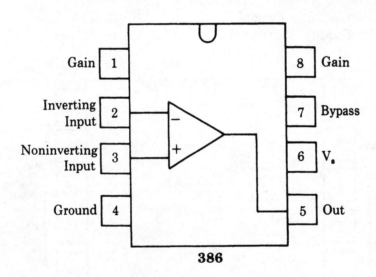

386

LM387 Low-noise dual preamplifier

The 387 is designed to amplify low-level signals in two channels (stereo), in applications where very low noise is required. The total input noise for this device is rated for a typical value of 1.0 μV.

To achieve this excellent noise performance, each of the two amplifiers on these chips is completely independent, with an internal power supply de-coupler-regulator. As a result, the supply rejection is 110 dB and the channel separation is 60 dB.

These preamplifiers offer high gain of 104 dB, and a large output voltage swing (up to 2 V under) the supply voltage, peak-to-peak. A single-ended supply voltage of 9 V to 30 V can be used to power this chip, so, for example, if a 15-V supply voltage is used, the output voltage can swing up to 13-V peak-to-peak. The amplifiers in the 387 are internally compensated for gains greater than 10.

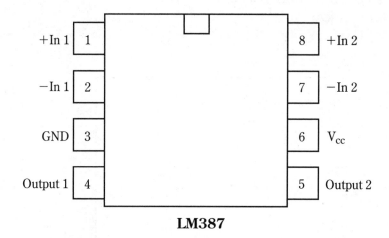

LM387

LM389 Low-voltage audio power amplifier with NPN transistor array

The 389 contains essentially the same amplifier as the 386, with the addition of an array of three NPN transistors on the same substrate. These transistors have high gain and excellent matching characteristics. They can operate from 1 μA to 25 mA at frequencies ranging from dc (0 Hz) to 100 MHz.

389 Specifications

	Minimum	Typical	Maximum
V+ Supply Voltage	12 V		28 V
Av Gain	40 V/V	50 V/V	60 V/V
Output Power	5 W	5.5 W	
Power Dissipation			1.67 W
Peak Current			1.3 A
THD (Total Harmonic Distortion)		0.25%	1.0%
Quiescent Output Voltage		11 V	
Quiescent Supply Current		8.5 mA	25 mA

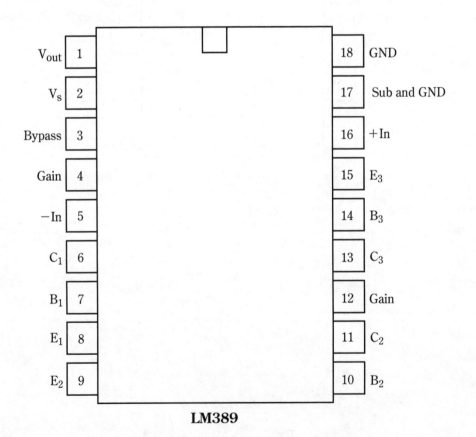

V_{out} — 1	18 — GND
V_s — 2	17 — Sub and GND
Bypass — 3	16 — +In
Gain — 4	15 — E_3
−In — 5	14 — B_3
C_1 — 6	13 — C_3
B_1 — 7	12 — Gain
E_1 — 8	11 — C_2
E_2 — 9	10 — B_2

LM389

LM390 1-Watt battery-operated audio power amplifier

The LM390 is a moderate-power audio amplifier designed specifically for battery-powered operation. It is designed to operate off one of the following standard battery voltages: 6 V, 7.5 V, or 9 V.

The amplifier's gain is internally set at 20. Accepting this "default" gain permits operation with a minimum of external parts. However, if the application requires, the gain can be increased to any value up to 200 by adding an external resistor and capacitor between pins 2 and 6. This 1-watt audio amplifier offers excellent supply rejection and low distortion.

390 Specifications

	Minimum	Typical	Maximum
V+ Supply Voltage	6 V		9 V
Output Power	0.8 W	1.0 W	
Package Dissipation (14-pin DIP)			8.3 W
Input Voltage			±0.4 V
Voltage Gain (pins 2 and 6 open)	23 dB	26 dB	30 dB
Voltage Gain (10 µF from pin 2 to 6)		46 dB	
Bandwidth (pins 2 and 6 open)		300 kHz	
THD (Total Harmonic Distortion)		0.2%	1.0%
Input Resistance	10 kΩ	50 kΩ	
Quiescent Current		10 mA	20 mA

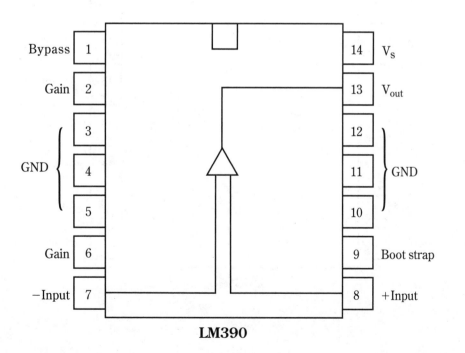

LM390

LM391 Audio power driver

The 391 is an audio amplifier driver designed to drive external power transistors from 10 watts to 100 watts. Notice that this level of output power does not (and should not) pass through the 391 itself, but through the external power transistors and their related circuitry. The 391 merely preconditions the signal for final amplification by the external transistors. The gain and bandwidth of this device are user selectable with appropriate external circuitry.

This chip is internally protected against thermal overloads and output faults, such as short circuits. An unusually wide range of supply voltages can be used with the 391, up to a maximum of ±50 V (+100 V). The input signal voltage can be as high as 5 volts under the actual supply voltage.

391 Specifications

Input Noise	3 μV
Supply Rejection	90 dB
THD (1 kHz)	0.01%
(20 kHz)	0.10%

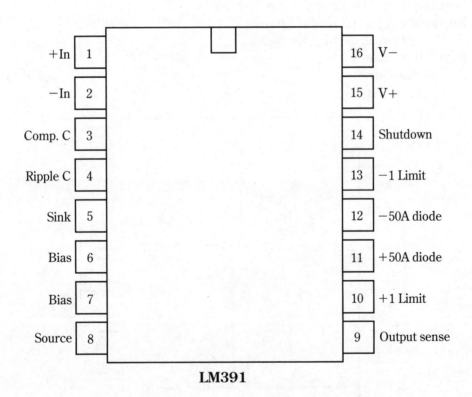

LM391

831 Low-voltage audio power amplifier

The 831 dual audio power amplifier is designed to be operated from low voltages, ranging from 1.8 V to 6.0 V. It contains two independent audio amplifiers for stereo applications, or the two amplifiers can be combined for higher-power bridged (BTL) operation. This chip can provide a power output of up to 440 mW into 8Ω in the BTL mode. Each single amplifier stage is rated for 220 mW.

A patented compensation technique reduces high-frequency radiation, making the 831 very well suited for AM radio applications. Lower distortion and less wide-band noise are additional advantages of this internal compensation.

Most audio amplifier ICs require an input coupling capacitor, but the 831 is designed for direct coupling. The voltage gain can be adjusted with a single external resistor.

831 Absolute Maximum Ratings

Supply Voltage	7.5 V	
Input Voltage	±0.4 V	
Power Dissipation	1.3 W	(M package)
	1.4 W	(N package)

831 Typical Ratings

Supply Voltage	3.0 V
Supply Current (V_{in} = 0)	
Dual Mode	5 mA
BTL Mode	6 mA
Voltage Gain	46 dB
Input resistance	25 kΩ
Channel separation	52 dB

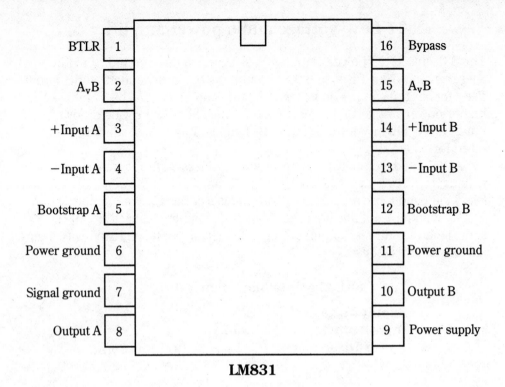

BTLR	1	16	Bypass
$A_v B$	2	15	$A_v B$
+Input A	3	14	+Input B
−Input A	4	13	−Input B
Bootstrap A	5	12	Bootstrap B
Power ground	6	11	Power ground
Signal ground	7	10	Output B
Output A	8	9	Power supply

LM831

LM833 Dual audio operation amplifier

The 833 contains a pair of op amps designed specifically for high-quality performance with audio-frequency signals. This chip is internally compensated for all closed loop gains, and is pin-for-pin compatible with industry-standard dual operational amplifiers.

811 Absolute Maximum Ratings

Supply Voltage	36 V
Input Voltage Range	±15 V
Differential Input Voltage	±30 V
Power Dissipation	500 mW

833 Typical Ratings

Power Bandwidth	120 kHz
High Gain-Bandwidth Product	15 MHz
(Minimum Rating)	10 MHz
Distortion	0.002%
Offset Voltage	0.3 mV
Slew Rate	7 V/µS
(Minimum Rating)	5 V/µS
Dynamic Range	>140 dB

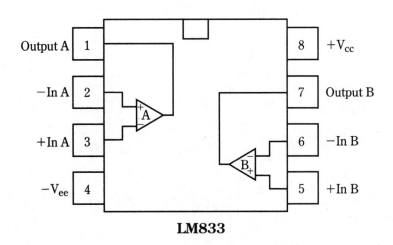

LM833

1875 20-Watt audio power amplifier

The 1875 is a fairly high-powered audio-amplifier IC, capable of outputting 20 watts into 4-Ω or 8-Ω loads, with very low distortion. The 1875 can provide high gain and a large output voltage swing. It is internally compensated and stable for gains of 10 or more. It also has a fast slew rate.

This chip features internal protection for ac or dc short circuits to ground, as well as internal current limiting and thermal shut-down protection. The 1875 is designed for use with a minimum of external components.

1875 Specifications

Supply Voltage	16 V to 60 V
Input Voltage	$-V_{ee}$ to V_{cc}
Supply current	
Typical	70 mA
Maximum	100 mA
Output power (THD = 1%)	25 Watts
THD (output power = 20 W)	
f = 1 kHz	0.015%
f = 20 kHz	0.05%
Power bandwidth	70 kHz
Gain	90 dB
Ripple rejection	94 dB
Current capability	4 A (maximum)

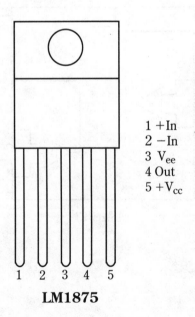

1 +In
2 −In
3 V_{ee}
4 Out
5 +V_{cc}

LM1875

LM1877 Dual audio power amplifier

The 1877 contains a pair of audio amplifiers that are capable of continuously delivering two watts into 8-Ω loads, with very few external components. Each of the amplifiers on this chip is biased from a common internal voltage regulator, which results in high power-supply rejection and output zero-point centering.

The 1877 is internally compensated for all gains of more than 10. These amplifiers also offer very low crossover distortion and low audio-band noise. The internal circuitry is protected against ac short circuits and includes thermal shutdown.

1877 Specifications

Supply Voltage		6 V to 24 V
Output Power (THD = 10%)		2 W
Ripple Rejection		
(Output Referred)		–65 dB
Channel Separation		
(Output Referred)		–65 dB
Total Supply Current		
(P_o = 0 W)	**Typical**	25 mA
	Maximum	50 mA
THD **(F = 1 kHz, Vs = 14 V)**		
50 mW/channel		0.075%
500 mW/channel		0.045%
1 W/channel		0.055%

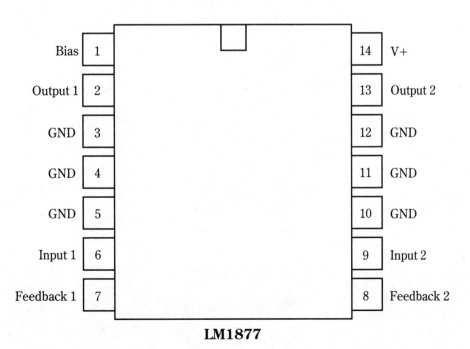

LM1877

LM1971 µPot digitally controlled 62-Db audio attenuator with mute

The 1971 is essentially a digitally controlled volume control or potentiometer in IC form. It uses CMOS circuitry to form a single-channel audio attenuator. At "full volume," the device essentially has no effect on the audio signal passing through it. The attenuation (or negative gain) at this setting, of course, is 0 dB. The attenuation can be increased in 1-dB steps down to –62 dB. The 1971 also features a mute function, which disconnects the input signal from the output. The minimum attenuation in the mute mode is –80 dB.

Because the attenuation is under digital control, specialized software can be written to create any desired attenuation curve to suit the intended application. A three-wire serial digital interface (pins 4, 5, and 6) is used to control the 1971. The device is compatible with either TTL or CMOS logic levels.

Control data to determine the desired attenuation setting is fed into the DATA line (pin 5) when the data input registers are enabled by the LOAD/SHIFT line (pin 4). The CLOCK line (pin 6) provides system timing.

The 1971 is designed to be as transparent as possible to audio signals, other than the desired attenuation, of course. Attenuation levels can be changed instantly, with no audible clicks or pops. The Total Harmonic Distortion plus Noise (THD+N) introduced by this device is rated at only 0.01%, and the signal-to-noise ratio is rated at a minimum of 80 dB.

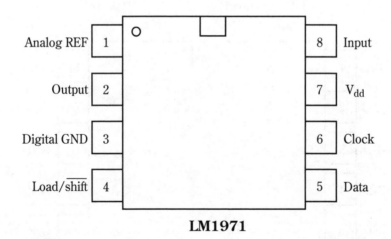

LM1971

MC13060 Mini-watt audio output amplifier

The 13060 is a small, but rugged and versatile, power amplifier IC. It has a typical maximum gain of 50, and can output up to 2 watts of output signal. The actual gain is externally determined by other components in the circuit. However, only a few external components are required in most applications for this device.

A rather wide range of supply voltages can be used to power this amplifier. Anything from 6 volts to 35 volts will do. Generally, the output is independent of the actual supply voltage used. The chip includes a built-in copper ground plane and self-protecting thermal shutdown.

13060 Specifications

	Minimum	Typical	Maximum
Audio Input			1.0 V_{p-p}
Power Supply Current			
(No Input Signal)		13 mA dc	
Quiescent Output Voltage			
(No Input Signal)		8.4 V dc	
Distortion: frequency = 1 kHz			
Output = 62.5 mW		0.2%	1.0%
Output = 900 mW		0.5%	3.0%
Input Resistance		28 k	

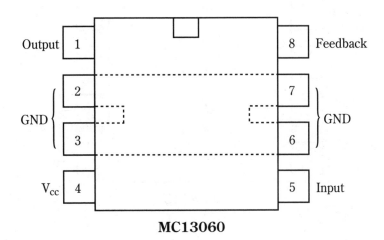

MC13060

MC34119 Low-power audio amplifier

The 34119 was designed with telephone applications in mind, though, of course, it can be used in many other applications as well. It can be powered from as little as 2.0 volts. The maximum supply voltage for use with this device is 16 volts. To maximize the output swing, especially when low supply voltages are used, differential speaker outputs are used in this device. Unlike many other similar audio amplifiers, the 34119 does not require coupling capacitors to the speaker.

The open loop gain is 80 dB. To reduce the gain, the loop can be closed with two external resistors. Their values will determine the closed loop gain.

Even though this is a fairly simple device in an 8-pin DIP (Dual-Inline Package) housing, it has a fairly unusual special feature. Pin 1 can be used to disable the chip, in effect, muting the amplifier.

	Minimum	Typical	Maximum
Supply Voltage	+2.0 V		+16 V dc
Load Impedance	8.0 Ω		100 Ω
Differential Gain			
Bandwidth = 5.0 kHz	0 dB		46 dB
Open Loop Gain	80 dB		
Peak Load Current			±200 mA

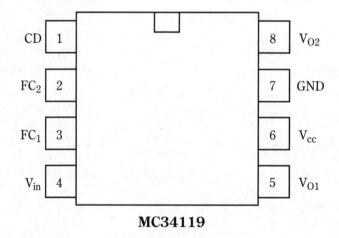

MC34119

6
SECTION

Timers, oscillators, and signal generators

Generally, electronic circuits are designed to do something to a signal from an outside source. In some cases, however, it is necessary to originate a signal. An oscillator produces an ac signal with a specific waveform and frequency.

In addition to oscillator and signal-generator ICs, this chapter also features timers. Essentially, a timer is a linear multivibrator. A multivibrator, like a digital gate, has just two possible output states, HIGH or LOW. There are three basic types of multivibrators.

The monostable multivibrator has just one stable output state. It can hold this state indefinitely (as long as power is continuously applied to the circuit). When an external trigger signal (typically, a short pulse) is received by the circuit, the output briefly switches to the opposite nonstable state for a specific, pre-determined period of time (determined by particular component values in the circuit), then the output reverts to its normal stable state. A monostable multivibrator circuit is sometimes referred to as a *timer*, but this can be a little confusing because a timer IC can usually do more than just the monostable multivibrator function.

The bistable multivibrator has two stable output states. The output can hold either state indefinitely (as long as power is continuously applied to the circuit). Each time an external trigger signal (typically, a short pulse) is received by the circuit, the output reverses its output stage (going from LOW to HIGH, or vice versa). The new output state is held until the next trigger pulse is detected (or power to the circuit is interrupted).

The astable multivibrator has no stable output states. It continuously switches back and forth between the LOW and HIGH states as long as power is applied to the circuit. Each output state is held for a time period that is determined by particular component values in the circuit. An astable multivibrator can also be referred to as a *rectangle wave generator*. In most astable multivibrators, the HIGH and LOW times are not equal, but if they are, the output signal is a square wave.

Most timer ICs can be operated as either monostable multivibrators or astable multivibrators. They usually can't function as bistable multivibrators, but this type of operation usually isn't appropriate to linear applications anyway.

LM1122/LM222/LM322 Precision timer

The LM122 series are precision timers that offer great versatility with high accuracy. They operate with unregulated supplies from 4.5 to 40 V while maintaining constant timing periods from microseconds to hours. Internal logic and regulator circuits complement the basic timing function, enabling the LM122 series to operate in many different applications with a minimum of external components.

The output of the timer is a floating transistor with built-in current limiting. It can drive either ground-referred or supply-referred loads up to 40 V and 50 mA. The floating nature of this output makes it ideal for interfacing, lamp or relay driving, and signal conditioning where an open collector or emitter is required. A logic-reverse circuit can be programmed by the user to make the output transistor either on or off during the timing period.

The trigger input to the LM122 series has a threshold of 1.6 V independent of supply voltage, but it is fully produced against inputs as high as (40 V, even when using a 5-V supply. The circuitry reacts only to the rising edge of the trigger signal, and is immune to any trigger voltage during the timing periods.

An internal 3.15-V regulator is included in the timer to reject supply voltage changes and to provide the user with a convenient reference for applications other than a basic timer. External loads up to 5 mA can be driven by the regulator. An internal 2-V divider between the reference and ground sets the timing period to 1 RC. The timing period can be voltage-controlled by driving this divider with an external source through the V_{ADJ} pin. Timing ratios of 50:1 can be easily achieved.

The comparator used in the LM122 utilizes high-gain PNP input transistors to achieve 300-pA typical input bias current over a common mode range of 0 to 3 V. A boost terminal allows the user to increase comparator operating current for timing periods less than 1 ms. This lets the timer operate over a 3 Ωs to multihour timing range with excellent repeatability.

The LM2905/LM3905 are identical to the LM122 series except that the boost and V_{REF} pin options are not available, limiting minimum timing period to 1 ms.

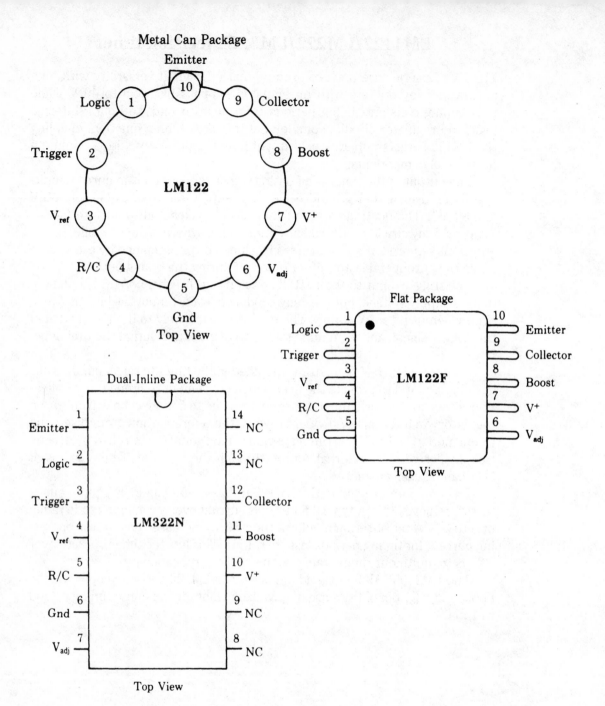

Metal Can Package

Emitter

LM122

Logic 1
Trigger 2
V_{ref} 3
R/C 4
10
9 Collector
8 Boost
7 V^+
6 V_{adj}
5

Gnd
Top View

Flat Package

LM122F

Logic 1
Trigger 2
V_{ref} 3
R/C 4
Gnd 5
10 Emitter
9 Collector
8 Boost
7 V^+
6 V_{adj}

Top View

Dual-Inline Package

LM322N

Emitter 1
Logic 2
Trigger 3
V_{ref} 4
R/C 5
Gnd 6
V_{adj} 7
14 NC
13 NC
12 Collector
11 Boost
10 V^+
9 NC
8 NC

Top View

BA222 Monolithic Timer

The BA222 is a timer IC, similar in concept to the popular 555. Like the 555, the BA222 can be used in either monostable-or astable-multivibrator circuits.

An external resistor and capacitor are used to set the RC time constant of the timer over a very wide range—from a few microseconds up to several hours.

The time period equation for the BA222 is the same as for the 555:

$$T = 1.1RC$$

The BA222 is contained within a 7-pin SIP (single-inline package).

Max. Supply Voltage	+18 V
Recommended Operating Range	+4.5 to 16.0 V
Max. Power Dissipation	500 mW
Quiescent Current	
V_{cc} = 5 V Infinite Load	3 mA typ.
V_{cc} = 15 V Infinite Load	7 mA max.
	15 mA max.
Monostable Operating Timing Regulation	1% typ.
Astable Operating Timing Regulation	2.5% typ.
Output Low Voltage	
V_{cc} = 5 V I_{sink} = 5m A	0.25 V typ.
	0.35 V max.

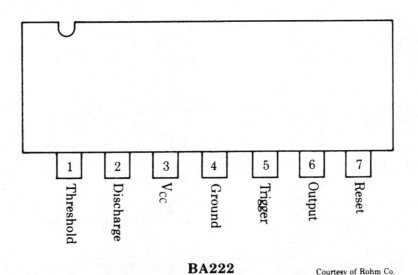

BA222

Courtesy of Rohm Co.

555 Timer

The 555 timer is probably one of the most-popular ICs ever. It has been around for years and is unlikely to be discontinued in the foreseeable future, despite the appearance of a number of various improved and more specialized versions. Despite being a relatively simple device, nominally designed for just two basic circuits (a monostable multivibrator or an astable multivibrator), the 555 is incredibly versatile, and has been put to work in countless circuits. It is very readily available, inexpensive, and quite easy to work with. For most basic applications, only a few external components are needed.

The timing period of the 555 is set by a straightforward external resistor/capacitor combination, according to a fairly simple formula:

$$T = 1.1RC$$

For reliable operation, the timing resistor should have a value between 10 kΩ and 14 MΩ, and the timing capacitor's value should be somewhere between 100 pF and 1000 μF. The larger either of these component values is, the longer the resulting timing period will be. The minimum and maximum component values given here produce a practical timing range from a low of 1.1 mS up to a high of 15,400 seconds (4 hours, 16 minutes, and 40 seconds). The timer certainly has quite an impressive range—especially for such an inexpensive and easy-to-use chip.

In astable multivibrator applications, two timing resistors are used, but the basic operating principles still apply. Two resistors are used in the astable circuit because different timing periods are needed for the LOW and HIGH portions of the cycle.

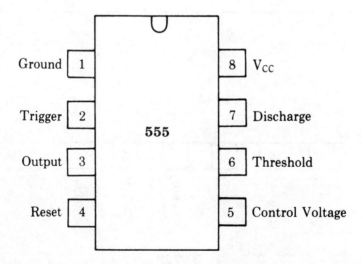

556 Dual timer

The 556 contains two 555-type timer sections. These two sections are functionally independent of one another, except for the power-supply connections. One timer on the 556 can be used in a monostable multivibrator circuit, and the other is used in an astable multivibrator circuit, or they can be used in the same operating mode.

Each of the timer stages in the 556 is the same as the 555 timer. They are directly interchangeable, except for the necessary changes in the pin-out configurations.

The timing period of the 556, like that of the 555, is set by a straightforward external resistor/capacitor combination, according to a fairly simple formula:

$$T = 1.1RC$$

For reliable operation, the timing resistor should have a value between 10 kΩ and 14 MΩ, and the timing capacitor's value should be somewhere between 100 pF and 1000 μF. The larger either of these component values is, the longer the resulting timing period will be. The minimum and maximum component values given here produce a practical timing range from a low of 1.1 mS up to a high of 15,400 seconds (4 hours, 16 minutes, and 40 seconds). The timer certainly has quite an impressive range—especially for such an inexpensive and easy-to-use chip.

In astable multivibrator applications, two timing resistors are used, but the basic operating principles still apply. Two resistors are used in the astable circuit because different timing periods are needed for the LOW and HIGH portions of the cycle.

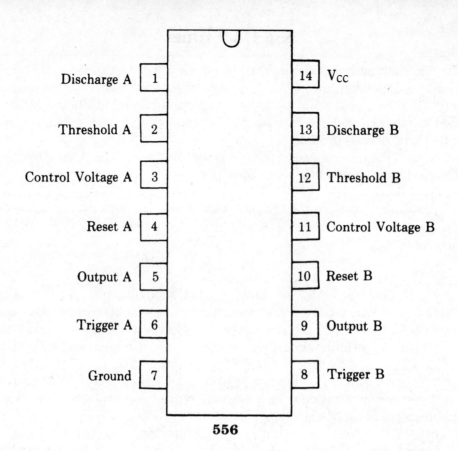

Discharge A — 1

Threshold A — 2

Control Voltage A — 3

Reset A — 4

Output A — 5

Trigger A — 6

Ground — 7

14 — V_{CC}

13 — Discharge B

12 — Threshold B

11 — Control Voltage B

10 — Reset B

9 — Output B

8 — Trigger B

556

558 Quad timer

The 558 contains four somewhat simplified 555-type timers in a single 16-pin package. Not all functions are brought out to the pins, so the 558 can not be substituted in all 555 applications. This device is intended primarily for operation in the monostable multivibrator, but it can be "tricked" into functioning as an astable multivibrator.

Functionally, each 558 timer stage is equivalent to a 555. Once again, the same simple timing period formula is used:

$$T = 1.1RC$$

For reliable operation, the timing resistor should have a value between 10 kΩ and 14 MΩ, and the timing capacitor's value should be somewhere between 100 pF and 1000 μF. The larger either of these component values is, the longer the resulting timing period will be. The minimum and maximum component values given here produce a practical timing range from a low of 1.1 mS up to a high of 15,400 seconds (4 hours, 16 minutes and 40 seconds). The timer certainly has quite an impressive range—especially for such an inexpensive and easy-to-use chip.

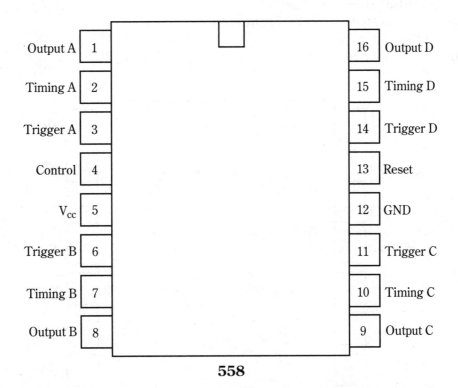

Output A	1	16	Output D
Timing A	2	15	Timing D
Trigger A	3	14	Trigger D
Control	4	13	Reset
V_{cc}	5	12	GND
Trigger B	6	11	Trigger C
Timing B	7	10	Timing C
Output B	8	9	Output C

558

LM565 Phase-locked loop

The 565 is an inexpensive general-purpose PLL (Phase-Locked Loop). The VCO (Voltage-Controlled Oscillator) frequency is set with an external resistor and capacitor. By varying the resistance, a 10:1 frequency tuning range can be achieved with a single capacitor. Similarly, an external resistor/capacitor pair set the circuit's bandwidth, response speed, capture and pull-in range over a wide range.

The internal circuitry of the 565 includes a stable, highly linear VCO, and a double-balanced phase detector with good carrier suppression. The supply voltage for this device can be anything from ±5 to ±12 Volts. Square-wave and linear triangle-wave outputs are available from this chip. The square-wave output is TTL and DTL compatible.

565 Specifications

	Minimum	Typical	Maximum
Supply Voltage	±5 V		±12 V
Power Dissipation			1.4 W
Power Supply Current		8.0 mA	12.5 mA
Differential Input Voltage			±1 V
VCO Maximum Operating Frequency			
(C_0 = 2.7 pF)	300 kHz	500 kHz	
Input Impedance	7 kΩ	10 kΩ	

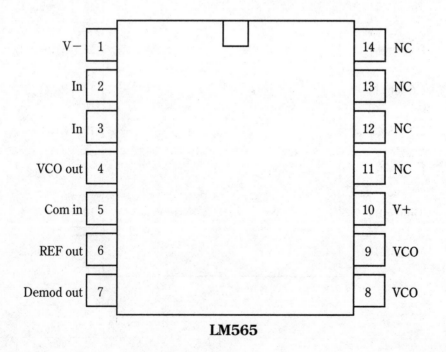

V−	1		14	NC
In	2		13	NC
In	3		12	NC
VCO out	4		11	NC
Com in	5		10	V+
REF out	6		9	VCO
Demod out	7		8	VCO

LM565

LM566 Voltage-controlled oscillator

The 566 is a general-purpose voltage-controlled oscillator (VCO). It has two outputs: one generates a square wave; the other, a triangle wave.

The control voltage's effect on the output signal's frequency is very linear, and a 10-to-1 frequency range can be achieved with a single fixed capacitor. The base signal frequency is set by an external resistor and capacitor. It can then be varied and controlled dynamically by an input voltage or current.

The 566 can be powered over a fairly wide range of supply voltages, extending from 10 to 24 volts.

	Minimum	Typical	Maximum
Supply Voltage	±10 V		±24 V
Power Dissipation			1.0 W
VCO Maximum Operating Frequency			
($R_0 = 2$ kΩ, $C_0 = 2.7$ pF)	0.5	1 MHz	
Input Voltage Range	0.75 Vdc	V_{cc}	

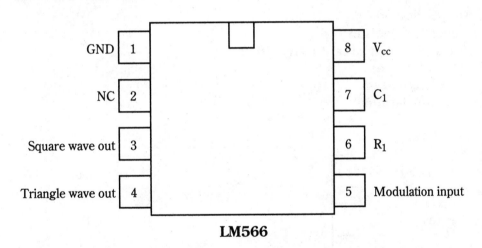

LM566

$$\text{Center frequency} = \frac{2(V_{cc} - \text{input volts})}{R_1 C_1 V_{cc}}$$

LM567 Tone decoder

The 567 is a general-purpose tone decoder. When its input signal has a frequency within the circuit's pre-programmed passband, an on-chip transistor is saturated, switching to ground. This effect can be easily detected by external circuitry to control almost any desired function.

An on-chip voltage-controlled oscillator (VCO) determines the center frequency of the decoder and drives an I and Q detector. A handful of external components are used to set up the key variables: center frequency, bandwidth, and output delay. An external resistor can cover a 20-to-1 frequency range. The center frequency, which is highly stable, can be adjusted over a very wide range, from 0.01 Hz up to 500 kHz. The bandwidth can be adjusted from 0% to 14%.

The 567 is quite good at rejecting noise and out-of-band signals, and offers good immunity to false signals. This device's output can sink up to 100 mA. It is compatible with TTL or CMOS logic circuits, when appropriate supply voltages are used.

	Minimum	Typical	Maximum
Supply Voltage	4.75 V	5.0 V	9.0 V
Power Dissipation			1.1 W
Power Supply Current (RL = 20 kΩ)			
Quiescent		6 mA	8 mA
Activated		11 mA	13 mA
Input Resistance	18 kΩ	20 kΩ	

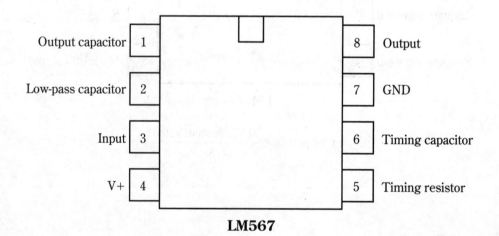

LM567

LM568 Low-power phase-locked loop

The 568 is an amplitude-linear phase-locked loop (PLL) designed for minimal power consumption. It can be powered by voltages ranging from 2 to 9 volts, with very low supply-current drain, as indicated in the specifications chart.

The internal circuitry of the 568 includes a linear voltage-controlled oscillator (VCO), fully balanced phase detectors, and a carrier detector output.

The linearized control range of the 568's VCO covers ±30%, permitting demodulation of FM or FSK signals with deviations up to ±15%. The input frequency can be as high as 500 kHz. When the PLL is locked to an appropriate input signal greater than 26 mVrms, carrier detect is indicated.

This device offers very low distortion. The Total Harmonic Distortion (THD) is typically just 0.5% for ±10% deviation.

	Minimum	Typical	Maximum
Supply Voltage	2 V		9 V
Input Voltage			$2\ V_{p\text{-}p}$
Power Dissipation			500 mW
Output Voltage			13 V
Output Current			30 mA
Input Resistance		40 kΩ	
Power Supply Current			
$V_s = 2$ V		0.35 mAdc	
$V_s = 5$ V		0.75	1.5 mAdc
$V_s = 9$ V		1.2	2.4 mAdc

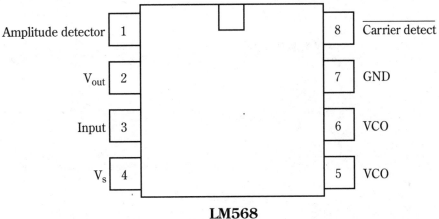

LM568

MC1455 Timer

The MC1455 is a monolithic timer, not entirely dissimilar to the popular 555 IC, discussed earlier in this section. In fact, the MC1455 is designed as a direct, improved replacement for the 555.

This chip can be used in both monostable multivibrator and astable multivibrator applications. In the monostable mode, the timing period is precisely set by one external resistor and one external capacitor. The timing periods (frequency and duty cycle) in the astable mode are precisely controlled by two external resistors and one external capacitor. The circuitry is similar to that used for the 555 timer. The same basic timing equation is used for the MC1455;

$$T = 1.1RC$$

Timing periods from a few microseconds up to several hours can be easily set up with this device. In the astable mode, the duty cycle is fully adjustable.

The output can be selected for either normally ON (HIGH) or normally OFF (LOW) to suit the individual application. The output of the MC1455 can source or sink up to 200 mA, or drive MTTL circuits.

Supply Voltage	+18 V max.
Recommended Operating Range	+4.5 V to +16 V
Supply Current	
V_{cc} = **5 V Infinite Load**	3.0 mA typ.
	6.0 mA max.
V_{cc} = **15 V Infinite Load**	10 mA typ.
	15 mA max.
Power Dissipation	680 mW max.
Discharge Current (Pin #7)	200 mA max.
Timing Error	1.0% typ.

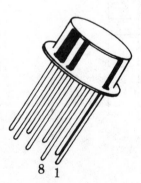

1. Ground
2. Trigger
3. Output
4. Reset
5. Control Voltage
6. Threshold
7. Discharge
8. V_{CC}

G Suffix

Metal Package
Case 601-04

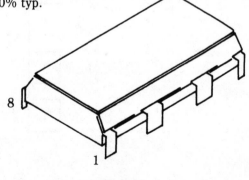

D Suffix
Plastic Package
Case 751-02
SO-8

MC3456, MC3556 Dual timer

The MC3456 and MC3556 each contain two independent timer stages. Each of these timers can be used in either monostable-multivibrator or astable-multivibrator circuits. Basically, these chips are improved versions of the 556 dual-timer IC, discussed earlier in this section.

Each of the timer sections in these chips are equivalent to Motorola's MC1455 timer IC, also discussed earlier in this section. The timing period is set by a simple external resistor/capacitor combination. The formula for the basic (monostable) timing period is:

$$T = 1.1RC$$

Timing periods from a few microseconds up to several hours can be set up with MC3456 and MC3556.

Supply Voltage	
MC3456	+4.5 V min.
	+16 V max.
MC3556	+4.5 V max.
	+ 18 V max.
Power Dissipation	1.0 W max.
Supply Current	
MC3456 (V_{cc} = 5-V Infinite Load)	6.0 mA typ.
	12 mA max.
(V_{cc} = 15-V Infinite Load)	20 mA typ.
	30 mA max.
MC3556 (V_{cc} = 5-V Infinite Load)	6.0 mA typ.
	10 mA max.
(V_{cc} = 15-V Infinite Load)	20 mA typ.
	24 mA max.
Discharge Current	200 mA max.

NOTE: Specifications are for both the MC3456 and the MC3556, except where noted.

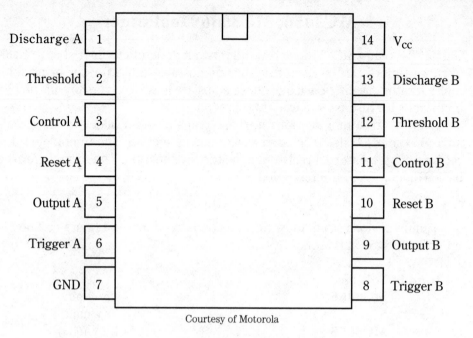

Discharge A	1	14	V$_{cc}$
Threshold	2	13	Discharge B
Control A	3	12	Threshold B
Reset A	4	11	Control B
Output A	5	10	Reset B
Trigger A	6	9	Output B
GND	7	8	Trigger B

Courtesy of Motorola

MC3456/MC3556

LM3905 Precision timer

The 3905 is part of the LM122 series of precision timers. These devices offer great versatility with high accuracy. They can operate with unregulated power supplies from 4.5 volts up to 40 volts, while maintaining constant timing periods from microseconds to hours. Internal logic and regulator circuits complement the basic timing, enabling the 3905 to operate in many different applications with a minimum of external components.

The output of the timer is a floating transistor with built-in current limiting. It can drive either ground-referred or supply referred loads up to 40 volts and 50 mA. The floating nature of this output makes it ideal for interfacing applications or lamp or relay driving. Another class of suitable applications for this device would be signal conditioning, where an open collector or emitter is required. A logic-reverse circuit can be programmed by the user to make the output transistor either on or off during the timing period, to suit the specific intended application.

The trigger input of the 3905 has a threshold of 1.6 volt, independent of the actual supply voltage, but it is fully protected against inputs as high as ±40 volts—even when using a 5-volt supply.

The timer's circuitry reacts only to the rising edge of the trigger signal, and is immune to any trigger voltage during the timing periods. The circuit must time-out before it can be re-triggered.

An internal 3.15-volt regulator is included in this timer to reject supply-voltage changes, and to provide the user with a convenient reference for applications other than a basic timer. Of course, this on-chip voltage regulation explains the 3905's tolerance to such a wide range of supply voltages. External loads up to 5 mA can be driven by the on-chip regulator. An internal 2-V divider between the reference and ground pins sets the timing period to 1RC. This is more convenient than the 1.1 constant in the timing equation for the 555 timer.

The comparator used in the 3905 utilizes high-gain PNP input transistors to achieve 300-pA typical input bias current over a common-mode range of 0 to 3 volts.

The 3905 is identical to the LM122 series, except that the boost and V_{ref} pin options are not available. This means the minimum timing period of the 3905 is 1 mS.

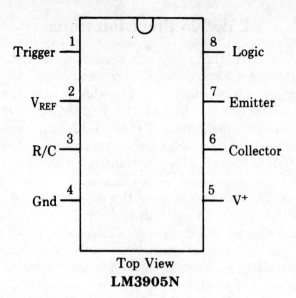

Top View
LM3905N

3909 LED Flasher/oscillator

The 3909 is a simple 8-pin oscillator IC. It is primarily intended for LED flasher circuits, but it can be used in many other low-frequency oscillator applications, too.

A simple external capacitor is used to set the frequency of the on-chip oscillator. The timing resistors are provided within the chip itself. This timing capacitor also provides a voltage boost, so pulses of over 2 volts are delivered to the external LED (or other load device), even with supply voltages as low as 1.5 volts.

Being inherently self-starting, the basic 3909 LED dasher circuit requires only the IC, a battery, the LED, and a timing capacitor.

A slow (pin 8) or a fast (pin 1) flash rate can be set up, depending on where the external timing capacitor is connected to the chip.

Supply Voltage	6.4 V max.
Power Dissipation	500 mW max.
Operating Current	75 mA max.
Peak Output Current	45 mA max.
Output Pulse Width	6 ms max.
Slow-Flash Frequency	1.3 Hz max.
Fast-Flash Frequency	1.1 kHz max.

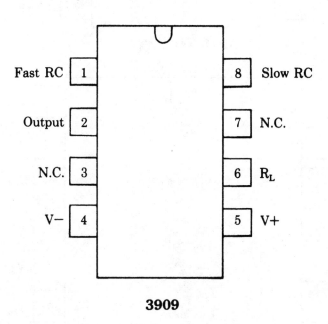

3909

MM5369 17-Stage oscillator/driver

The 5369 uses a quartz crystal to set its precise base frequency. Then, a 17-stage binary divider drops the base frequency to a precise reference of a lower value. Both the original base frequency and the divider output can be tapped off from this chip.

CMOS circuitry is used in the 5369. This device can be used in either analog or digital circuits. Of course, the output signals are in the form of square waves.

The supply voltage can be anything from 3 volts to 15 volts, and current consumption is LOW. Assuming that the supply voltage is 10 volts, the 5369 can handle signal frequencies up to 4 MHz.

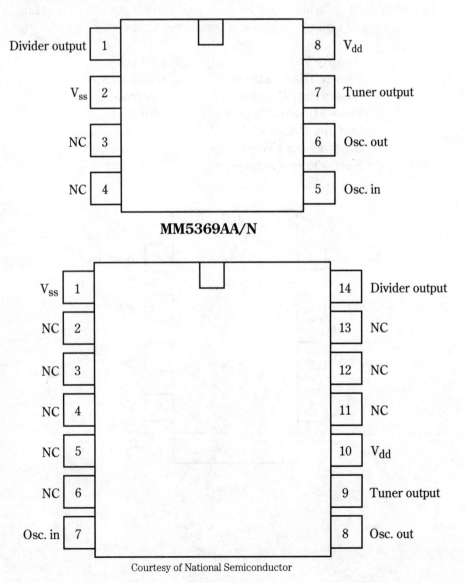

MM5369AA/N

Courtesy of National Semiconductor

7
SECTION

RF amplifiers and related devices

Probably the most commonly used type of electronic circuit is the amplifier. Almost all electronic systems of any complexity include at least one amplifier stage. Audio amplifiers typically only need to handle frequencies up to about 20 kHz, and generally involve few inherent problems to the circuit designer. In radio frequency (RF) applications, however, the high-frequency signals introduce a number of special considerations and problems. RF signals can have frequencies of several hundred kilohertz or even megahertz. Such high-frequency signals behave somewhat differently than lower frequency signals. For example, two adjacent conductors can act as a small capacitance, which is ignored by AF signals, but a RF signal might see it as a partial or complete short circuit.

This section covers ICs designed for the amplification and other modification of signals within the radio-frequency range. A number of related devices for RF applications are also covered here. This chapter is not restricted just to amplifiers. All of these ICs are intended for use in radio and television receivers and communications systems, as well as related applications.

Because RF circuits operate at much higher signal frequencies than audio or dc circuits, greater care must be taken when working on circuit layouts, lead lengths, and heatsinks.

403

TDA1190P TV sound system

The TDA1109P could have been included in Section 5 (Audio Amplifiers) because it is basically just an audio amplification system, but it is included here because it designed specifically for TV sound-system applications. It even includes an on-chip FM detector and related circuitry to demodulate television audio signals.

The functional stages within this device include:

- IF limiter
- IF amplifier
- Low-pass filter
- FM detector
- Dc volume control
- Audio preamplifier'
- Audio power amplifier

The dc volume-control input permits an external dc voltage to set the amplifier gain or output volume. This feature can be particularly handy in remote control systems. The volume control response is linear.

The TDA1109P is basically a low power version of the TDA3190P.

	Minimum	**Maximum**
Supply Voltage	9.0 V	22 V
Output Power	(V_{cc} = 18 V, R_1 = 32 Ω)	1.3 W

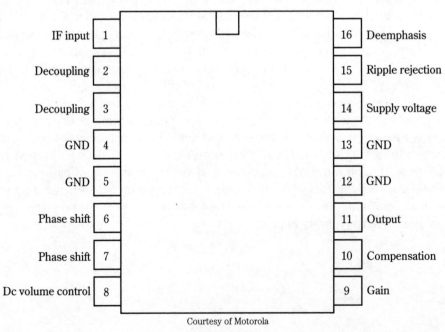

Courtesy of Motorola

TDA1190P

LM1201 Video amplifier system

The 1201 is an essentially self-contained wideband video amplifier system. It can be used in high-resolution monochrome (BandW) or RGB (color) monitor applications. This chip features a gated differential input black-level clamp comparator for brightness control. There is also an attenuator circuit for contrast control. The attenuation can be operated over a 40-dB range. An on-chip voltage reference for the video input is also provided.

The wideband video amplifier in the 1201 can operate over a 200-MHz range, at –3 dB. The absolute maximum supply voltage for this chip is 13.5 Volts. A nominal 12-V supply is preferred.

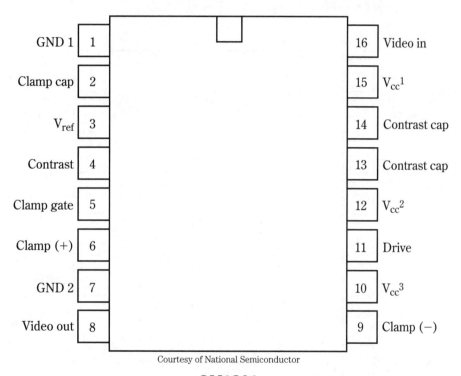

Courtesy of National Semiconductor

LM1201

LM1202 230-MHz video amplifier system

The 1202 is a high-frequency video amplifier system, designed to operate over a 230-MHz bandwidth in high-resolution monochrome (BandW) or RGB (color) monitor applications. This chip features a gated differential input black-level clamp comparator for brightness control. There is also a dc-controlled subcontrast attenuator circuit for contrast control. The attenuation can be operated over a 40-dB range. This dc control is brought out to its own independent pins to provide a more-accurate control system in RGB monitor applications. All of the dc controls on this chip have a high input impedance and operate over a 0- to 4-volt range.

The 1202 is designed for operation from a 12-volt power supply, but it can be operated reliably off voltages as low as 8 volts.

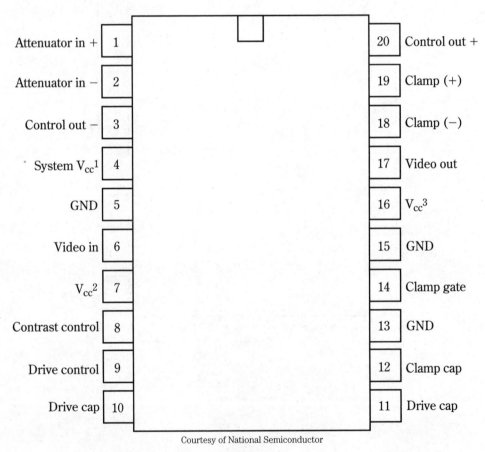

Attenuator in +	1	20	Control out +
Attenuator in −	2	19	Clamp (+)
Control out −	3	18	Clamp (−)
System V_{cc}^1	4	17	Video out
GND	5	16	V_{cc}^3
Video in	6	15	GND
V_{cc}^2	7	14	Clamp gate
Contrast control	8	13	GND
Drive control	9	12	Clamp cap
Drive cap	10	11	Drive cap

Courtesy of National Semiconductor

LM1202

LM1205 130-MHz RGB
video amplifier system with blanking

The 1205 is a very high frequency amplifier designed for use in high-resolution RGB video monitors. Actually, it contains three matched amplifiers, one for each of the primary colors (red, green, and blue). This chip's circuitry also includes three matched dc-controlled attenuators for contrast control, three dc-controlled subcontrast attenuators for trimming white balance, and three single-ended input black-level clamp comparators for brightness control. All of these dc control inputs operate with signals from 0 to 4 volts and have a high input impedance. The inputs are easily interfaced to bus-controlled alignment systems.

The video output voltage is blanked within 0.1 V above ground via an on-chip blanking circuit. This permits blanking at the cathodes of the CRT. The 1305 also contains a spot killer to protect the CRT's phosphor coating during power-down.

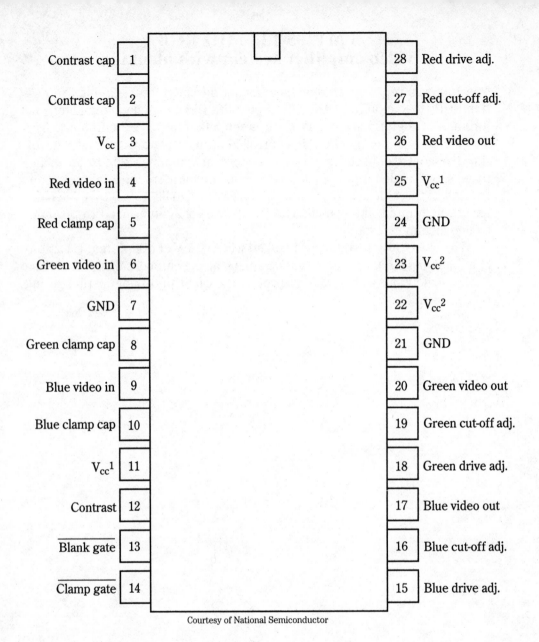

Contrast cap	1	28	Red drive adj.
Contrast cap	2	27	Red cut-off adj.
V_{cc}	3	26	Red video out
Red video in	4	25	$V_{cc}1$
Red clamp cap	5	24	GND
Green video in	6	23	$V_{cc}2$
GND	7	22	$V_{cc}2$
Green clamp cap	8	21	GND
Blue video in	9	20	Green video out
Blue clamp cap	10	19	Green cut-off adj.
$V_{cc}1$	11	18	Green drive adj.
Contrast	12	17	Blue video out
$\overline{\text{Blank gate}}$	13	16	Blue cut-off adj.
$\overline{\text{Clamp gate}}$	14	15	Blue drive adj.

Courtesy of National Semiconductor

LM1205

LM1310 Phase-locked-loop FM stereo demodulator

The LM1310 is an integrated FM stereo demodulator using phase-locked-loop techniques to regenerate the 38-kHz subcarrier. A second version also available is the LM1800, which adds superb power-supply rejection and buffered (emitter-follower) outputs to the basic phase-locked decoder circuit. The features available in these integrated circuits make possible a system that can deliver high-fidelity sound within the cost restraints of inexpensive stereo receivers. Features include:

- Automatic stereo/monaural switching
- No coils, all tuning performed with single potentiometer
- Wide supply operating voltage range
- Excellent channel separation

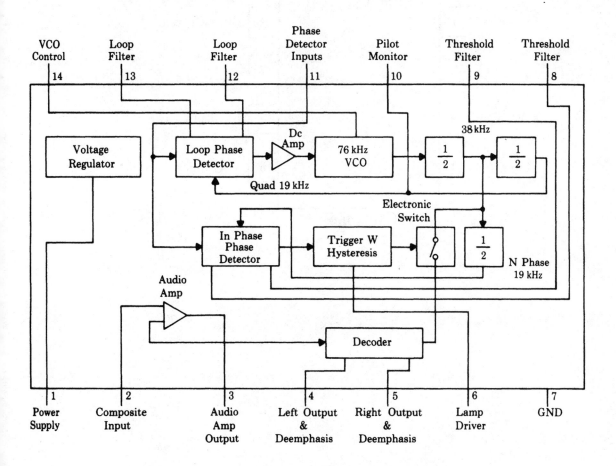

MC1350 IF amplifier

The MC1350 is a monolithic IF amplifier, featuring a wide-range AGC (automatic gain control). It is intended for use with video IF signals in a television receiver.

Max. Supply Voltage	18V
Output Supply Voltage	18V
Differential Input Voltage	5.0 V
Power Gain 45 MHz	50 dB
58 MHz	48 dB
Min. AGC Range	60 dB (0 to 45 MHz)

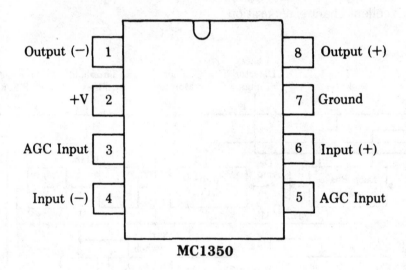

MC1350

MC1374 TV Modulator circuit

The MC1374 is designed to generate a TV signal from audio and video inputs. The internal circuits within this chip include:

- FM audio modulator
- Sound carrier oscillator
- RF oscillator
- RF dual-input modulator

The MC1374 features a wide dynamic range and low-distortion audio. The gain of the RF modulator can be externally adjusted. The output signal can use either positive or negative sync, depending upon the requirements of the specific application.

A single-ended power-supply voltage is used with the 1374. This voltage can be anything from +5 to +14 volts, although it is recommended that the supply voltage not exceed +12 volts in normal operation.

The output signal can be set for TV channel 3 or 4. It is best to select a channel that is not used by a local broadcaster in your area.

Typical applications for the 1374 include:

- Video tape players
- Video disc players
- Computer monitors
- Video games
- Cable or satellite subscription decoders

	Minimum	Typical	Maximum
Supply Voltage	+5 V		+14 V
Power Dissipation			1250 mW
RF Output		170 mV p-p	
(pin 9 resistor = 75 Ω, no external load)			
Minimum Video Bandwidth		30 MHz	
(75-W input source)			
AM Oscillator Frequency Range		105 MHz	
Frequency Range of Modulator		4.5 MHz	

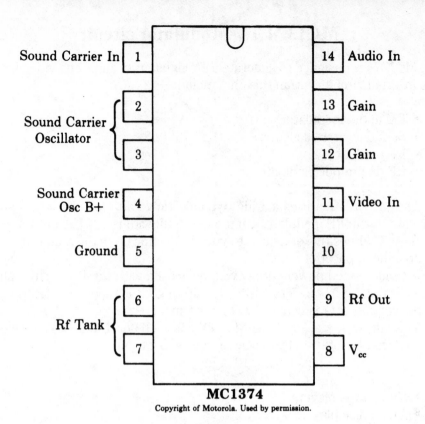

MC1374

Copyright of Motorola. Used by permission.

MC1377 Color television RGB-to-PAL/NTSC encoder

The 1377 accepts baseband red, blue, and green, and sync inputs, and generates a composite video signal. Either the PAL or NTSC standard can be easily selected, simply by applying an appropriate control signal to pin 20.

The external circuitry used with this chip does not have to be extensive. A complete circuit can include only a handful of external parts. On the other hand, the 1377 does have sufficient pin-outs for more sophisticated external circuitry, and a fully implemented top-quality composite signal, if that suits the intended application.

The internal circuits within this device include:

- Color subcarrier oscillator
- Voltage-controlled 90° phase shifter
- Into DSB-suppressed-carrier chroma modulators
- RGB-input matrices
- Blanking-level clamps

An external reference oscillator can be used, or the on chip oscillator can be selected.

Typical applications for the MC1377 include color video cameras and graphics generators.

Max. Supply Voltage	+15 V
Preferred Range	+10.8 to 13.2 V
8.2-V Regulator Output Current	10 mA max.
Supply Current	32 mA typ.
	40 mA max.
Power Dissipation	1250 mW max.

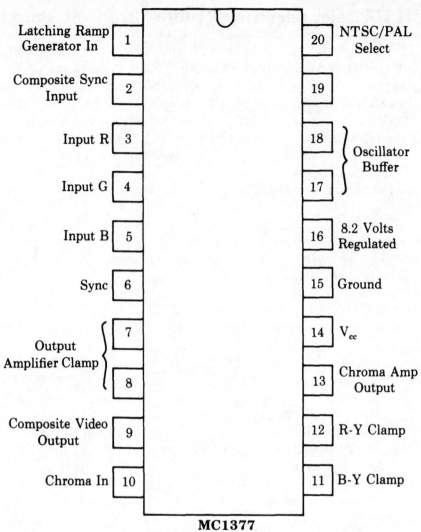

Latching Ramp Generator In	1	20	NTSC/PAL Select
Composite Sync Input	2	19	
Input R	3	18	Oscillator Buffer
Input G	4	17	
Input B	5	16	8.2 Volts Regulated
Sync	6	15	Ground
Output Amplifier Clamp	7	14	V_{cc}
	8	13	Chroma Amp Output
Composite Video Output	9	12	R-Y Clamp
Chroma In	10	11	B-Y Clamp

MC1377

Copyright of Motorola. Used by permission.

MC1378 Color-television composite-video-overlay synchronizer

The MC1378 is a bipolar composite video overlay encoder and microcomputer synchronizer. It's internal circuitry includes quadrature color modulators, RGB matrices, blanking-level clamps, and a complete complement of synchronizers to lock a microcomputer-based video source to any remote video source. Except for the microcomputer synchronizers, the MC1378 is quite similar to the MC1377, discussed previously.

All necessary reference oscillators are included on chip. The device can be operated in either the PAL or NTSC modes, with 625 or 525 lines. The MC1378 was specifically designed for use with the Motorola RMS (Raster Memory System), but it can be applied to other controllable video sources. It will work with nonstandard video.

The MC1378 can be operated from a standard, single-handed 5.0-volt power supply.

Max. Supply Voltage	+6.0 V
Preferred Range	+4.75 to 5.25 V
Max. Power Dissipation	1250 mW
Typ. Supply Current	100 mA

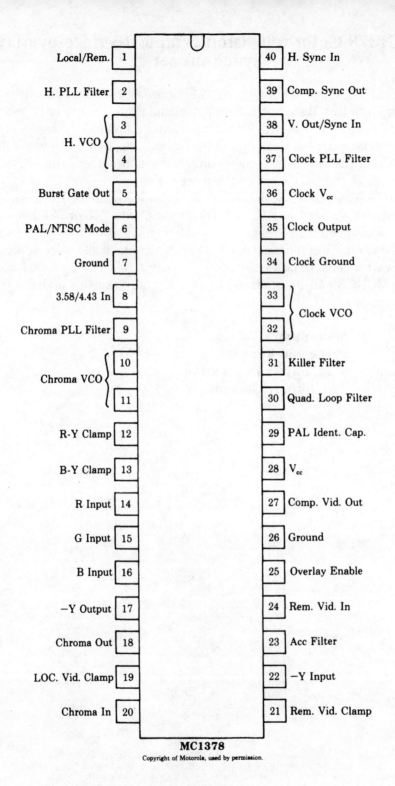

Local/Rem.	1	40	H. Sync In
H. PLL Filter	2	39	Comp. Sync Out
H. VCO	3	38	V. Out/Sync In
	4	37	Clock PLL Filter
Burst Gate Out	5	36	Clock V_{cc}
PAL/NTSC Mode	6	35	Clock Output
Ground	7	34	Clock Ground
3.58/4.43 In	8	33	Clock VCO
Chroma PLL Filter	9	32	
Chroma VCO	10	31	Killer Filter
	11	30	Quad. Loop Filter
R-Y Clamp	12	29	PAL Ident. Cap.
B-Y Clamp	13	28	V_{cc}
R Input	14	27	Comp. Vid. Out
G Input	15	26	Ground
B Input	16	25	Overlay Enable
−Y Output	17	24	Rem. Vid. In
Chroma Out	18	23	Acc Filter
LOC. Vid. Clamp	19	22	−Y Input
Chroma In	20	21	Rem. Vid. Clamp

MC1378

LM1391 Phase-locked loop

The 1391 is a phase-locked loop (PLL) circuit designed specifically for use in the horizontal section of video monitors and TV receivers. Because the PLL function is rather generalized, this device can also be used in many other applications as well, particularly those involving low-frequency signal processing.

The circuitry within this chip includes a stable VCO, linear pulse phase detector, and variable duty-cycle output driver. The output signal's duty cycle can be controlled by an external dc voltage.

The collector of the output transistor is internally uncommitted for maximum circuit design flexibility. This transistor features a high-voltage swing and low saturation.

The 1391 also contains an on-chip active voltage regulator for improved power-supply rejection. The thermal frequency drift and static phase error of this device are small. The dc loop gain is externally adjustable. Typical pull-in for the 1391 PLL is ±300 Hz.

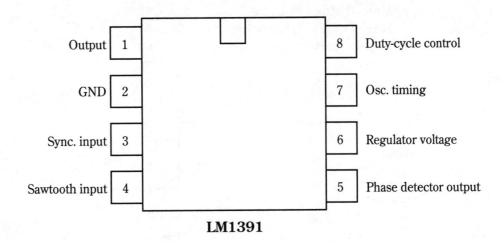

LM1391

MC1391P TV Horizontal processor

The MC1391P contains the horizontal circuitry for all types of television receivers. The on-chip circuits include:

- Phase detector
- Oscillator
- Predriver

This device features low thermal frequency drift and ±300-Hz typical pull-in capability. The hold control capability can be externally preset.

The MC1391P can be used for driving either transistor or tube television circuits, because the output duty cycle is variable.

Supply Current	40 mA max.
	20 mA typ.
Max. Output Current	30 mA
Max. Output Voltage	40 V
Max. Power Dissipation	625 mW
Regulated Voltage (pin 6)	8.0 V min.
	8.6 V typ.
	9.4 V max.

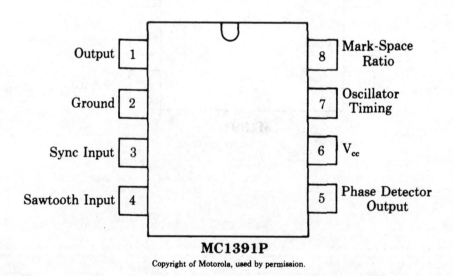

MC1391P

LM1496/LM1596 Balanced modulator/demodulator

The LM1496/LM1596 are double-balanced modulator/demodulators that produce an output voltage proportional to the product of an input (signal) voltage and a switching (carrier) signal. Typical applications include suppressed carrier modulation, amplitude modulation, synchronous detection, FM or PM detection, and broadband frequency doubling and chopping.

The LM1596 is specified for operation over the −55°C to +125°C military temperature range. The LM1496 is specified for operation over the 0°C to +70°C temperature range. Features include:

- Excellent carrier suppression
  65 dB typical at 9.5 MHz
  50 dB typical at 10 MHz
- Adjustable gain and signal handling
- Fully balanced inputs and outputs
- Low offset and drift
- Wide frequency response up to 100 MHz

Dual-In-Line Package

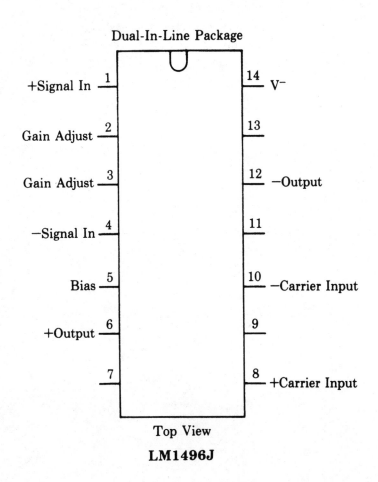

Top View
LM1496J

Metal Can Package

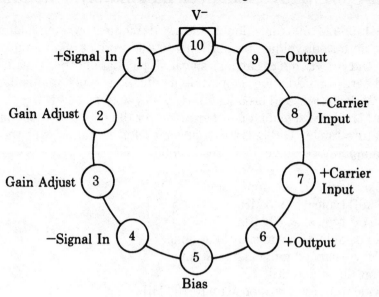

Top View

Note: Pin 10 is connected electrically to the case through the device substrate.

LM1496H

LM1823 Video IF amplifier/PLL detector system

The 1823 contains virtually a complete video IF signal-processing system. Very few external components are required in most applications. The on-chip circuitry includes a five-stage gain-controlled IF amplifier, a phase-locked loop (PLL) synchronous amplitude detector, a gated automatic gain control (AGC), and a switchable automatic frequency control (AFC) detector. This video amplifier has a 9-MHz video bandwidth and reverse tuner AGC output. Typical operating frequencies for this device are 38.9 MHz, 45.75 MHz, 58.75 MHz, and 61.25 MHz.

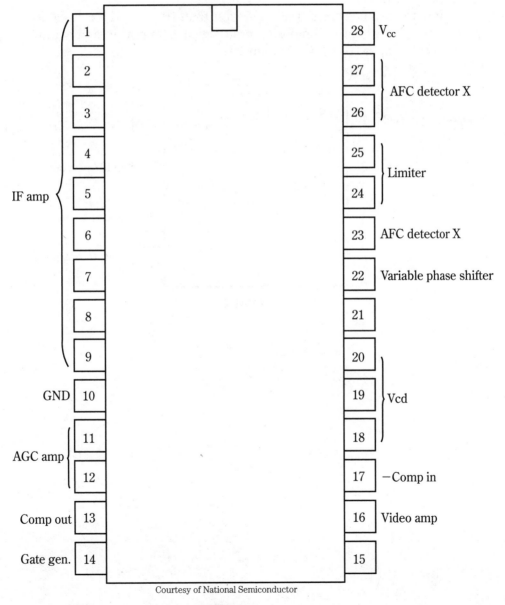

Courtesy of National Semiconductor

LM1823

421

LM1881 Video sync separator

The 1881 is a fairly simple video sync separator. This device is designed to extract vital timing information from video signals, including composite and vertical sync, burst/back-porch timing, and odd/even field information. The input signal is assumed to be a standard negative going sync signal with amplitude ranging from 0.5 V to 2 V peak-to-peak. The video input signal may use the NTSC, PAL, or SECAM standards.

The 1881 is versatile and can also provide sync separation for nonstandard faster horizontal-rate video signals. This device can handle horizontal scan rates up to 150 kHz.

The input resistance of the 1881 is greater than 10 kΩ, and it is designed for use with an ac-coupled composite input signal. The power supply drain current of this chip is typically less than 10 mA.

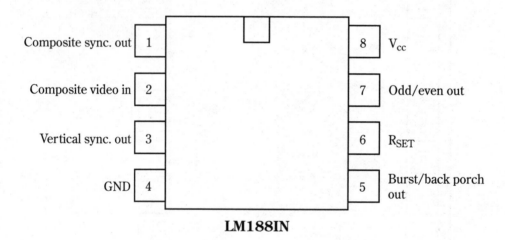

LM188IN

LM2416 Triple 50-MHz CRT driver

The 2416 consists of three wide-bandwidth, large-signal amplifiers capable of wide voltage swings. Each of these amplifiers has a gain of −13. Using this device in color CRT monitors permits conformation to VGA, Super VGA, or IBM 8514 graphics standards at relatively low cost.

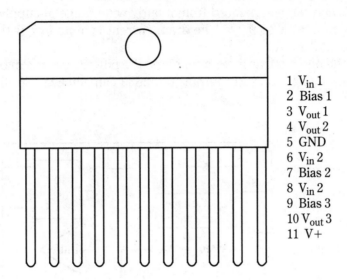

1 V_{in} 1
2 Bias 1
3 V_{out} 1
4 V_{out} 2
5 GND
6 V_{in} 2
7 Bias 2
8 V_{in} 2
9 Bias 3
10 V_{out} 3
11 V+

MC2833 Low-power FM transmitter system

The 2833's internal circuitry includes a built-in microphone amplifier, VCO, and two auxiliary transistors to form the basis of a one-chip FM transmitter subsystem, particularly well-suited for such applications as cordless telephones, wireless microphones, and FM communication equipment. Only a very few external components are required for a functional FM transmitter circuit.

This device can be operated from a fairly wide range of supply voltages, extending from 2.8 V to 9.0 V. The drain current is quite low, with a typical value of about 2.9 mA.

Using the on-chip transistor amplifiers, the power output can be boosted to +10 dBm. Using direct RF output, the 2833 can achieve –30-dBm power output to 60 MHz.

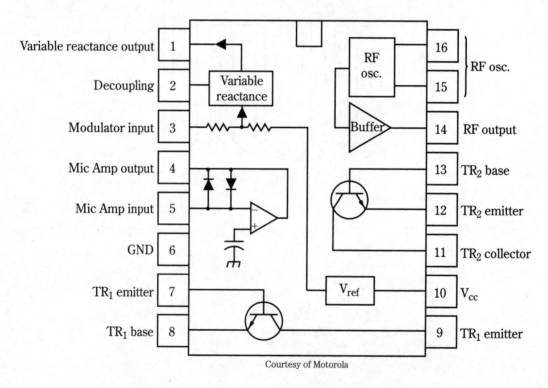

Courtesy of Motorola

MC2833

LM2889 TV video modulator

The 2889 adapts video and audio signals from an unmodulated source (such as a computer, video game, video-disc or video-tape player, etc.) and modulates it into a suitable form to be applied to the antenna terminals of a black-and-white or color television set. The TV "thinks" it is receiving a broadcast program over the air. The circuitry in this chip includes a sound subcarrier oscillator and FM modulator, video clamp, and RF oscillators and modulators for two low-VHF channels. (Television channels 3 and 4 are the most commonly used for this purpose.)

This device can operate on supply voltages from 10 to 16 volts. It offers excellent oscillator stability and low intermodulation products.

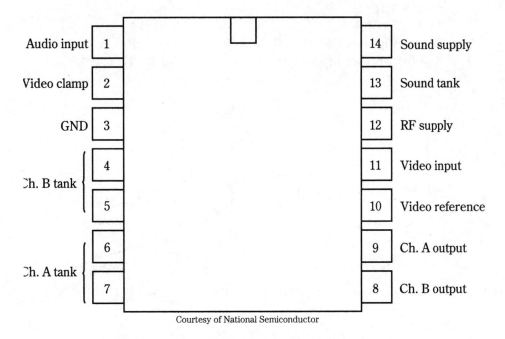

Courtesy of National Semiconductor

LM2889

CA3011 FM IF amplifier

The CA3011 is a special-purpose amplifier used in IF amplifiers for FM broadcast and TV-sound applications. It comes in a 10-lead TO-5 package.

Max. Positive dc Supply Voltage	+10 V
Max. Recommended Minimum dc	
Supply Voltage (V_{cc})	5.5 V
Max. Input Signal Voltage (single-ended)	±3 V
Max. Total Device Dissipation	300 mW
Typ. Device Dissipation	120 mW
Typ. Voltage Gain:	
f = 1 MHz	70 dB
f = 4.5 MHz	67 dB
f = 10.7 MHz	61 dB
Typ. Noise Figure (f = 4.5 MHz)	8.7 dB
Typ. Useful Frequency Range	100 kHz to > 20 MHz

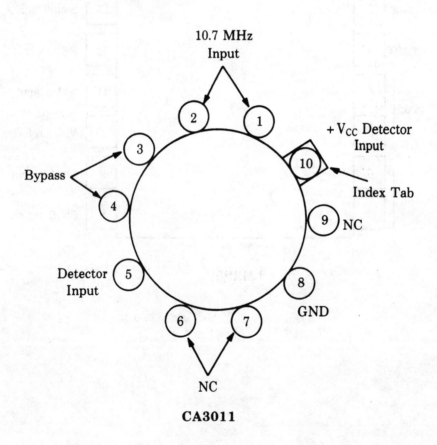

CA3011

CA3013 FM IF amplifier/discriminator/AF amplifier

The CA3013 is a special-purpose amplifier used in IF amplifier, AM, and noise-limiter, FM-detector, and AF-applications. It is available in a 10-lead TO-5 package.

Max. Positive Dc Supply Voltage	+10 V
Max. Recommended Minimum Dc Supply Voltage (V_{CC})	5.5 V
Max. Input Signal Voltage (Between terminals 1 and 2)	±3 V
Max. Total Device Dissipation	300 mW
Typ. Device Dissipation	120 mW
Typ. Voltage Gain:	
f = 1 MHz	70 dB
f = 4.5 MHz	67 dB
f = 10.7 MHz	60 dB
Typ. Noise Figure (f = 4.5 MHz)	8.7 dB
Typ. Useful Frequency Range	100 kHz to > 20 MHz

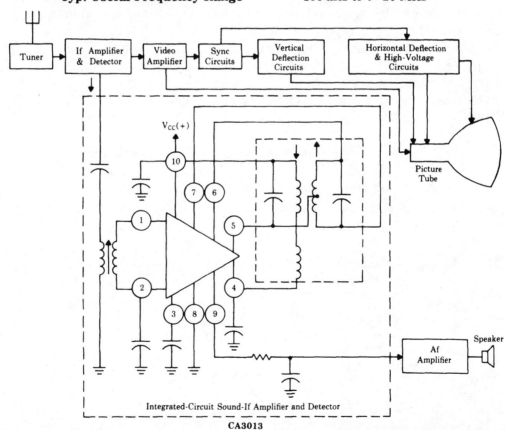

CA3013

TDA3190P TV sound system

The TDA3109P could have been included in Section 5 (Audio Amplifiers) because it is basically just an audio amplification system, but it is included here because it is designed specifically for TV sound-system applications. It even includes an on-chip FM detector and related circuitry to demodulate television audio signals.

The functional stages within this device include:

- IF limiter
- IF amplifier
- Low-pass filter
- FM detector
- Dc volume control
- Audio preamplifier
- Audio power amplifier

The audio power-amplifier stage in the TDA3109P can put out up to 4.2 watts (assuming the supply voltage is 24 volts, and the output load impedance is 16 Ω).

The dc volume control input permits an external dc voltage to set the amplifier gain (output volume). This feature can be particularly handy in remote-control systems. The volume-control response is linear.

A related device is the TDA1190P, which is a low-power version of the TDA3109P.

	Minimum	**Maximum**
Supply Voltage	9.0 V	22 V
Output Power	(V_{cc} = 24 V, R_1 = 16 Ω)	4.2 W

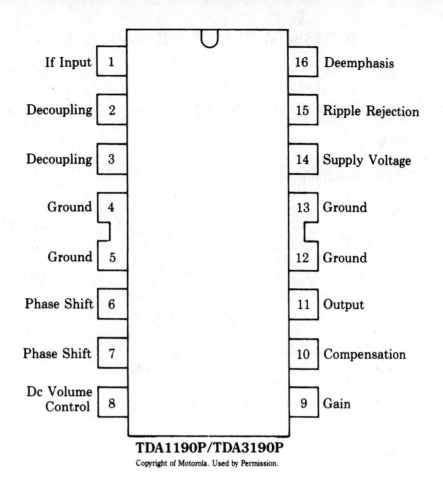

If Input	1		16	Deemphasis
Decoupling	2		15	Ripple Rejection
Decoupling	3		14	Supply Voltage
Ground	4		13	Ground
Ground	5		12	Ground
Phase Shift	6		11	Output
Phase Shift	7		10	Compensation
Dc Volume Control	8		9	Gain

TDA1190P/TDA3190P

MC3356 Wideband FSK receiver

The MC3356 is designed for use in a receiver in a communication system using FSK (frequency shift keying).

The circuitry within this chip includes:

- Oscillator
- Mixer
- Limiting IF amplifier
- Quadrature detector
- Audio buffer
- Squelch
- Meter drive
- Squelch status output
- Data shaper comparator

The MC3356 can reliably handle data rates up to 500 kilobaud.

Max. Supply Voltage	+15 V
Operating Power Supply Voltage Range	
Pins 6 and 10	3 to 9 V
Operating RF Supply Voltage Range	
Pin 4	3 to 12 V
Max. Power Dissipation	125 mW
Sensitivity	
30 µV rms at 100 MHz	3 dB

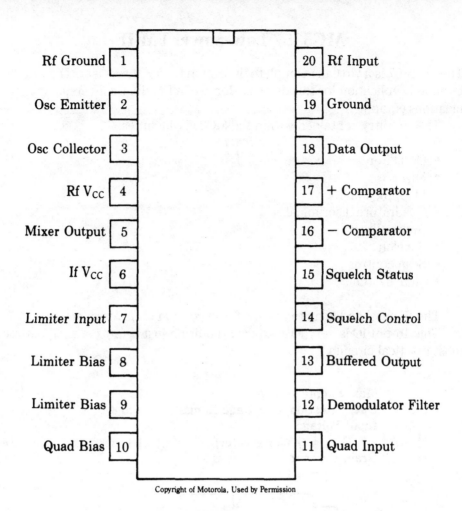

Rf Ground — 1

Osc Emitter — 2

Osc Collector — 3

Rf V$_{CC}$ — 4

Mixer Output — 5

If V$_{CC}$ — 6

Limiter Input — 7

Limiter Bias — 8

Limiter Bias — 9

Quad Bias — 10

20 — Rf Input

19 — Ground

18 — Data Output

17 — + Comparator

16 — − Comparator

15 — Squelch Status

14 — Squelch Control

13 — Buffered Output

12 — Demodulator Filter

11 — Quad Input

MC3356

MC3357 Low-power FM IF

The MC3357 is a virtually complete IF section for a FM receiver. The primary intended application for this device is for use in FM dual-conversion communications equipment.

The circuitry contained within the MC3357 includes:

- Oscillator
- Mixer
- Limiting amplifier
- Quadrature discriminator
- Active filter
- Squelch
- Scan control
- Mute switch

The IF passband is quite narrow for maximum receiver selectivity.

This 16-pin IC is very easy to use, requiring just a few external parts in most practical circuits.

Max. Supply Voltage	12 V
Operating Supply Voltage Range	4 to 8 V
Input Voltage	
(V$_{CC}$ = 6 V or greater)	1 V rms max.
Drain Current	
(V$_{CC}$ = 6 V)	3 mA typ.

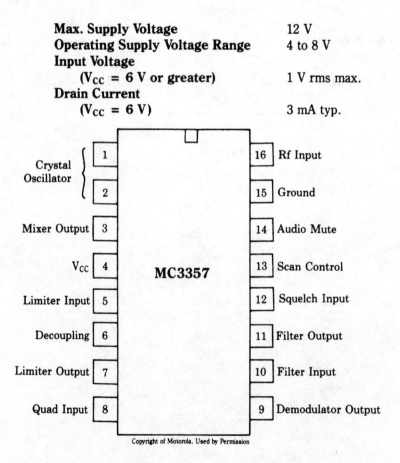

MC3359 High-gain low-power narrowband FM IF

The 3359 is designed to detect narrowband FM signals using a 455-kHz ceramic filter for use in FM dual-conversion communications systems. The on-chip circuitry of this device includes an oscillator, mixer, limiting amplifier, AFC, quadrature discriminator, operational amplifier, squelch, scan control, and mute switch. Very few external components are required for most practical FM IF applications.

	Minimum	Typical	Maximum
Supply Voltage	6	9	12 Vdc
Input Voltage			1.0 Vrms
Drain Current			
Squelch Off		3.6	6.0 mA
Squelch On		5.4	7.0 mA

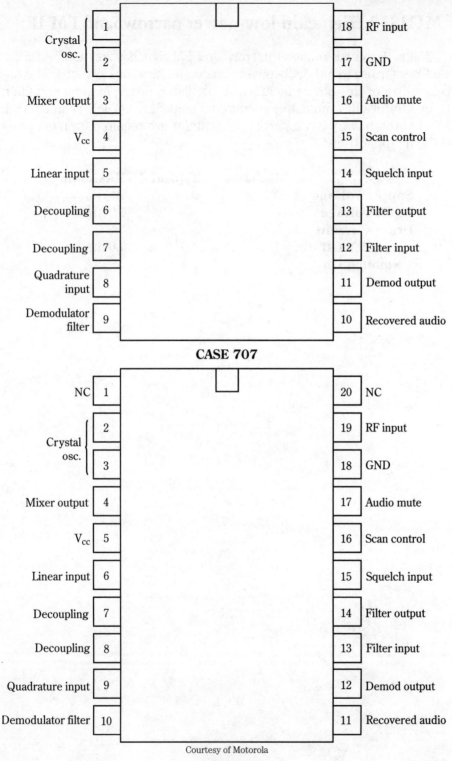

CASE 707

Courtesy of Motorola

CASE 7510
MC3359

434

MC3362 Low-power dual-conversion FM receiver

The MC3362 is just about a complete dual-conversion FM receiver in a single 24-pin package. Only a few external parts are required in most practical circuits.

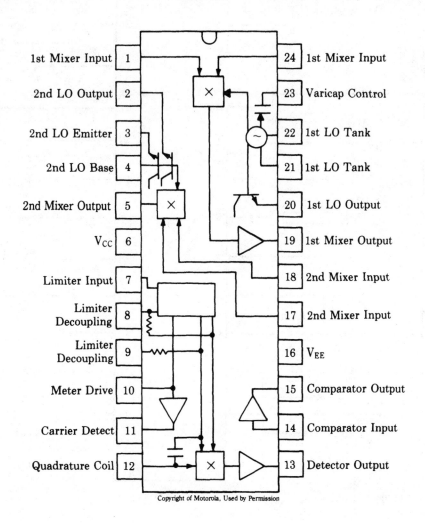

1st Mixer Input — 1
2nd LO Output — 2
2nd LO Emitter — 3
2nd LO Base — 4
2nd Mixer Output — 5
V_{CC} — 6
Limiter Input — 7
Limiter Decoupling — 8
Limiter Decoupling — 9
Meter Drive — 10
Carrier Detect — 11
Quadrature Coil — 12

24 — 1st Mixer Input
23 — Varicap Control
22 — 1st LO Tank
21 — 1st LO Tank
20 — 1st LO Output
19 — 1st Mixer Output
18 — 2nd Mixer Input
17 — 2nd Mixer Input
16 — V_{EE}
15 — Comparator Output
14 — Comparator Input
13 — Detector Output

435

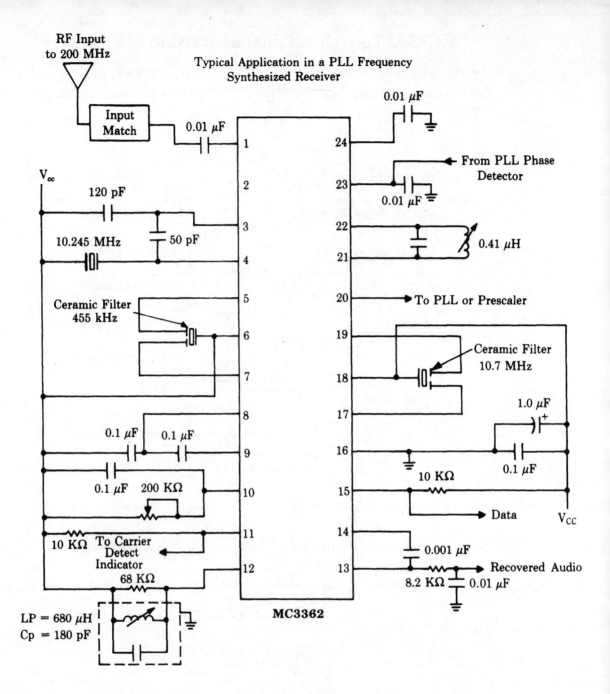

RF Input
to 200 MHz

Typical Application in a PLL Frequency
Synthesized Receiver

Input
Match

0.01 μF

0.01 μF

From PLL Phase
Detector

0.01 μF

V_{cc}

120 pF

50 pF

10.245 MHz

0.41 μH

Ceramic Filter
455 kHz

To PLL or Prescaler

Ceramic Filter
10.7 MHz

1.0 μF

0.1 μF   0.1 μF

0.1 μF

0.1 μF   200 KΩ

10 KΩ

Data

V_{CC}

10 KΩ   To Carrier
Detect
Indicator

0.001 μF

68 KΩ

Recovered Audio

8.2 KΩ   0.01 μF

LP = 680 μH
Cp = 180 pF

MC3362

MC3363 Low-power dual-conversion FM receiver

The MC3363 contains a nearly complete narrowband-FM receiver in a single chip designed to operate in the VHF band. The circuitry is a dual-conversion receiver.

The subcircuits contained within this chip include:

- RF amplifier transistor
- Oscillators
- Mixers
- Quadrature detector
- Meter drive/carrier detect
- Mute circuitry

In addition, the MC3363 features a buffered first-local-oscillator output. This output signal is useful in frequency synthesizers. The MC3363 also contains a data-slicing comparator for FSK (frequency shift keying) detection.

This receiver has an impressively wide input bandwidth. If the internal local oscillator is used, the input bandwidth is 200 MHz. In some applications, it might be preferable to use an external local oscillator. Depending on the oscillator circuitry employed, the input bandwidth can be increased to 450 MHz.

Max. Supply Voltage	8 V
Recommended Operating Voltage Range	2 V to 7 V
Input Voltage	
(V_{cc} = 5 V)	1 V rms max.
Drain Current (Carrier Detect Low)	4.5 mA typ.
	8.0 mA max.
Carrier Detect Threshold (Below V_{cc})	0.53 V min.
	0.64 V typ.
	0.77 V max.
Mute Output Impedance	
Typ. High	10 Ω
Typ. Low	25 Ω

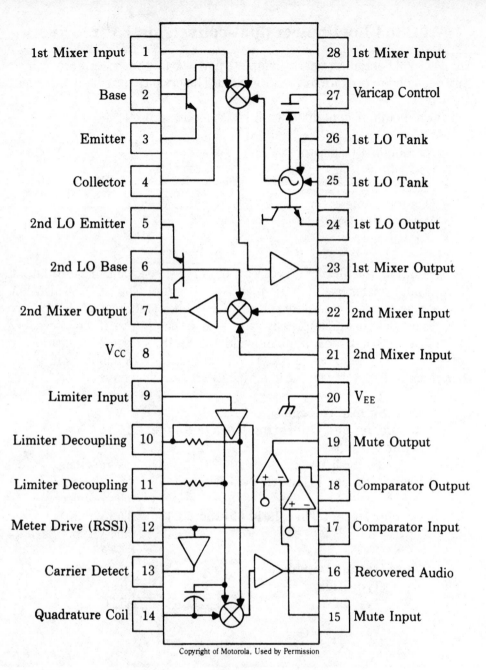

1st Mixer Input	1	28	1st Mixer Input
Base	2	27	Varicap Control
Emitter	3	26	1st LO Tank
Collector	4	25	1st LO Tank
2nd LO Emitter	5	24	1st LO Output
2nd LO Base	6	23	1st Mixer Output
2nd Mixer Output	7	22	2nd Mixer Input
V_{CC}	8	21	2nd Mixer Input
Limiter Input	9	20	V_{EE}
Limiter Decoupling	10	19	Mute Output
Limiter Decoupling	11	18	Comparator Output
Meter Drive (RSSI)	12	17	Comparator Input
Carrier Detect	13	16	Recovered Audio
Quadrature Coil	14	15	Mute Input

MC3363

MC3371 Low-power narrowband FM IF

The 3371 is a narrowband FM IF circuit, designed to perform dual-conversion FM reception. Only a few external components are required to construct a functional circuit. On-chip circuitry includes an oscillator, mixer, limiting amplifier, IF amplifier, quadrature discriminator, active filter, meter drive, and squelch switch.

This device can be operated from a fairly wide variety of supply voltages, from 2.0 to 9.0 volts, with a low drain current, typically 3.2 mA, assuming that the supply voltage is 4.0 V and the squelch is off. There is only a minimal increase in the drain current when the squelch is switched on.

The 3371 is designed for use with parallel LC components.

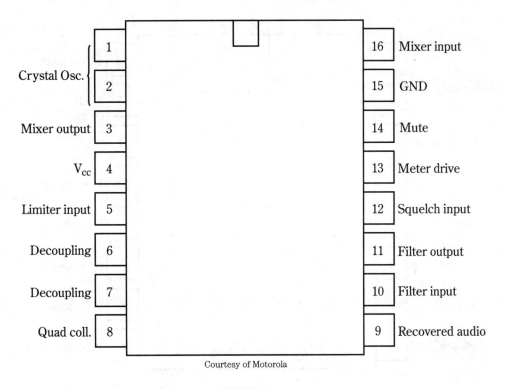

Courtesy of Motorola

MC3371

MC3372 Low-power narrowband FM IF

The 3372 is a narrowband FM IF circuit, designed to perform dual-conversion FM reception. Only a few external components are required to construct a functional circuit. On-chip circuitry includes an oscillator, mixer, limiting amplifier, IF amplifier, quadrature discriminator, active filter, meter drive, and squelch switch.

This device can be operated from a fairly wide variety of supply voltages, from 2.0 to 9.0 volts, with a low drain current, typically 3.2 mA, assuming that the supply voltage is 4.0 V and the squelch is off. There is only a minimal increase in the drain current when the squelch is switched on.

The 3372 is designed for use with either parallel LC components or a 455-kHz ceramic discriminator.

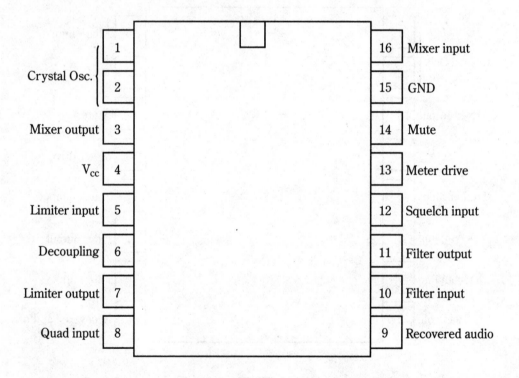

MC3372

LM6171 High-speed low-power low-distortion voltage feedback amplifier

The 6171 is a high-speed unity-gain RF voltage-feedback amplifier with very good stability. It performs very well on either ac or dc signals. Its high-speed capability makes it well suited for the high-frequency requirements of video applications.

This device is designed for use with a ±15-volt power supply. This relatively large supply voltage permits greater output signal swings, a larger dynamic range, and an improved signal-to-noise ratio. However, this chip can also be used for ±5 volts in applications where a lower voltage would be more practical, such as in portable (battery powered) equipment.

6171 Typical Ratings

Supply Current	2.5 mA
Open Loop Gain	90 dB
CMRR	110 dB
Slew Rate	3600 V/μS
Unity-Gain Bandwidth Product	100 MHz
–3-dB Frequency (*Gain* = +2)	62 MHz

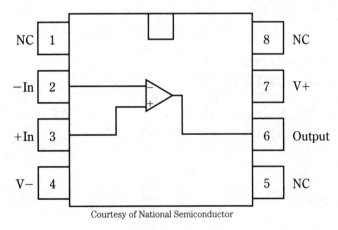

Courtesy of National Semiconductor

LM6171

MC13001XP MONOMAX®
black and white TV subsystem

The MC13001XP is part of Motorola's MONOMANIA® series. It is virtually a complete, full-performance black-and-white television receiver in a single 28-pin IC.

The video IF detector is fully on-chip. No external coils are required. In fact, this part of the device has no pins, except for the inputs. The MC13001XP features a minimum of pins and external circuitry requirements. The on-chip oscillator does not require external precision capacitors.

On-chip features of the MC13001XP include:

- Noise and video processing
- Black-level clamp
- Dc contrast
- Beam limiter
- High-performance vertical countdown
- 2-Loop horizontal system with low-power start-up mode
- Noise-protected sync and gated-AGC system

The 13001 is designed to work with Motorola's TDA1109P or TDA3109P sound-IF and audio output devices. This chip is also designed to operate with the 525 line NTSC (American) video system. A similar device (the MC13002XP) is available for use with the 625 line CCIR system.

MAX. SUPPLY VOLTAGE	+16 V
MAX. POWER DISSIPATION	1 W
MAX. POWER SUPPLY CURRENT	76 mA
REGULATOR VOLTAGE	7.2 V (min.)
	8.2 V (typ.)
	8.8 V (max.)

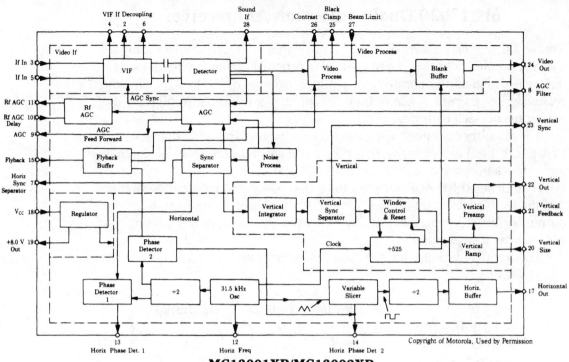

MC13001XP/MC13002XP

MC13030 Dual-conversion AM receiver

The 13030 is a dual-conversion AM receiver that is designed specifically for car radio applications, but it can be used in other AM radio applications. This device is designed to operate from a supply voltage of 7.5 to 9.0 volts with relatively low current consumption, though it might be a little high for use with some dry-cell batteries.

The circuitry contained within the 13030 includes a high dynamic-range first mixer, local oscillator, second mixer and second oscillator, IF amplifier with high-gain automatic gain control (AGC), detector, signal strength output, two delayed RF AGC outputs for a cascode FET/bipolar RF amplifier and diode attenuator, a buffered IF output stage, and a first local oscillator output buffer for driving a synthesizer. In other words, it is a nearly complete AM radio on a single chip.

The frequency range for the first mixer and first oscillator covers 100 kHz to 50 MHz.

	Minimum	Typical	Maximum
Supply Voltage	7.5 V	8.0 V	10.0 V
Supply Current	26 mA	32 mA	44 mA
Audio S/N Ratio	48 dB	52 dB	

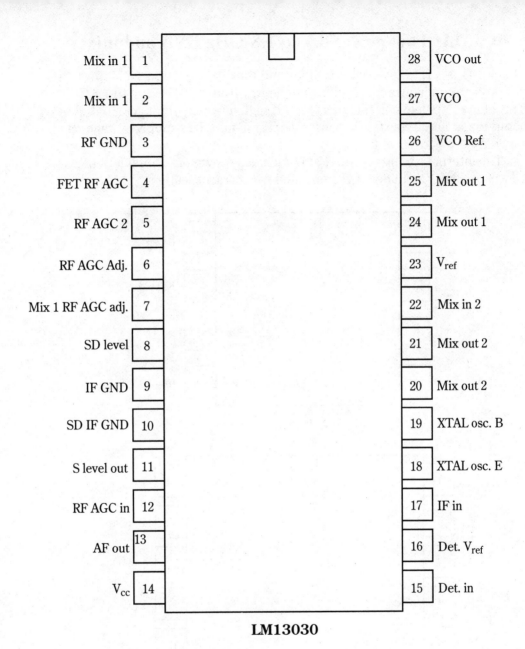

LM13030

MC13141 Low-power dc- to 1.8-GHz LNA and mixer

The 13141 is a RF first-stage amplifier and down converter, which covers a very wide frequency range, running all the way from dc (0 Hz) up to 1.8 GHz. This chip offers low noise, high gain, and high linearity, yet its current consumption is low. Typically, it consumes just 7 mA. The supply voltage can range from 2.7 to 6.5 volts.

The internal circuitry of the 13141 includes a low-noise amplifier (LNA), a local oscillator amplifier (LO amp), and a dc control section.

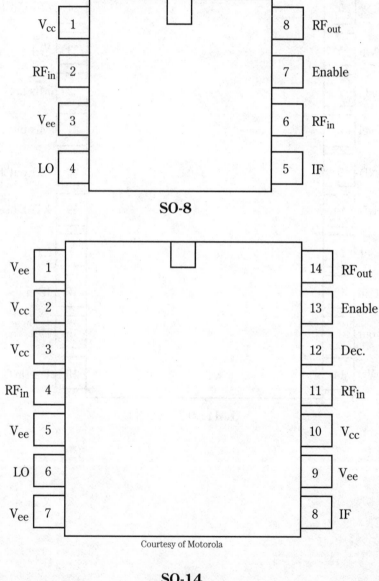

SO-8

Courtesy of Motorola

SO-14
MC13141

MC13142 Low-power dc- to 1.8-GHz LNA, Mixer, and VCO

The 13142 is somewhat similar to the 13141, except that this chip's internal circuitry also includes a voltage-controlled oscillator (VCO) stage. This device is designed to serve as a RF first-stage amplifier and down converter, which covers a very wide frequency range, running all the way from dc (0 Hz) up to 1.8 GHz. This chip offers low noise, high gain, and high linearity, yet its current consumption is low. Typically, it consumes just 7 mA. The supply voltage can range from 2.7 to 6.5 volts.

The internal circuitry of the 13141 includes a low-noise amplifier (LNA), a VCO, a buffered oscillator output, a mixer, an intermediate frequency amplifier (IF amp), and a dc control section.

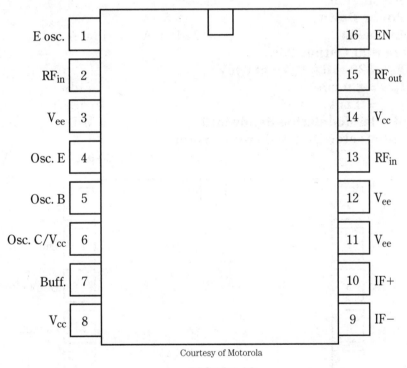

Courtesy of Motorola

MC13142

447

MC13175 UHF FM/AM transmitter

The 13175 is a single-chip FM/AM transmitter subsystem. It is designed primarily for use in communications systems using either amplitude or frequency modulation. This chip can handle either. It is designed to operate in the UHF band, which ranges from 260 to 470 MHz.

The internal circuitry contained on this chip includes a Colpitts crystal reference oscillator, UHF oscillator, divide-by-eight prescaler, and phase detector. Together, these subcircuits form a versatile phase-locked loop (PLL) system.

The 13175 is similar to the 13176, except for the prescaler stage.

	Minimum	Typical	Maximum
Supply Voltage	1.8	5.0	7.0 Vdc
Supply Current			
(Power Down)	$-0.5\ \mu A$		
(Enabled)	$-18\ mA$	-14 mA	
Differential Output Power			
(F_0 = 320 MHz, V_{ref} = 500 mV)			
(I_{mod} = 2.0 mA)	2.0 dBm	+4.7 dBm	
(I_{mod} = 0 mA)		-45 dBm	
Amplitude-Modulation Bandwidth		25 MHz	
Phase-Detector Output-Error Current			
		$20\ \mu A$	$25\ \mu A$

Osc. 1	1	16	I_{mod}
NC	2	15	Out GND
NC	3	14	Output 2
Osc. 4	4	13	Output 1
V_{ee}	5	12	V_{cc}
I_{cont}	6	11	Enable
RD out	7	10	Reg. GND
Xtal E	8	9	Xtal B

Courtesy of Motorola

MC13175

MC13176 UHF FM/AM transmitter

The 13176 is a single-chip FM/AM transmitter subsystem. It is designed primarily for use in communications systems using either amplitude or frequency modulation. This chip can handle either. It is designed to operate in the UHF band, which ranges from 260 to 470 MHz.

The internal circuitry contained on this chip includes a Colpitts crystal reference oscillator, UHF oscillator, divide-by-32 prescaler, and phase detector. Together, these subcircuits form a versatile phase-locked loop (PLL) system.

The 13176 is similar to the 13175, except for the prescaler stage.

	Minimum	Typical	Maximum
Supply Voltage	1.8	5.0	7.0 Vdc
Supply Current			
(Power Down)	$-0.5\ \Omega A$		
(Enabled)	-18 mA	-14 mA	
Differential Output Power			
(F_o = 320 MHz, V_{ref} = 500 mV)			
(I_{mod} = 2.0 mA)	2.0 dBm	+4.7 dBm	
(I_{mod} = 0 mA)		-45 dBm	
Amplitude-Modulation Bandwidth		25 MHz	
Phase-Detector			
Output-Error Current	22 μA	27 μA	

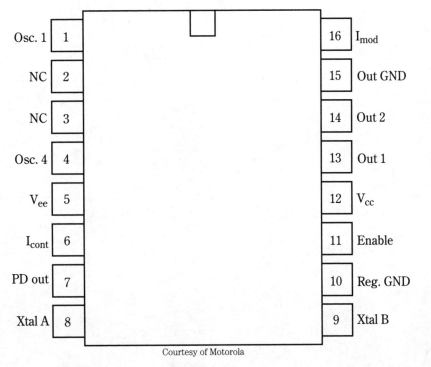

Courtesy of Motorola

MC13176

MC44002 Chroma 4 video processor

The 44002 is designed to perform most of the basic functions required for a color TV. Naturally, it is a fairly sophisticated and complex IC. It has a larger pin-out than most of the devices in this book.

In addition to offering many features and functions, it is processor-controlled via an I2C bus. This permits the system designer to eliminate standard potentiometer controls entirely, if so desired. The number of required external components can thereby be reduced considerably, reducing system cost and bulk. It is also very convenient for remote-control applications.

The 44002 offers four selectable matrix modes (primarily for NTSC), fast beam-current limiting, and 16:9 display. It can also support PAL or SECAM video signals. This device can use either dual composite video or S-VHS inputs. This chip also offers auxiliary Y, R-Y, and B-Y inputs.

Only a single 5.0-volt power supply is required to operate the 44002. Current consumption is also quite low, typically only 120 mA.

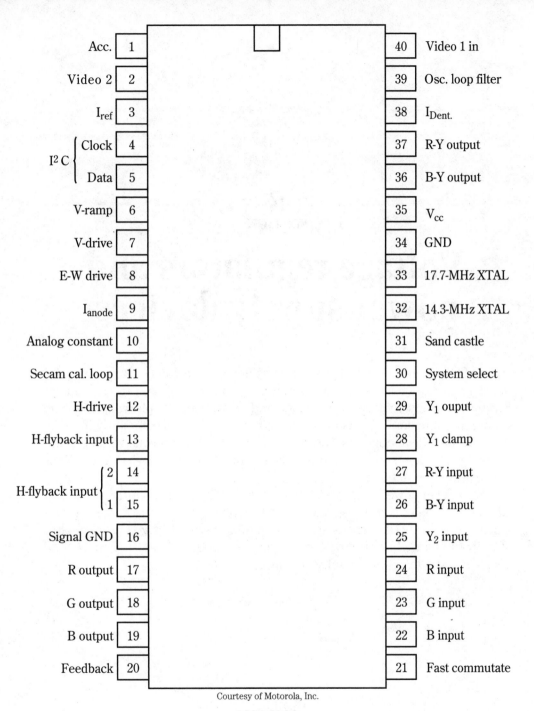

Acc.	1		40	Video 1 in
Video 2	2		39	Osc. loop filter
I_{ref}	3		38	$I_{Dent.}$
$I^2 C$ { Clock	4		37	R-Y output
Data	5		36	B-Y output
V-ramp	6		35	V_{cc}
V-drive	7		34	GND
E-W drive	8		33	17.7-MHz XTAL
I_{anode}	9		32	14.3-MHz XTAL
Analog constant	10		31	Sand castle
Secam cal. loop	11		30	System select
H-drive	12		29	Y_1 ouput
H-flyback input	13		28	Y_1 clamp
H-flyback input { 2	14		27	R-Y input
1	15		26	B-Y input
Signal GND	16		25	Y_2 input
R output	17		24	R input
G output	18		23	G input
B output	19		22	B input
Feedback	20		21	Fast commutate

Courtesy of Motorola, Inc.

MC44002

8
SECTION

Voltage regulators and power-supply devices

Except for a very few, simple passive circuits, all electronics circuits require a power supply of some sort. Most digital circuitry and high-precision linear circuitry usually requires very exact and consistent supply voltages (and sometimes currents). To meet this sort of need, which is constantly increasing with new developments in the electronics field, numerous voltage-regulator ICs have been created, along with a number of other devices for various power-supply functions.

By itself, a power supply is rather mundane. Yet, it is essential for almost any other type of electronics circuit to function. Despite the apparent simplicity of the function, there are an incredible number of voltage-regulator ICs on the market, with new devices appearing every month. This section looks at just a few representative examples.

Voltage Regulators

current-limit sense voltage The voltage across the current-limit terminals required to cause the regulator to current-limit with a short-circuited output. This voltage is used to determine the value of the external current-limit resistor when external booster transistors are used.

dropout voltage The input-output voltage differential at which the circuit ceases to regulate against further reductions in input voltage.

feedback sense voltage The voltage, referred to ground, on the feedback terminal of the regulator while it is operating in regulation.

input voltage range The range of dc input voltages over which the regulator will operate within specifications.

line regulation The change in output voltage for a change in the input voltage. The measurement is made under conditions of low dissipation or by using pulse techniques such that the average chip temperature is not significantly affected.

load regulation The change in output voltage for a change in load current at constant chip temperature.

long-term stability Output voltage stability under accelerated life-test conditions at 125°C with maximum rated voltages and power dissipation for 1000 hours.

maximum power dissipation The maximum total device dissipation for which the regulator will operate within specifications.

output-input voltage differential The voltage difference between the unregulated input voltage and the regulated output voltage for which the regulator will operate within specifications.

output noise voltage The rms ac voltage at the output with constant load and no input ripple, measured over a specified frequency range.

output voltage range The range of regulated output voltages over which the specifications apply.

output voltage scale factor The output voltage obtained for a unit value of resistance between the adjustment terminal and ground.

quiescent current That part of input current to the regulator that is not delivered to the load.

ripple rejection The one regulation for ac input signals at or above a given frequency with a specified value of bypass capacitor on the reference bypass terminal.

standby current drain That part of the operating current of the regulator which does not contribute to the load current.

temperature stability The percentage change in output voltage for a thermal variation from room temperature to either temperature extreme.

LM117/LM217/LM317 3-Terminal adjustable regulator

The LM117/LM217/LM317 are adjustable 3-terminal positive voltage regulators capable of supplying in excess of 1.5 A over a 1.2 to 37 V output range. They are exceptionally easy to use and require only two external resistors to set the output voltage. Further, both line and load regulation are better than standard fixed regulators. Also, the LM117 is packaged in standard transistor packages which are easily mounted and handled.

In addition to performing better than fixed regulators, the LM117 series offers full overload protection available only in ICs. Included on the chip are current-limit, thermal-overload protection and safe-area protection. All overload-protection circuitry remains fully functional, even if the adjustment terminal is disconnected. Features include:

- Adjustable output down to 1.2 V
- Guaranteed 1.5 A output current
- Line regulation typically 0.01 percent/V
- Load regulation typically 0.1 percent
- Current limit constant with temperature
- 100-percent electrical burn-in
- Eliminates the need to stock many voltages
- Standard 3-lead transistor package
- 80-dB ripple rejection

Normally, no capacitors are needed unless the device is situated far from the input filter capacitors, in which case an input bypass is needed. An optional output capacitor can be added to improve transient response. The adjustment terminal can be bypassed to achieve very high ripple-rejection ratios that are difficult to achieve with standard 3-terminal regulators.

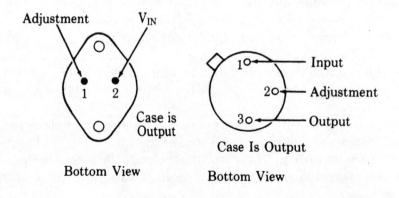

LM117

LM120/LM220/LM320 3-Terminal negative regulator

The LM120 series are 3-terminal negative regulators with a fixed output voltage of –5 V, –5.2 V, –6 V, –8 V, –9 V, –12 V, –15 V, –18 V, and –24 V with up to 1.5 A load current capability (LM320-5, LM320-5.2, LM320-6, etc.).

These devices need only one external component: a compensation capacitor at the output, making them easy to apply. Worst-case guarantees on output voltage deviation due to any combination of line, load, or temperature variation ensure satisfactory system operation.

Exceptional effort has been made to make the LM120 series immune to overload conditions. The regulators have current limiting, which is independent of temperature, combined with thermal-overload protection. Internal current limiting protects against momentary faults and thermal shutdown prevents junction temperatures from exceeding safe limits during prolonged overloads.

Although primarily intended for fixed output voltage applications, the LM120 series can be programmed for higher output voltages with a simple resistive divider. The low quiescent drain current of the devices allows this technique to be used with good regulation. They feature:

- Preset output voltage error less than ±3 percent
- Preset current limit
- Internal thermal shutdown
- Operates with input-output voltage differential down to 1 V
- Excellent ripple rejection
- Low temperature drift
- Easily adjustable to higher output voltage

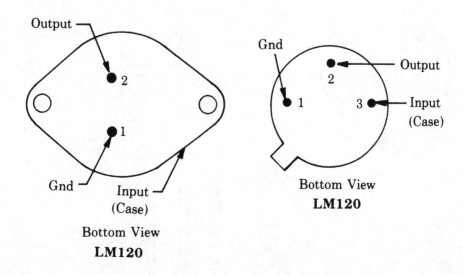

Bottom View
LM120

Bottom View
LM120

LM123/LM223/LM323 3-Amp,
5-volt positive voltage regulator

The 123 (and its related devices) is a three-terminal positive voltage regulator with a preset 5 volt output and a load driving capability of a full three amperes. Special circuit design and processing techniques are used to provide the high output current without sacrificing the regulation characteristics of lower-current devices. This 3-amp regulator is virtually blow-out proof. Current limiting, power limiting, and thermal shutdown provide a high level of reliability.

No external components are required for basic operation of the 123. If the device is more than 4 inches from the filter capacitor, however, a 1-µF solid tantalum capacitor should be used on the input. It is also advisable to use a 0.1-µF (or larger) capacitor across the output to swamp out any stray load capacitance or to reduce any load transient spikes created by fast-switching digital logic.

An overall worst-case specification for the combined effects of input voltage, load currents, ambient temperature, and power dissipation ensure that the 123 will perform satisfactorily as a system element in almost any practical application.

Features of this device include:

- 3-A output current
- Internal current and thermal limiting
- 0.01-Ω typical output impedance
- 7.5-volts minimum input voltage
- 30 watts power dissipation

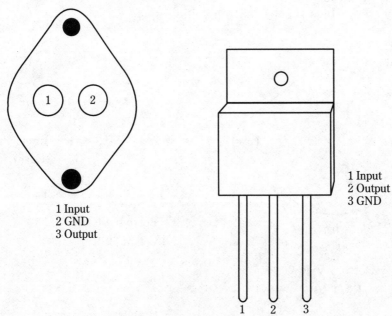

1 Input
2 GND
3 Output

1 Input
2 Output
3 GND

LM123/LM223/LM323

LM285/385 Micropower voltage-reference diode

The LM285 is a very low-power two-terminal band-gap voltage reference diode, which can operate over a wide current range of 10 µA to 20 mA. On-chip trimming permits tight voltage tolerances. The device also features low noise, low dynamic impedance, and good stability over time and temperature.

The 285 is available in two package styles (designated "D" and "Z") and two voltages (1.2325 V and 2.500 V). These variations are identified in the device number suffix, as outlined in the table.

The 285 is designed to operate over the standard commercial temperature range. The 385 is identical to the 285 in all respects, except that it meets the wider military temperature range standards.

Device Number	Package	Reverse Breakdown Voltage	Tolerance
LM285D-1.2	8-pin DIP	1.235 V	±1.0%
LM285Z-1.2	3-pin	1.235 V	±1.0%
LM285D-2.5	8-pin DIP	2.500 V	±1.5%
LM285Z-2.5	3-pin	2.500 V	±1.5%

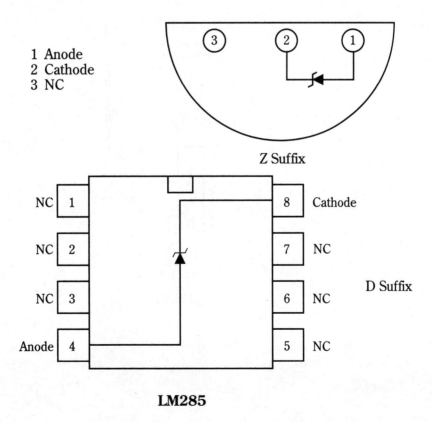

1 Anode
2 Cathode
3 NC

Z Suffix

D Suffix

LM285

LM337 Three-terminal
adjustable negative-voltage regulator

The 337 is a three-terminal voltage regulator that can put out an externally adjustable negative voltage anywhere between −1.2 V to −37 V. Only two external resistors are required to set the desired output voltage of this device. It can supply a hefty amount of current, as well. It is rated for output currents exceeding 1.5 A. The 337 is also designed for maximum reliability, with internal current limiting, thermal shutdown, and safe area compensation.

By connecting a resistor between the adjustment and output pins, this IC can also be used in precision current-regulator applications.

	Typical	Maximum
Input/Output Voltage Differential		40 V
Line Regulation	0.01% V	0.04% V
Load Regulation		
Vo < 5 volts	15 mV	50 mV
Vo > 5 volts	0.3 mV	1.0% Vo

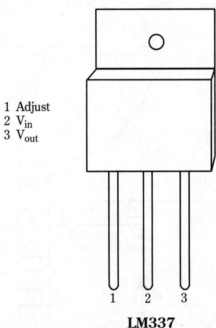

1 Adjust
2 V_{in}
3 V_{out}

LM337

LM320L Series 3-terminal negative regulator

The LM320L-XX series of 3-terminal negative voltage regulators features several selected fixed output voltages from –5 to –24 V with load current capabilities to 100 mA. Internal protective circuitry includes safe operating area for the output transistor, short-circuit current limit, and thermal shutdown. Features include:

- Preset output voltage error less than ±5 percent over temperature
- 100 mA output current capability
- Internal thermal overload protection
- Input-output voltage differential down to 2 V
- Internal current limit
- Maximum load regulation –0.15 percent/mA
- Maximum line regulation –0.1 percent/V
- Output transistor safe-area protection

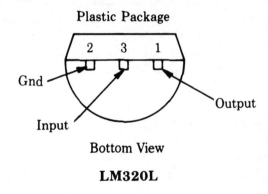

Plastic Package

Bottom View

LM320L

LM340 Series voltage regulator

The LM340-XX series of 3-terminal regulators is available with several fixed output voltages, making them useful in a wide range of applications. One of these is local on-card regulation, eliminating the distribution problems associated with single-point regulation. The voltages available allow these regulators to be used in logic systems, instrumentation, stereo, and other solid-state electronic equipment. Although designed primarily as fixed voltage regulators, these devices can be used with external components to obtain adjustable voltages and currents.

The LM340-XX series is available in two power packages. Both the plastic TO-220 and metal TO-3 packages allow these regulators to deliver over 1.5 A, if adequate heat sinking is provided. Current limiting is included to limit the peak output current to a safe value. Safe-area protection for the output transistor is provided to limit internal power dissipation. If internal power dissipation becomes too high for the heat sinking provided, the thermal-shutdown circuit takes over, preventing the IC from overheating.

Considerable effort was expended to make the LM340-XX series of regulators easy to use, with a minimal number of external components. It is not necessary to bypass the output, although this does improve transient response. Input bypassing is needed only if the regulator is located far from the alter capacitor of the power supply. Features include:

- Output current in excess of 1.5 A
- Internal thermal-overload protection
- No external components required
- Output transistor safe-area protection
- Internal short-circuit current limit

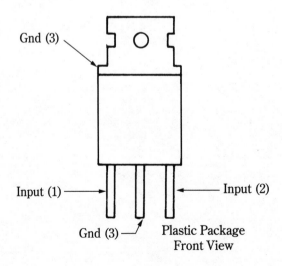

LM340

LM350 Three-terminal
adjustable positive-voltage regulator

The 350 is a three-terminal voltage regulator that can put out an externally adjustable positive voltage anywhere between +1.2 V to +33 V. Only two external resistors are required to set the desired output voltage of this device. It can supply a hefty amount of current. It is rated for output currents exceeding 3.0 A. The 350 is also designed for maximum reliability, with internal current limiting, thermal shutdown, and safe-area compensation.

By connecting a resistor between the adjustment and output pins, this IC can also be used in precision current-regulator applications.

	Typical	Maximum
Input/Output Voltage Differential		35 V
Line Regulation	0.0005% V	0.03% V
Load Regulation		
Vo < 5 volts	5.0 mV	25 mV
Vo > 5 volts	0.1 mV	0.5% Vo

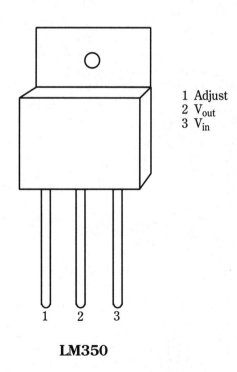

1 Adjust
2 V_{out}
3 V_{in}

LM350

LM377 Three-terminal adjustable negative-voltage regulator

The LM377 can be externally adjusted to regulate voltages ranging from -1.2 V to -37 V, with current capability of more than 1.5 A (under certain conditions). To some degree, it is a negative-voltage equivalent to the LM317.

Only two external resistors are required to set the output voltage, making the LM337 very easy to use. Because it features internal current limiting, thermal shutdown, and safe-area compensation, this device is also very reliable, and virtually blow-out proof. It is unlikely to be damaged—even under most short circuit conditions. Because it is designed to shut itself down under such emergency conditions, the LM377 will also help prevent damage to other components in the event of a short circuit.

Because the regulated output voltage is externally adjustable, there is no need to stock a number of different devices for various voltage applications. This one chip can do the job in most cases. It is even designed to permit floating operation for high-voltage applications.

LM337 Maximum Ratings

Input-Output Voltage Differential	40 Vdc
Power Dissipation	
Case 221A Ta = 24 C	*
Thermal Resistance, Junction to Ambient	65 C/W
Thermal Resistance, Junction to Case	5.0 C/W
Case 936 (D2PAK) Ta = 24 C	*
Thermal Resistance, Junction to Ambient	70 C/W
Thermal Resistance, Junction to Case	5.0 C/W
Maximum Output Current	
(Vi-Vo) < 15 V	2.2 A
(Vi-Vo) < 40 V	0.4 A

* = Internally limited

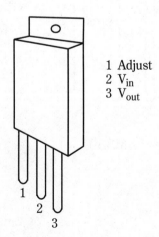

1 Adjust
2 V_{in}
3 V_{out}

LM350 Three-terminal
adjustable positive-voltage regulator

The LM350 can be externally adjusted to regulate positive voltages ranging from 1.2 V to 33 V, with current capability of 3.0 A (guaranteed for output voltages less than or equal to 10 volts.

Only two external resistors are required to set the output voltage, making the LM350 very easy to use. Because it features internal current limiting, thermal shutdown, and safe-area compensation, this device is also very reliable, and virtually indestructible. It is unlikely to be damaged—even under most short-circuit conditions. Because it is designed to shut itself down under such emergency conditions, the LM377 will also help prevent damage to other components in the event of a short circuit.

Typical applications for the LM350 include local on-card voltage regulation, simple adjustable switching regulators, or programmable output regulators. If a fixed resistor is connected between the Adjust (pin 1) and Output (pin 2) pins of the LM350, this device can also function as a precision current regulator.

The load regulation for the LM350 is rated for a typical value of 0.1%, and the line regulation is typically 0.005%/V. Because the regulated output voltage is externally adjustable, there is no need to stock a number of different devices for various voltage applications.

LM350 Maximum Ratings

Input-Output Voltage Differential	35 Vdc
Power Dissipation	Internally Limited
Output Current	
V_i-V_o < 10 V	4.5 A
V_i-V_o = 30 V	1.0 A

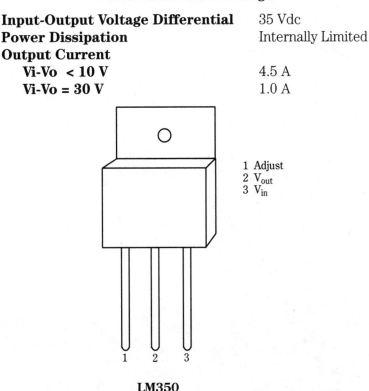

1 Adjust
2 V_{out}
3 V_{in}

LM350

TL431 Programmable precision reference

The 431 is a three-terminal programmable shunt regulator diode. It is available in three different package styles. Notice that there are only three active connections, even in the 8-pin packages: anode, cathode, and reference. A stable 2.5-volt reference voltage is available from this last pin.

The 431 can be user-programmed for any voltage ranging from V_{ref} (2.5 V) up to 36 V. Just two external resistors are required to program the device's output voltage. It can operate with currents ranging from 1.0 mA to 100 mA, and has a typical dynamic impedance of a just 0.22 Ω.

	Minimum	Maximum
Cathode to Anode Voltage	2.5 V	36 V
Cathode Current	1.0 mA	100 mA
Cathode Current Range		
(Continuous)	–100 mA	+150 mA
Reference Input Current Range		
(Continuous)	–0.05 mA	+10 mA
Power Dissipation		
D Package		0.70 W
P Package		1.10 W
LP Package		0.70 W

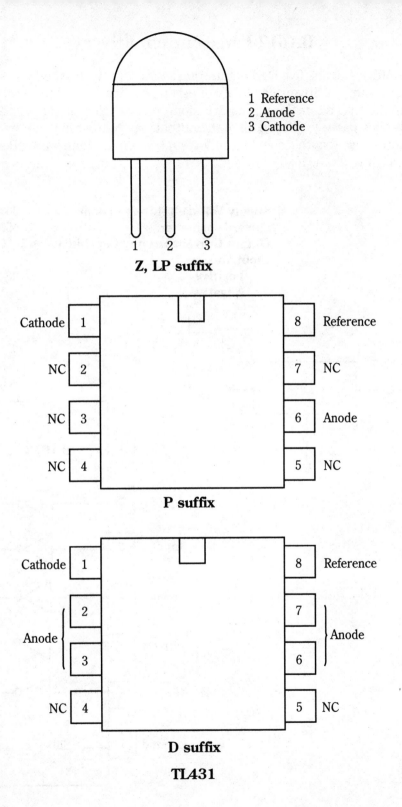

1 Reference
2 Anode
3 Cathode

Z, LP suffix

Cathode 1   8 Reference

NC 2   7 NC

NC 3   6 Anode

NC 4   5 NC

P suffix

Cathode 1   8 Reference

2   7

Anode { 3   6 } Anode

NC 4   5 NC

D suffix

TL431

BA612 Large current driver

The BA612 contains five Darlington transistor arrays with input resistors to boost the current-driving capability of other circuitry.

Notice that all of the inputs are along the left side of the 14-pin DIP (dual-inline package) IC and all of the outputs are to the right.

Notice also that this chip has no V_{CC} supply-voltage connection. Power is "stolen" through the load.

Supply Voltage (Through Load)	30 V max.
Power Dissipation	550 mW max.
Output Current Driving Capability	400 mA Max.
Input Voltage	
Positive	30 V Max.
Negative	−0.5 V Max.

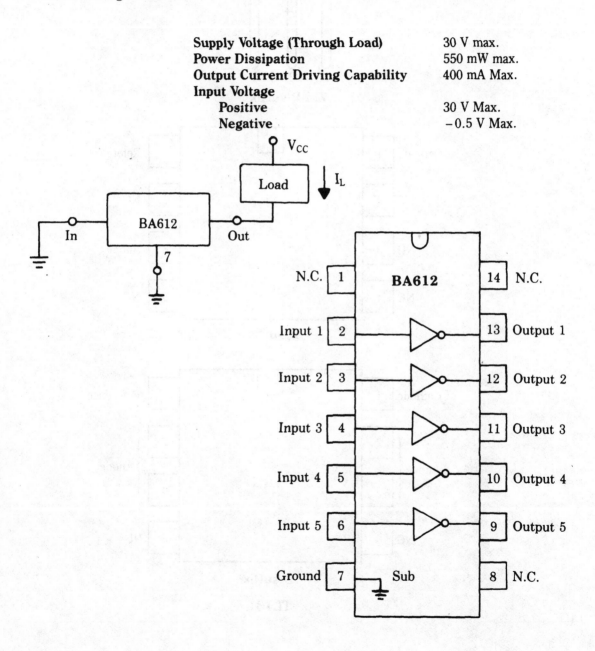

LM723/LM723C Voltage regulator

The LM723/LM723C is a voltage regulator designed primarily for series regulator applications. By itself, it will supply output currents up to 150 mA, but external transistors can be added to provide any desired load current. The circuit features extremely low standby current drain, and provisions are made for either linear or foldback current limiting. Important characteristics are:

- 150-mA output current without an external-pass transistor
- Output currents in excess of 10 A possible by adding external transistors
- Input voltage 40 V max
- Output voltage adjustable from 2 to 37 V
- Can be used as either a linear or a switching regulator

The LM723/LM723C is also useful in a wide range of other applications such as a shunt regulator, a current regulator, or a temperature controller.

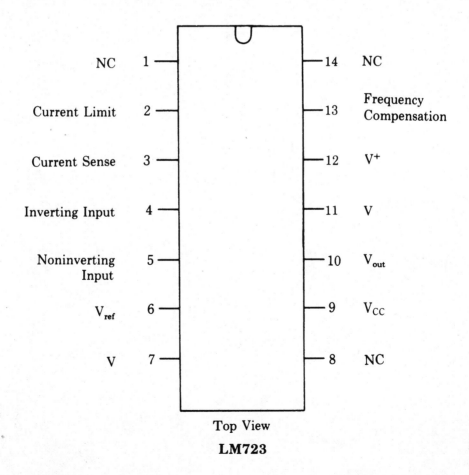

Top View

LM723

TL780 Series three-terminal
positive fixed-voltage regulators

The TL780 series is a group of precision fixed voltage regulators that can drive loads over 1.5 A. These devices are available in several standard voltages: 5 V, 12 V, and 15 V. In each case, the input voltage can be as high as 35 V. At 25°C, the output voltage tolerance is ±1%. Over the device's entire operating temperature range (0 to 125 C), the output voltage tolerance is still within ±2%.

The TL780 series of voltage regulators include internal thermal overload protection, and short-circuit current limiting. In basic voltage regulation applications, no external components are required.

THE TL780 SERIES

Device	Output Voltage
TL780-05CKC	5.0 V
TL780-12CKC	12 V
TL780-15CKC	15 V

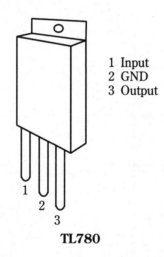

1 Input
2 GND
3 Output

TL780

MC1403 Precision low-voltage reference

The 1403 is a 2.5-V precision band-gap voltage reference. It can accept input voltages from 4.5 V up to 40 V, yet the output voltage maintains accuracy within ±25 mV. The 1403 is equivalent to the AD580. This chip is designed for critical instrumentation and 8-to-12 bit D/A converter applications.

1403 Specifications

Input Voltage Range	4.5 V to 40 V
Output Voltage	2.5 V (±25 mV)
Quiescent Current	1.2 mA
Output Current	10 mA

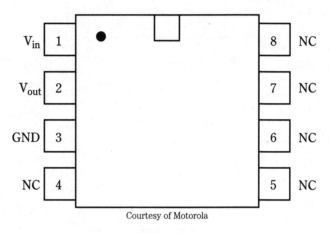

Courtesy of Motorola

MC1403

MC1404 Precision low-drift voltage reference

The 1404 is a temperature-compensated voltage reference, available in three standard voltages: 5.0 V, 6.25 V, and 10 V. These devices are stable, and accurate to 12 bits over a wide range of temperatures. They also offer fine long-term stability, and low noise. The input voltage for all three versions can range from +2.5 V to +40 V. The output voltage is trimmable ±6%.

Very little power is consumed by the 1404. Typically, the quiescent current is just 1.25 mA.

Device	Output Voltage
MC1404P5	5.0 V
MC1404P6	6.25 V
MC1404P10	10 V

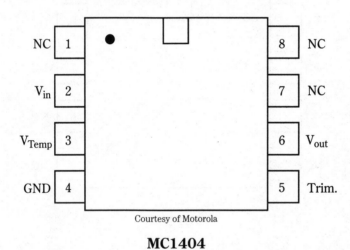

Courtesy of Motorola

MC1404

LM2931 Series low-dropout voltage regulators

The 2931 is available either as a fixed 5.0-V or an adjustable output voltage regulator. It is offered in a variety of package styles. The fixed-voltage version is available in with either ±5.0% or ±3.8% tolerances. The adjustable output devices are rated for a ±5.0% tolerance.

The 2931 can handle output currents higher than 100 mV and input voltages as high as 40 volts. The adjustable output voltage version can be set up for output voltages ranging from 3 to 24 volts.

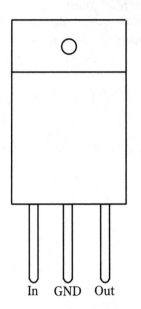

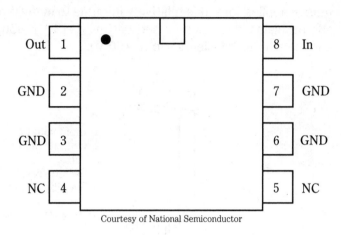

Courtesy of National Semiconductor

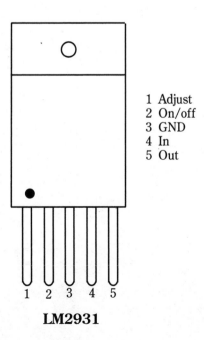

1 Adjust
2 On/off
3 GND
4 In
5 Out

LM2931

LM2935 Low-dropout dual-voltage regulator

The 2935 is a dual +5.0-V low-dropout voltage regulator intended primarily for use in standby power systems. The main output of this device can supply up to 750 mA of current. This output can be turned on and off via the SWITCH/RESET pin (pin 4). The second output is not switchable. It is intended to supply power to volatile memory. It can drive loads up to 10 mA. When supplying 10 mA from the standby output, the quiescent current of the 2935 is no greater than 3.0 mA.

The 2935 was designed for use in harsh environments, such as in automotive applications. It is relatively immune to many typical input supply voltage problems, such as reversed battery (−12 V), doubled battery (+24 V), and load dump transients (up to +60 V).

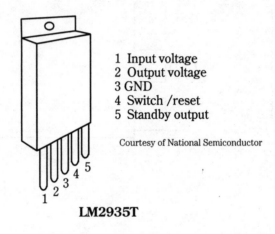

1 Input voltage
2 Output voltage
3 GND
4 Switch /reset
5 Standby output

Courtesy of National Semiconductor

LM2935T

LP2950 Low-dropout micropower voltage regulators

In some applications, only low levels of power need to be regulated. In such cases, it is overkill to use a heavy-duty voltage regulator that can carry several amps of current. The LP2950 is a micropower voltage regulator designed to maintain tight regulation with an extremely low input-to-output voltage differential. It can supply currents to over 100 mA. Internal current- and thermal-limiting protection is included in this chip.

In any voltage regulator, some of the input power is inevitably consumed by the voltage regulator itself. In the case of the LP2950, the quiescent bias current is a mere 75 μA. The low power consumption and input-to-output voltage differential makes the LP2950 very well suited for applications in battery powered equipment.

The output voltage of the LP2950 is preset to a nominal 5.0 volts. Even under extreme operating conditions, the actual output voltage will be between 4.880 V and 5.120 V. The input voltage can range from 6.0 to 30 volts.

The LP2950 features a low input-to-output voltage differential of 50 mV at 100 μA, and 380 mV at 100 mA. This voltage-regulator chip is exceptionally easy to use, requiring only a 1.0-μF output capacitor for stability in its most basic application circuit.

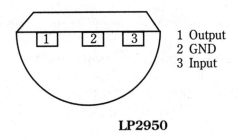

1 Output
2 GND
3 Input

LP2950

LP2951 Low-dropout micropower voltage regulators

In some applications, only low levels of power need to be regulated. In such cases, it is overkill to use a heavy-duty voltage regulator that can carry several amps of current. The LP2951 is a micropower voltage regulator designed to maintain tight regulation with an extremely low input-to-output voltage differential. It can supply currents to over 100 mA. Internal current- and thermal-limiting protection is included in this chip.

In any voltage regulator, some of the input power is inevitably consumed by the voltage regulator itself. In the case of the LP2951, the quiescent bias current is a mere 75 μA. The low power consumption and input-to-output voltage differential makes the LP2951 very well suited for applications in battery-powered equipment.

The LP2951 is a somewhat more versatile version of the LP2950. It is supplied in an 8-pin DIP package, rather than the three-lead TO-226AA or TO-92 case of the LP2950. The extra pins permit additional functions, including external adjustment of the output voltage. The LP2951 can be used as a tight 5.0-V fixed regulator or it can be programmed for output voltages ranging from 1.25 V to 29 V. This programmability is simply accomplished via a pinned out resistor divider, along with direct access to the regulator's Error Amplifier feedback input. The LP2951 also features a special Shutdown input, permitting the regulated voltage output to be switched on or off under external digital logical control. There are many possibilities for remote-control and automation applications, thanks to this rather unusual feature.

The nominal, preprogrammed output voltage of the LP2951 is preset to a nominal 5.0 volts. Even under extreme operating conditions, the actual output voltage will be between 4.880 V and 5.120 V. The input voltage can range from 6.0 to 30 volts. The LP2951 features low input-to-output voltage differential of 50 mV at 100 μA, and 380 mV at 100 mA.

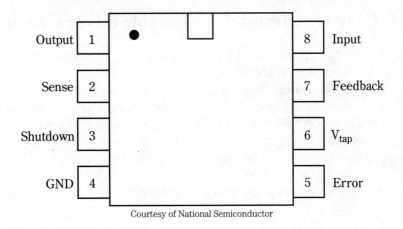

Courtesy of National Semiconductor

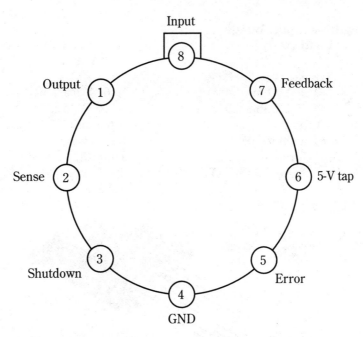

LR2951H

MC3399T Automotive high-side driver switch

The MC3399T is a high-side driver switch that is designed to drive loads from the positive side of the power supply.

A TTL-compatible enable pin is used to control the output of this device. In the ON state, the chip exhibits very low saturation voltages for load currents up to (at least) 750 mA. The MC3399T protects the load from high-voltage transients by switching into an open-circuit state for the length of the transient. When the applied voltage drops back to an acceptable level, the device automatically switches back to its normal ON state.

As the name of this chip suggests, it is intended primarily for use in automotive applications.

Ignition Input Voltage	
Continuous	+25 V max.
	−12 V max.
Transient	
T = 100 ms	± 60 V max.
T = 1.0 ms	± 100 V max.
Input Voltage	−0.3 to +7.0 V max.
Power Dissipation	2.0 W max.
Output Current	Internally Limited
Output Current Limit (V_o = 0 V)	1.6 A typ.
	2.5 A max.

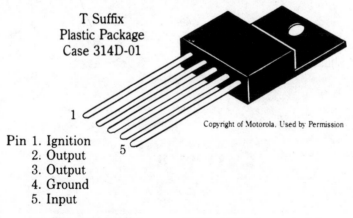

T Suffix
Plastic Package
Case 314D-01

1

5

Copyright of Motorola. Used by Permission

Pin 1. Ignition
2. Output
3. Output
4. Ground
5. Input

Heatsink surface connected to Pin 2.

MC3399T

7805 5-Volt voltage regulator

The 7805 is a popular and easy-to-use voltage-regulator IC. It has just three pins—input (1), output (2), and common or ground (3).

The 7805 is designed for a well-regulated output voltage of 5 volts. This voltage-regulator chip can handle currents over 1.5 ampere if adequate heatsinking is provided.

Input Voltage	+35 V max.
Output Voltage	+5 V
Power Dissipation	Internally Limited

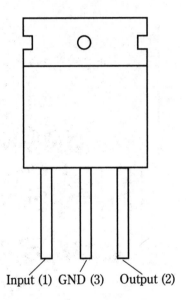

Input (1) GND (3)   Output (2)

7812 12-Volt voltage regulator

The 7812 is a popular and easy-to-use voltage-regulator IC. It has just three pins-input (1), output (2), and common or ground (3).

The 7812 is designed for a well-regulated output voltage of 12 volts. This voltage-regulator chip can handle currents over 1.5 ampere if adequate heatsinking is provided.

Input Voltage	35 V max.
Output Voltage	12 V
Power Dissipation	Internally Limited

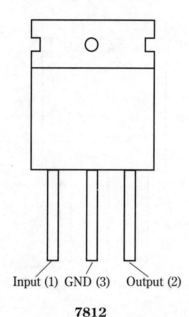

Input (1) GND (3)   Output (2)

7812

7815 15-V voltage regulator

The 7815 is a popular and easy-to-use voltage regulator IC. It has just three pins: input (pin 1), output (pin 2), and ground (pin 3).

The 7815 is designed for a well-regulated output of 15 volts. This voltage-regulator chip can handle currents over 1.5 ampere, if adequate heatsinking is provided.

Input Voltage	+35 V max.
Output Voltage	+15 V
Power Dissipation	Internally Limited

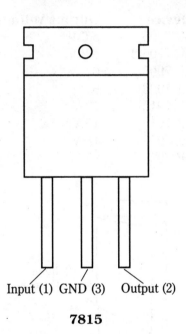

Input (1) GND (3)   Output (2)

7815

MC7900 Series negative fixed-voltage regulators

The 7900 series of fixed-voltage regulators are essentially negative-voltage versions of the popular 7800 series. These devices are simple to use, with three pins (input, output, and common). They are available in a variety of standard output voltages and can deliver output currents greater than 1.0 A.

The 7900-series voltage regulators feature internal short-circuit protection and thermal overload protection. They are so complete that no external components are required for basic operation.

The 7900 Series

Device	Output Voltage
MC7905	5.0 V
MC7905.2	5.2 V
MC7906	6.0 V
MC7908	8.0 V
MC7912	12 V
MC7915	15 V
MC7918	18 V
MC7924	24 V

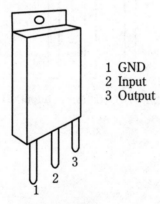

1 GND
2 Input
3 Output

Courtesy of Motorola, Inc.

7900

MC33128 Power management controller

The MC33128 power management controller is designed primarily for use in battery-powered cellular telephones and pagers. It provides three positive regulated outputs with low dropout voltage, and a negative regulated output for full depletion of GaAs MESFETs, as well as battery latch and low-battery shutdown.

This device also features the active functions needed for interfacing to the system electronics with a microprocessor. A power up reset function for the MPU is provided, as well as a pinned-out reference for the MPU's A/D converter.

Regulator Output 1: Pin 15

Output Voltage	**Min.**	**Typ.**	**Max.**
$V_{cc} = 3.15 - 4.5$ V	2.9 V	3.0 V	3.1 V
Output Current			150 mA

Regulator Output 2: Pin 1

Output Voltage	**Min.**	**Typ.**	**Max.**
$V_{cc} = 3.15 - 4.5$ V	2.9 V	3.0 V	3.1 V
Output Current			250 mA

Regulator Output 3: Pin 14

Output Voltage	**Min.**	**Typ.**	**Max.**
$V_{cc} = 3.15 - 4.5$ V	2.9 V	3.0 V	3.1 V
Output Current			50 mA

Regulator Output 4: Pin 5

Output Voltage	**Min.**	**Typ.**	**Max.**
$V_{cc} = 3.15 - 4.5$ V	–2.35 V	–2.5 V	–2.65 V
Output Current			10 mA

Reference Output: Pin 12

Output Voltage	**Min.**	**Typ.**	**Max.**
	1.46 V	1.5 V	1.54 V
Output Current			40 mA

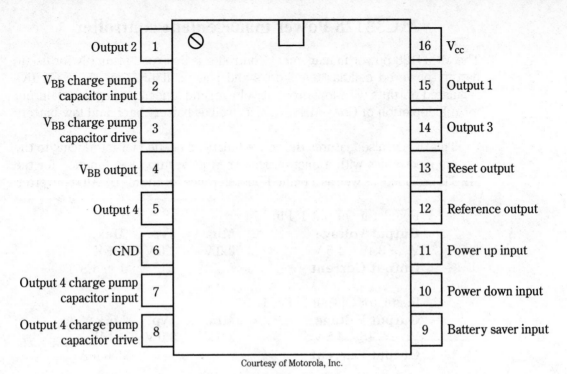

Output 2	1		16	V_{cc}
V_{BB} charge pump capacitor input	2		15	Output 1
V_{BB} charge pump capacitor drive	3		14	Output 3
V_{BB} output	4		13	Reset output
Output 4	5		12	Reference output
GND	6		11	Power up input
Output 4 charge pump capacitor input	7		10	Power down input
Output 4 charge pump capacitor drive	8		9	Battery saver input

Courtesy of Motorola, Inc.

MC33128

MC33023/MC34023
High-speed single-ended PWM controller

The MC33023/34023 is more sophisticated and more complex than most standard voltage-regulator ICs. It uses pulse-width modulation (PWM) controllers optimized for high-frequency operation. Instead of the standard three-terminal (or occasionally, 8-pin) package appropriate for most voltage regulators, the MC33023/34023 is a 16-pin device. It is specifically designed for off-line and dc-to-dc converter applications, rather than basic ac-to-dc voltage control. Such circuits are normally fairly complex, with a high parts count, but this chip can be used with a minimum of external components.

The MC33023 and the MC34023 are essentially the same, except that the MC33023 is designed to operate over a somewhat wider temperature range. The MC33023 is rated for –40 to +105 C, and the MC34023 is designed for operation at temperatures between 0 and 70 C.

Either chip can operate in either a current or voltage mode at frequencies up to 1.0 MHz. It is functionally similar to the UC3823. There is only 50 nS of propagation delay to the output. Other features include a wide-bandwidth-error amplifier, fully latched logic with double pulse suppression, latching PWM for cycle-by-cycle current limiting, and a precision trimmed on-chip oscillator.

The output-pulse duty cycle is externally adjustable up to a maximum of 90%. The MC33023 has a soft-start control with latched overcurrent reset, and the start-up current is quite low, typically about 500 μA.

MC33023/MC34023 Maximum Ratings

Power Supply Voltage	30 V
Output Driver Supply Voltage	20 V
Output Current, Source or Sink	
Dc	0.5 A
Pulsed (0.5 μS)	2.0 A
Output Short-Circuit Current	–100 mA
Oscillator Frequency Change	
With Voltage (V_{cc} = 10 to 30 V)	1.0%
With Temperature	2.0% (typical)

Output Voltage	Min.	Typ.	Max.
Low State			
Isink = 20 mA	–	0.25 V	0.4 V
Isink = 200 mA	–	1.2 V	2.2 V
High State			
Isink = 20 mA	13 V	13.5 V	–
Isink = 200 mA	12 V	13 V	–

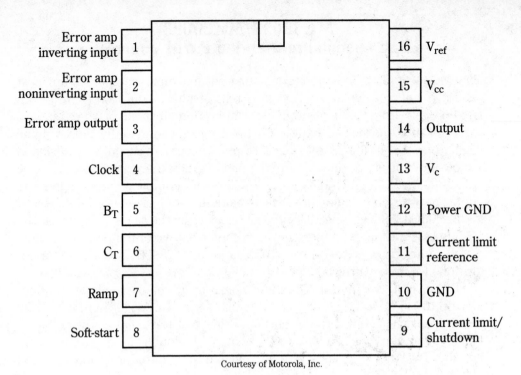

Error amp inverting input	1	16 V_{ref}
Error amp noninverting input	2	15 V_{cc}
Error amp output	3	14 Output
Clock	4	13 V_c
B_T	5	12 Power GND
C_T	6	11 Current limit reference
Ramp	7	10 GND
Soft-start	8	9 Current limit/shutdown

Courtesy of Motorola, Inc.

MC33023

MC33063A/MC34063A
Dc-to-dc converter control circuits

The MC33063A/MC34063A dc-to-dc converters can operate with input voltages that range from 3 to 40 volts. The MC33063 and the MC34063 are essentially the same, except that the MC33063 is designed to operate over a somewhat wider temperature range. The MC33063 is rated for –40 to +105 C, and the MC34063 is designed for operation at temperatures between 0 and 70 C. For convenience, from here on, only the MC334063A is referred to.

This chip was designed with step-up, step-down, and voltage-inverting applications in mind. Its output can switch up to 1.5 A. The output voltage is externally adjustable. It can operate at frequencies up to 100 kHz.

Internal features of the MC34063A include a comparator and a controlled duty-cycle oscillator with an active current-limit circuit, driver, and output switch. The MC34063A also includes an internal temperature-compensated precision 2% reference.

MC34063A Maximum Ratings

Power Supply Voltage	40 Vdc
Comparator Input Voltage Range	–0.3 to +40 Vdc
Switch Collector Voltage	!
Switch Emitter Voltage (Pin 1 = 40 V)	!
Switch Collector to Emitter Voltage	!
Driver Collector Voltage	!
Driver Collector Current	100 mA
Switch Current	1.5 A
Power Dissipation	
TA = 25 C	
Platic Package (P Suffix)	1.25 W
SOIC Package (D Suffix)	625 mW

Switch collector	1	8	Driver collector
Switch emitter	2	7	I_{PK} sense
Timing capacitor	3	6	V_{cc}
GND	4	5	Comparator inverting input

Courtesy of Motorola, Inc.

MC33063

MC33161/MC34161 Universal voltage monitor

The MC33161 and MC34161 are designed for use in a wide variety of voltage-sensing applications. These two devices are functionally identical, except that the 33161 is designed to operate over a wider temperature range than the MC34161. For convenience, from here on, only the MC34161 is referred to.

The MC34161 is versatile enough to sense either positive or negative voltages. The chip can be programmed in several operating modes. It can sense if the detected voltage is under the reference voltage, over the reference voltage, or within a specific user-determined window range of voltages. Channel programming is also possible via a unique mode-select input.

The MC34161 is fully functional at 2.0 V for positive-voltage sensing, and 4.0 V for negative-voltage sensing. It has a built-in 2.54 reference voltage, which is available at pin 1. Because this chip features open collector outputs, its flexibility is maximized.

MC34161 Absolute Maximum Ratings

Power Supply Input Voltage	40 V
Comparator Input Voltage Range	–1.0 to 40 V
Comparator Output Voltage	40 V
Comparator Output Sink Current	
(pins 5 and 6)	20 mA
Power Dissipation	
P Suffix Plastic Package	
Case 626	800 mW
D Suffix Plastic Package	
Case 751	450 mW

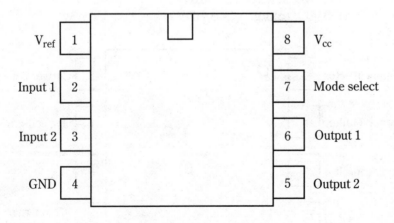

MC33161

MC33166/MC34166 Power switching regulator

The MC33166 and MC34166 are high-performance fixed-frequency switching regulators. These two devices are functionally identical, except that the 33166 is designed to operate over a wider temperature range than the MC34166. For convenience, from here on, only to the MC34166 is referred to.

The 34166 contains the primary functions required for a dc-to-dc converter. It is intended for use in step-down and voltage-inverting applications, and it generally requires a minimum number of external components. With just a little extra external circuitry, the MC34166 can also be used in step-up applications.

The output of the MC34166 can safely switch more than 3.0 A. If no external resistor voltage divider is used, the output voltage is 5.05 V. The MC34166 can be operated from supply voltages ranging from 7.5 V to 40 V. In the standby mode, the power-supply current is reduced to a mere 36 µA. The chip's fixed-frequency oscillator operates at 72 kHz with on-chip timing, and the duty cycle can be adjusted from 0% to 95%.

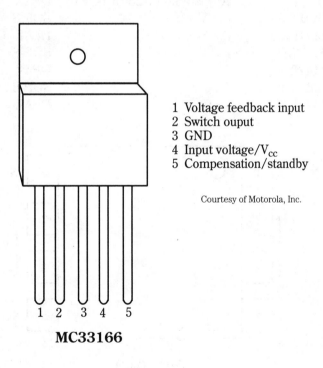

1 Voltage feedback input
2 Switch ouput
3 GND
4 Input voltage/V_{cc}
5 Compensation/standby

Courtesy of Motorola, Inc.

MC33166

MC33269 Low-dropout positive-voltage regulator

The MC3369 is a three-terminal voltage regulator, which is available in adjustable and several fixed voltage versions. It designed for medium-current applications, up to about 800 mA. The MC33269 features a 1.0-V PNP-NPN pass transistor, internal current limiting and thermal shutdown. It is not internally compensated, however. An external output capacitor is required for stability in most applications. This capacitor should be at least 10 µF, with an equivalent series resistance no greater than 10 Ω.

The fixed-output-voltage versions have no minimum-load requirement. The adjustable version has a minimum required load current of 8.0 mA. The input-output voltage for the adjustable versions of the MC33269 can range from 1.35 V to 10 V at 800 mA or 1.25 V to 15 V at 500 mA.

Device	Nominal Output Voltage	Package Style
MC33269D	Adjustable (1.25 to 15 V)	SOP-8
MC33269DT	Adjustable (1.25 to 15 V)	DPAK
MC33269D-3.3	3.3 V	SOP-8
MC33269DT-3.3	3.3 V	DPAK
MC33269D-5.0	5.0 V	SOP-8
MC33269DT-5.0	5.0 V	DPAK
MC33269D-12	12.0 V	SOP-8
MC33269DT-12	12.0 V	DPAK

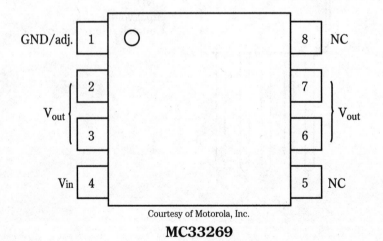

Courtesy of Motorola, Inc.

MC33269

Other linear devices

The last few sections have examined various common types of linear ICs, but there are many other, less common linear functions, as well. This section contains a miscellaneous selection of linear devices that don't quite fit into any of the other section headings, including, but not limited to, voltage comparators, multipliers, and arrays.

Voltage Comparator

bandwidth That frequency at which the voltage gain is reduced to $1/\sqrt{2}$ times the low-frequency value.

common-mode rejection ratio The ratio of the input common-mode voltage range to the peak-to-peak change in input offset voltage over this range.

harmonic distortion The percentage of harmonic distortion being defined as one-hundred times the ratio of the root-mean-square (rms) sum of the harmonics to the fundamental. % harmonic distortion =

$$\frac{(V_2^2 + V_3^2 + V_4^2 + \ldots)^{1/2}}{V_1}(100\%)$$

where V1 is the rms amplitude of the fundamental and $V_2, V_3, V_4, \ldots$ are the rms amplitudes of the individual harmonics.

input bias current The average of the two input currents.

input common-mode voltage range The range of voltages on the input terminals for which the amplifier is operational. Note that the specifications are not guaranteed over the full common-mode voltage range unless specifically stated.

input impedance The ratio of input voltage to input current under the stated conditions for source resistance (R_S) and load resistance (R_L).

input offset current The difference in the currents into the two input terminals when the output is at zero.

input offset voltage That voltage which must be applied between the input terminals through two equal resistances to obtain zero output voltage.

input resistance The ratio of the change in input voltage to the change in input current on either input with the other grounded.

input voltage range The range of voltages on the input terminals for which the amplifier operates within specifications.

large-signal voltage gain The ratio of the output voltage swing to the change in input voltage required to drive the output from zero to this voltage.

output impedance The ratio of output voltage to output current under the stated conditions for source resistance (R_S) and load resistance (R_L).

output resistance The small signal resistance seen at the output when the output voltage is near zero.

output voltage swing The peak output voltage swing, referred to zero, that can be obtained without clipping.

offset voltage temperature drift The average drift rate of offset voltage for a thermal variation from room temperature to the indicated temperature extreme.

power supply rejection The ratio of the change between input offset voltage to the change in power-supply voltages that produced it.

setting time The time between the initiation of the input-step function and the time when the output voltage has settled to within a specified error band of the final output voltage.

slew rate The internally limited rate of change in output voltage with a large amplitude step function applied to the input.

supply current The current required from the power supply to operate the amplifier with no load and the output midway between the supply voltages.

transient response The closed-loop step function response of the amplifier under small-signal conditions.

unity-gain bandwidth The frequency range from dc to the frequency where the amplifier open-loop gain rolls off to one.

voltage gain The ratio of output voltage to input voltage under the stated conditions for source resistance (R_s) and load resistance (R_L).

LM139/LM139A Low-power low-offset quad voltage comparator

The LM139 series consists of four independent precision voltage comparators, with an offset voltage specification as low as 2 mV maximum for all four comparator stages.

This device was designed specifically to operate from a single-ended power supply over a wide range of voltages. Operation from split power supplies is also possible, and is preferred for the A-suffix versions. The low power-supply current drain of the 139 is independent of the magnitude of the power-supply voltage. These comparators also have a unique characteristic in that the input common-mode voltage range includes ground—even when the chip is operated from a single-ended supply voltage.

Application areas include limit comparators, simple analog-to-digital converters, pulse generators, square-wave generators, time-delay generators, wide-range VCOs, window comparators, MOS clock timers, multivibrators, and high-voltage digital logic gates. The 139 series is designed to directly interface with TTL and CMOS circuitry. When operated from a split power supply, this device will directly interface with MOS logic.

Supply Voltage Range	2 to 36 Vdc
	or ±1 to ±18 Vdc
Supply Current Drain	0.8 mA
Input Biasing Current	25 nA
Input Offset Current	±5 nA
Offset Voltage	±3 mV
Output Saturation Voltage	250 mV at 4 mA

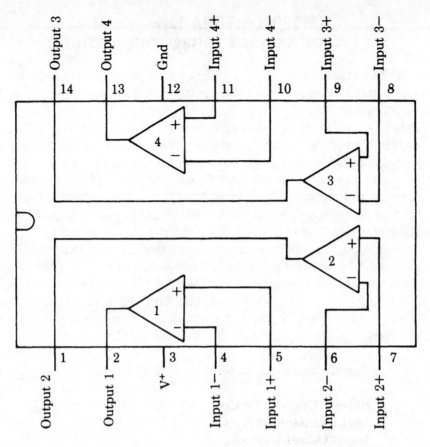

Top View

LM139

LM239/LM239A Low-power low-offset quad voltage comparator

The LM239 series consists of four independent precision voltage comparators, with an offset voltage specification as low as 2 mV maximum for all four comparator stages.

This device was designed specifically to operate from a single-ended power supply over a wide range of voltages. Operation from split power supplies is also possible, and is preferred for the A-suffix versions. The low power-supply current drain of the 239 is independent of the magnitude of the power-supply voltage. These comparators also have a unique characteristic in that the input common-mode voltage range includes ground—even when the chip is operated from a single-ended supply voltage.

Application areas include limit comparators, simple analog-to-digital converters, pulse generators, square-wave generators, time-delay generators, wide-range VCOs, window comparators, MOS clock timers, multivibrators, and high-voltage digital logic gates. The 239 series is designed to directly interface with TTL and CMOS circuitry. When operated from a split power supply, this device will directly interface with MOS logic.

Supply Voltage Range	2 to 36 Vdc
	or ±1 to ±18 Vdc
Supply Current Drain	0.8 mA
Input Biasing Current	25 nA
Input Offset Current	±5 nA
Offset Voltage	±3 mV
Output Saturation Voltage	250 mV at 4 mA

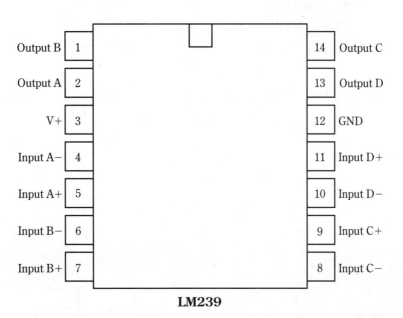

LM239

LM339/LM339A Low-power
low-offset quad voltage comparator

The LM339 series consists of four independent precision voltage comparators, with an offset voltage specification as low as 2 mV maximum for all four comparator stages.

This device was designed specifically to operate from a single-ended power supply over a wide range of voltages. Operation from split power supplies is also possible, and is preferred for the A-suffix versions. The low power-supply current drain of the 339 is independent of the magnitude of the power-supply voltage. These comparators also have a unique characteristic in that the input common-mode voltage range includes ground—even when the chip is operated from a single-ended supply voltage.

Application areas include limit comparators, simple analog-to-digital converters, pulse generators, square-wave generators, time-delay generators, wide-range VCOs, window comparators, MOS clock timers, multivibrators, and high-voltage digital logic gates. The 339 series is designed to directly interface with TTL and CMOS circuitry. When operated from a split power supply, this device will directly interface with MOS logic, where the low power drain of the LM339 is a distinct advantage over most standard voltage-comparator devices.

Supply Voltage Range	2 to 36 Vdc
	or ±1 to ±18 Vdc
Supply Current Drain	0.8 mA
Input Biasing Current	25 nA
Input Offset Current	±5 nA
Offset Voltage	±3 mV
Output Saturation Voltage	250 mV at 4 mA

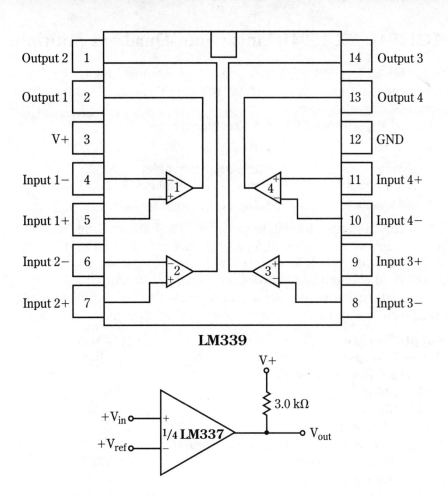

LM339

MC1494L MC1594L Linear Four-Quadrant Multiplier

The MC1494L and MC1594L are linear four-quadrant multiplier ICs—the output voltage is the linear product of two input voltages. This chip offers excellent linearity in its operation.

Typical applications for this device include:

- Multiply
- Divide
- Square root
- Mean square

- Phase detector
- Frequency doubler
- Balanced modulator/demodulator
- Electronic Gain Control

In technical terms, the MC1494 and MC1594 are variable transconductance multipliers with internal level-shift circuitry and an on-chip voltage regulator. A pair of complementary-regulated voltages are provided to simplify offset adjustment and improve power-supply rejection.

Four external potentiometers can be added to the circuit to allow full manual adjustment over scale factor, input offset, and output offset.

Supply Voltage	±18 V Max.
Power Dissipation	750 mW Max.
Maximum Error (X or Y)	
MC1494L	±1.0% Max.
MC1594L	±0.5% Max.
Input Voltage Range	±10 V
Frequency Response (3-dB Small Signal)	1.0 MHz
Power Supply Sensitivity	30 mV/V typ.
Power Supply Current	
MC1494L	±6.0 mA typ.
	±12 mA max.
	+6.0 mA, –6.5 mA typ.
MC1494L	±9.0 mA max.

MC1594L

16

1

L Suffix
Ceramic Package
Case 620-10

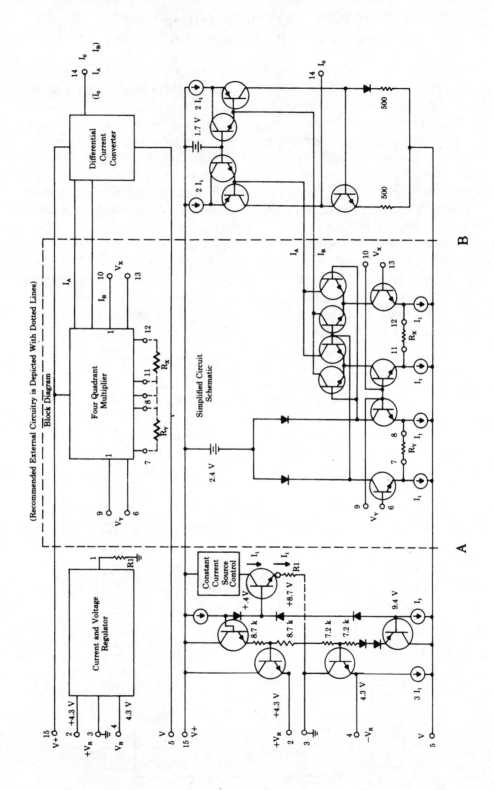

LM2901/LM2901A Low-power low-offset quad voltage comparator

The LM2901 series consists of four independent precision voltage comparators, with an offset voltage specification as low as 2 mV maximum for all four comparator stages.

This device was designed specifically to operate from a single-ended power supply over a wide range of voltages. Operation from split power supplies is also possible, and is preferred for the A-suffix versions. The low power-supply current drain of the 2901 is independent of the magnitude of the power-supply voltage. These comparators also have a unique characteristic in that the input common-mode voltage range includes ground—even when the chip is operated from a single-ended supply voltage.

Application areas include limit comparators, simple analog-to-digital converters, pulse generators, square-wave generators, time-delay generators, wide-range VCOs, window comparators, MOS clock timers, multivibrators, and high-voltage digital logic gates. The 2901 series is designed to directly interface with TTL and CMOS circuitry. When operated from a split power supply, this device will directly interface with MOS logic.

Supply Voltage Range	±1 to ±18 Vdc
Supply Current Drain	0.8 mA
Input Biasing Current	25 nA
Input Offset Current	±5 nA
Offset Voltage	±3 mV
Output Saturation Voltage	250 mV at 4 mA

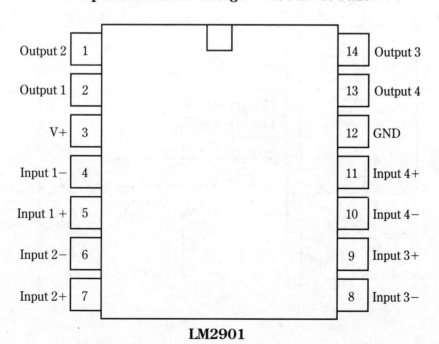

LM2901

CA3000 Dc amplifier

The CA3000 is a general-purpose amplifier used in Schmitt trigger, RC-coupled feedback amplifier, mixer, comparator, crystal oscillator, sense amplifier, and modulator applications. It comes in a 10-lead TO-5 package.

Max. Positive Dc Supply Voltage	$+10$ V
Max. Negative Dc Supply Voltage	-10 V
Max. Input Signal Voltage:	
Single-ended	± 2 V
Common mode	± 2 V
Max. Total Device Dissipation	300 mW
Typ. Input Offset Voltage	1.4 mV
Typ. Input Offset Current	1.2 μA
Typ. Input Bias Current	23 μA

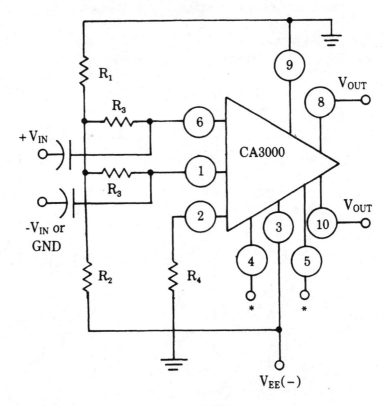

CA3000

CA3054 General Purpose Transistor Array

As integrated circuits go, the CA3054 is a pretty simple device. It basically contains six NPN transistors arranged for a number of general-purpose applications.

The six transistors on this chip are divided into two groups of three each. Each three-transistor combination is a simple differential amplifier with a constant-current transistor. The two differential amplifiers are electrically independent of one another, except for the chip's common-ground connection.

Each differential-amplifier stage can handle signals from dc (0 Hz) up to 120 MHz.

Collector-Emitter Voltage	15 V max.
Collector-Base Voltage	20 V max.
Emitter-Base Voltage	5.0 V max.
Collector-Substrate Voltage	20 V max.
Collector Current	
Continuous	50 mA max.
Input Offset Voltage	± 5 mV max.

Pin Connections

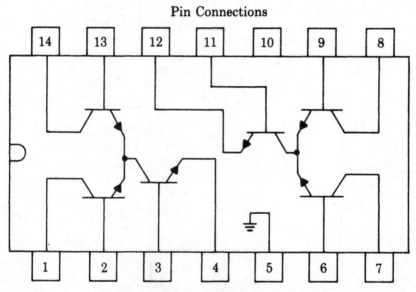

Pin 5 is connected to substrate
and must remain at the lowest circuit potential

CA3054

LM3302/LM3302A Low-power low-offset quad voltage comparator

The LM3302 series consists of four independent precision voltage comparators, with an offset voltage specification as low as 2 mV maximum for all four comparator stages.

This device was designed specifically to operate from a single-ended power supply over a wide range of voltages. Operation from split power supplies is also possible, and is preferred for the A-suffix versions. The low power-supply current drain of the 3302 is independent of the magnitude of the power-supply voltage. These comparators also have a unique characteristic in that the input common-mode voltage range includes ground—even when the chip is operated from a single-ended supply voltage.

Application areas include limit comparators, simple analog-to-digital converters, pulse generators, square-wave generators, time-delay generators, wide-range VCOs, window comparators, MOS clock timers, multivibrators, and high-voltage digital logic gates. The 3302 series is designed to directly interface with TTL and CMOS circuitry. When operated from a split power supply, this device will directly interface with MOS logic.

Supply Voltage Range	2 to 28 Vdc
	or ±1 to ±14 Vdc
Supply Current Drain	0.8 mA
Input Biasing Current	25 nA
Input Offset Current	±5 nA
Offset Voltage	±3 mV
Output Saturation Voltage	250 mV at 4 mA

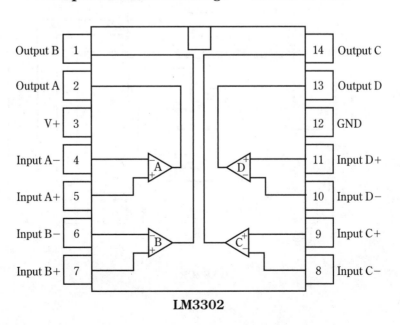

LM3302

BA6209 Reversible motor driver

The BA6209 is an IC designed primarily for use in VCRs to control reversible motors, such as the loading motor, the capstan motor, and the reel motor.

This chip features an internal surge suppressor, making it capable of withstanding brief bursts of current up to 1.6 A. Such current surges can frequently occur when a motor is stopping or changing direction.

The BA6209 automatically brakes the motor when the control inputs are both set HIGH or LOW. The control pin (V_r, pin 4) accepts a control voltage, adjusting the motor speed over a wide range.

This chip is available in a 10 pin-SIP (single-inline package).

Supply Voltage
V_{cc1}	18 V max.
	6.0 V min.
V_{cc2}	18 V max.
Power Dissipation	2200 mW 2.2 W max.
Output Voltage	7.2 V typ.
	6.6 V min.
Output Current	1.6 A max.

Block Diagram

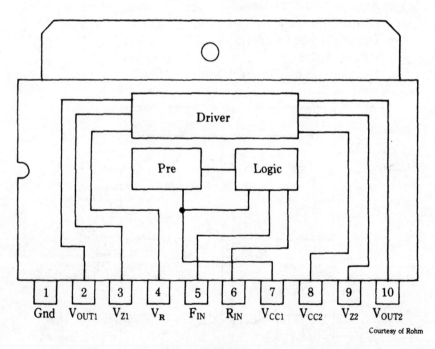

Courtesy of Rohm

BA6209

10
SECTION

Digital/analog hybrids

Ordinarily, digital and linear (or analog) circuitry are two separate things. A digital circuitry can recognize only two voltage levels: LOW and HIGH, but an analog circuit responds to a continuous range of intermediate voltage levels. In some applications, however, it might be desirable to combine digital and analog functions in a single circuit or system.

Some circuits, by their nature can function either way. For example, the output of a voltage comparator can be used as input to a digital gate. Multivibrators, which produce square wave outputs, can also be used in either linear or digital applications. This section looks at ICs that use digital means to perform what are normally analog functions. It also covers digital-to-analog (D/A) converters, which transform a multi-bit digital value to a proportional linear signal. An analog-to-digital (A/D) converter works in the exact opposite manner.

MC1408/MC1508 Eight-bit multiplying D/A converter

The MC1408/MC1508 is a digital-to-analog converter that accepts an eight-bit digital word at its inputs, and converts it into a proportional linear output current. The unusual feature of this device is that it also has an additional analog input. The voltage applied to this input serves as a constant multiplier. The output current of the MC1408/MC1508 is the linear product of the eight-bit digital input and the analog input voltage. This chip is very fast, and very accurate in its operation. Its noninverting digital inputs are both TTL and CMOS compatible.

Supply Voltage
V_{cc} +4.5 to +5.5 V
V_{ee} −5.0 to −15 V
Output voltage swing   +0.4 to −5.0 V
Settling time   300 nS typ.
Multiplying input slew rate

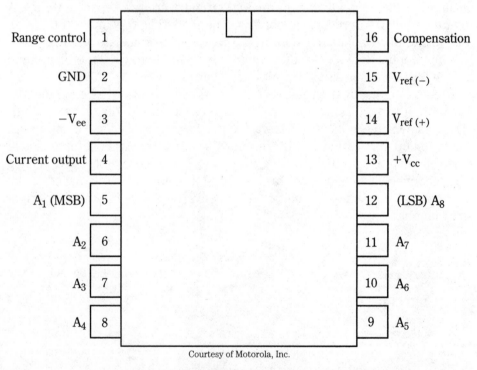

Range control — 1	16 — Compensation
GND — 2	15 — $V_{ref\,(-)}$
$-V_{ee}$ — 3	14 — $V_{ref\,(+)}$
Current output — 4	13 — $+V_{cc}$
A_1 (MSB) — 5	12 — (LSB) A_8
A_2 — 6	11 — A_7
A_3 — 7	10 — A_6
A_4 — 8	9 — A_5

Courtesy of Motorola, Inc.

MC1408/MC1508

4016 Quad bilateral switch

The 4016 is a quad bilateral switch with an extremely high OFF-resistance and low ON-resistance. The switch will pass analog or digital signals in either direction, and is extremely useful in digital switching. In effect, it is the equivalent of an ordinary mechanical switch, except that it is operated by logic signals on the appropriate control input pin, rather than physical motion or pressure.

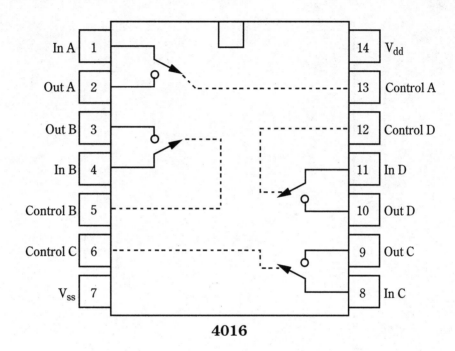

4016

4046 Micropower phase-locked loop

The 4046 micropower phase-locked loop (PLL) consists of a low-power, linear voltage-controlled oscillator (VCO), a source follower, a zener diode, and two phase comparators. The two phase comparators have a common signal input and a common comparator input. The signal input can be directly coupled for a large-voltage signal, or capacitively coupled to the self-biasing amplifier at the signal input for a small-voltage signal.

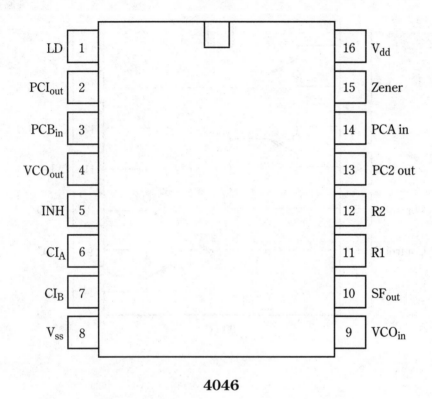

LD	1		16	V_{dd}
PCI_{out}	2		15	Zener
PCB_{in}	3		14	PCA in
VCO_{out}	4		13	PC2 out
INH	5		12	R2
CI_A	6		11	R1
CI_B	7		10	SF_{out}
V_{ss}	8		9	VCO_{in}

4046

4053 Triple two-channel analog multiplexer/demultiplexer

The 4053 is a triple two-channel multiplexer/demultiplexer having three separate digital control inputs (A, B, and C), and an inhibit input. Each control input selects one of a pair of channels that are connected in a single-pole, double-throw (SPDT) configuration.

The 4053 analog multiplexers/demultiplexers are digitally controlled analog switches, having low ON-impedance, and very low OFF-leakage currents. Three two-channel units are included on this chip.

Control of analog signals up to 15 volts (peak-to-peak) can be achieved by digital signal amplitudes of 3 to 15 V. For example, if $V_{dd} = 5$ V, $V_{ss} = 0$ V, and $V_{ee} = 5$ V, analog signals from –5 to +5 V can be controlled by digital inputs of 0 to 5 V.

The multiplexer circuits dissipate extremely low quiescent power over the full V_{dd}-to-V_{ss} and V_{dd}-to-V_{ee} supply voltage ranges, independent of the logic state of the control signals. When the inhibit input terminal is made HIGH, all channels are turned off. The 4052 is similar, except it contains two analog multiplexer/demultiplexers of four channels each.

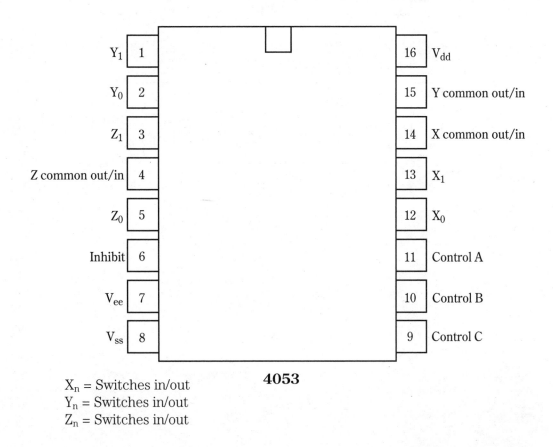

4053

X_n = Switches in/out
Y_n = Switches in/out
Z_n = Switches in/out

4066 Quad bilateral switch

The 4066 is a quad bilateral switch with an extremely high OFF resistance and low ON resistance. The switch will pass analog or digital signals in either direction, and it is extremely useful in digital switching. In effect, it is the equivalent of an ordinary mechanical switch, except it is operated by logic signals on the appropriate control input pin, rather than physical motion or pressure.

The 4066 is pin-for-pin compatible with the 4016, but it has a much lower ON resistance, and its ON resistance is relatively constant over the input-signal range. Essentially, the 4066 is an improved version of the 4016.

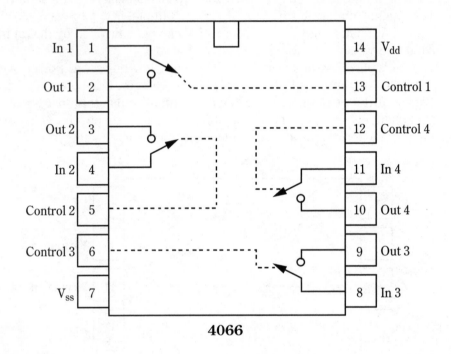

4066

4067 16-Channel analog multiplexer/demultiplexer

The 4067 consists of 16 digitally controlled analog switches, which can be used in either analog or digital applications. These switches have very low ON resistance and low OFF leakage current.

The 4067 has four binary control inputs (A, B, C, and D), which select one-of-16 channels by turning on the appropriate switch. Only one switch is ON (closed) at any given time. In effect, this device permits the selection of any of 16 separate input signals. It also has an inhibit input, which turns off all 16 switches so that no signal is output.

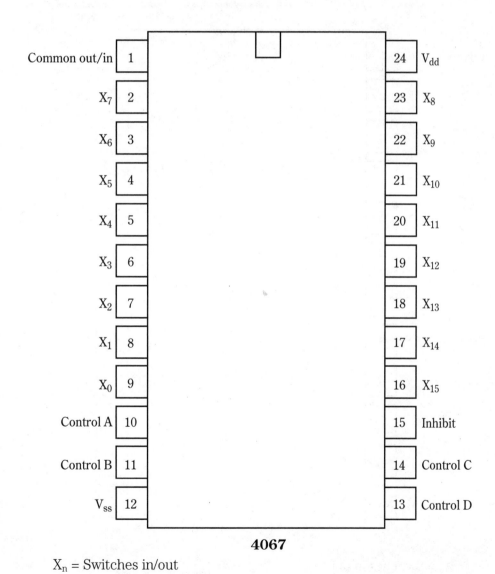

4067

X_n = Switches in/out

4097 8-Channel analog multiplexer/demultiplexer

The 4097 consists of eight digitally controlled analog switches, which can be used in either analog or digital applications. These switches have very low ON resistance and low OFF leakage current.

The 4097 has three binary control inputs (A, B, and C), which select one-of-eight channels by turning on the appropriate switch. Only one switch is ON (closed) at any given time. In effect, this device permits the selection of any of eight separate input signals. There is also an inhibit input, which turns off all eight switches so that no signal is output.

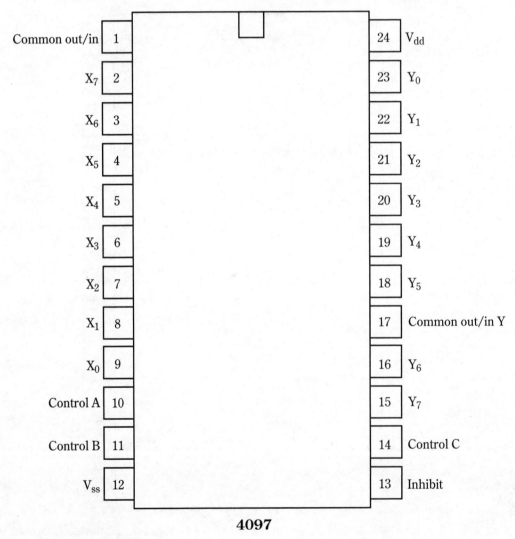

4097

X_n = Switches in/out
Y_n = Switches in/out

4415 Quad precision timer/driver

The 4415 consists of four very precise timer/driver stages. Each stage's output pulse has a width determined by the frequency of the input clock signal. Once the output buffer is set (turned on) by the appropriate input sequence, it will remain on for 100 clock pulses, then it will be reset (turned off). The inputs to this device can be set up to respond to either LOW-to-HIGH or HIGH-to-LOW transitions to suit the specific intended application.

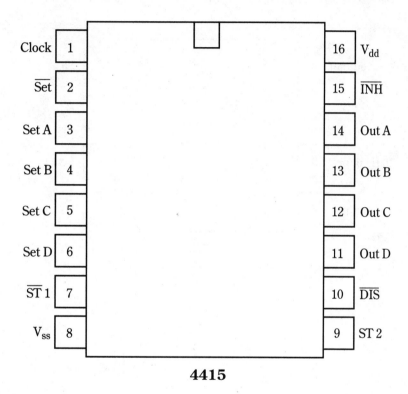

Clock	1		16	V_{dd}
$\overline{Set}$	2		15	$\overline{INH}$
Set A	3		14	Out A
Set B	4		13	Out B
Set C	5		12	Out C
Set D	6		11	Out D
$\overline{ST\,1}$	7		10	$\overline{DIS}$
V_{ss}	8		9	ST 2

4415

MM5368 CMOS oscillator divider

The MM5368 accepts a square-wave input signal from an oscillator (or other signal source), and divides the signal frequency to three much lower output frequencies. The input signal at pin 6 should be a floating 16-kHz square wave. The same signal is buffered by the MM5368 and can be tapped off at pin 5. A 10-kHz signal is available at pin 3, and pin 4 offers a 1-kHz signal. The signal frequency at pin 1 can be either 50 Hz or 60 Hz, depending on the logic state at the select input (pin 7).

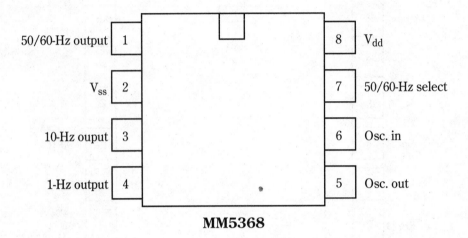

50/60-Hz output	1		8	V_{dd}
V_{ss}	2		7	50/60-Hz select
10-Hz ouput	3		6	Osc. in
1-Hz output	4		5	Osc. out

MM5368

7555 CMOS timer

The 7555 is a CMOS version of the popular 555 timer IC, which is described in section 8 of this book.

The advantages of the CMOS version include a wider range of acceptable supply voltages and lower power consumption. The CMOS 7555 is more stable than the standard 555, permitting longer timing cycles.

Like the ordinary 555 timer, the 7555 CMOS timer can be used in both monostable-multivibrator and astable-multivibrator applications.

Supply Voltage   2 to 18 V

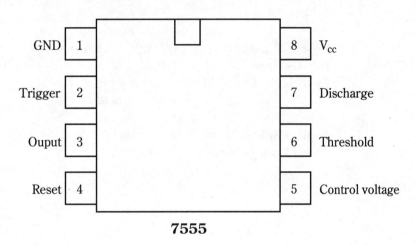

7555

BA9201/BA9201F 8-Bit D/A converter with latch

The BA9201 and BA9201F are 8-bit digital-to-analog (D/A) converter ICs, complete with internal-reference voltage-source and input-data latch circuitry. An external reference voltage can be substituted for the internal reference voltage source, if desired in a specific application.

The input data latch permits the use of multiple D/A converters in appropriate systems.

Supply Voltage	
Pin 18	+ 6 V max.
	+4.5 V min.
Pin 1	–8.5 V max.
	–6.3 V min.
Power Dissipation	500 mW max.
Resolution	8 bits
Full-Scale Current	1.992 mA typ.
	2.10 mA max.
	1.90 mA min.
Input High Voltage	2.3 V min.
Input Low Voltage	0.8 V min.

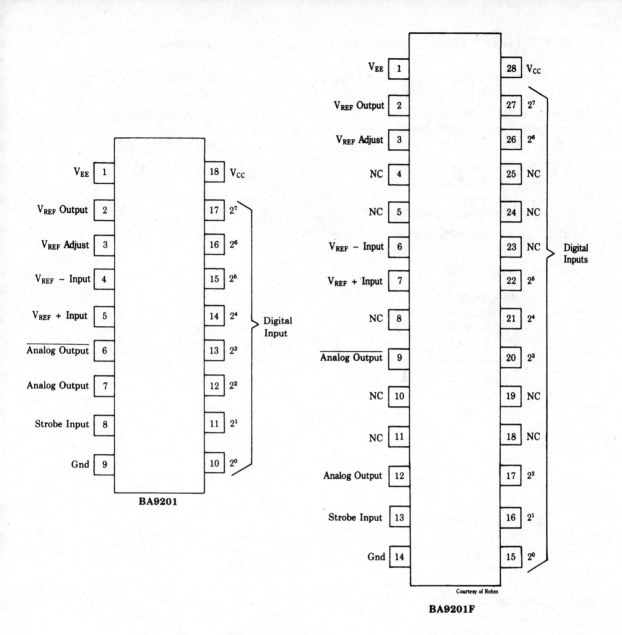

BA9201

BA9201F

Courtesy of Rohm

MC10319 High-speed 8-bit A/D flash converter

The 10319 is a high-speed parallel flash analog-to-digital converter with an eight-bit output. It can easily be interconnected for nine-bit conversion, too. An internal Grey Code structure is utilized within this chip to eliminate large output errors on rapidly changing input signals.

The 10319 is fully TTL compatible, and uses tri-state logic. A +5.0-volt power supply is required, along with a more flexible negative supply voltage, which can range from –3 to –6 volts.

The 10319 contains 256 parallel comparators across a precision-input reference network. The input signal is sampled at a rate of 25 MHz. The comparator outputs are latched, then fed to an encoder network to produce an eight-bit digital output, plus an overrange bit.

Supply Voltage	
V_{cc} **Pin #15**	+7.0 V absolute max.
V_{cc}**(D) Pin #11 and 17**	+7.0 V absolute max.
V_{ee} **Pin #13**	–7.0 V absolute max.
Recommended Supply Voltages	
V_{cc}**(A)**	+5.5 V max.
	+5.0 V typ.
	+4.5 V min.
V_{cc}**(D)**	+5.5 V max.
	+5.0 V typ.
	+4.5 V min.
V_{ee}	–6.0 V max.
	–5.0 V typ.
	–3.0 V min.
Positive Supply-Voltage Differential	
V_{cc}**(D)**—V_{cc}**(A)**	–0.3 to +0.3 V
Power Dissipation	618 mW max.
Clock Frequency	25 MHz max.
	0 MHz min.
Resolution	8 Bits max.
Accuracy	9 Bits typ.
Analog Input Voltage	+2.5 V max.
	–2.5 V min.
Input Capacitance	50 pF

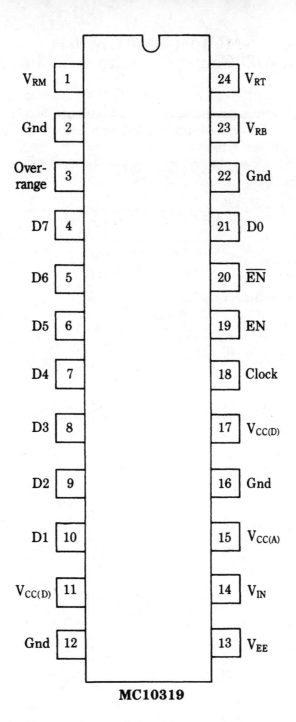

MC10319

MC145040/MC145041
8-Bit A/D Converters with serial interface

The MC145040 and MC145041 perform fast and accurate analog-to-digital conversions. The output is in the form of an 8-bit digital word.

A single-ended power supply is used with both chips, and no external trimming circuitry is required.

With the MC145040, an external clock input (A/D CLK) is used to operate the dynamic A/D conversion sequence.

The MC145041 features an internal clock, and the clock input pin is replaced with an end-of-conversion (EOF) output signal.

Supply Voltage   4.5 to 5.5 V
Power Consumption   11 mW

Successive Approximation Conversion Time
  MC145040   10 μs (with 2 MHz A/D CLK)
  MC145041   20 μs max. (Internal Clock)

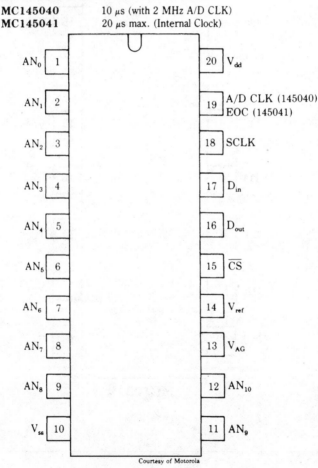

Courtesy of Motorola

MC145040/MC145041

A
APPENDIX

Symbols and definitions

Dc voltages

All voltages are referenced to ground. Negative voltages limits are specified as absolute values (i.e., 10 V is greater than –1.0 V).

Vcc Supply voltage   The range of power supply voltage over which the device is guaranteed to operate within the specified limits.

Vcd (Max) Input clamp diode voltage   The most negative voltage at an input when the specified current is forced out of that input terminal. This parameter guaranteed the integrity of the input diode intended to clamp negative ringing at the input terminal.

Vih Input HIGH voltage   The range of input voltages recognized by the device as a logic HIGH.

Vih (Min) Minimum input HIGH voltage   This value is the guaranteed input HIGH threshold for the device. The minimum allowed input HIGH in a logic system.

Vil Input LOW voltage   The range of input voltages recognized by the device as a logic LOW.

Vil (Max) Maximum Input LOW voltage   This value is the guaranteed input LOW threshold for the device. The maximum allowed input LOW in a logic system.

Vm Measurement voltage   The reference voltage level on ac waveforms for determining ac performance. Usually specified as 1.5 V for most TTL families, but it is 1.3 V for the low-power Schottky 74LS family.

Voh (Min) Output HIGH voltage The minimum guaranteed HIGH voltage at an output terminal for the specified output current (*Ioh*) and at the minimum *Vcc* value.

Vol (Max) Output LOW voltage The minimum guaranteed LOW voltage at an output terminal sinking the specified load current, *Iol*.

Vt+ Positive-going threshold voltage The input voltage of a variable threshold device, which causes operation according to the specification as the input transition rises from below *Vt-* (Min).

V_t- Negative-going threshold voltage The input voltage of a variable threshold device which causes operation according to specification as the input transition falls from above VT+ (Max).

DC Currents

Positive current is defined as conventional current flow into a device. Negative current is defined as conventional current flow out of a device. All current limits are specified as absolute values.

I_{CC} Supply current The current flowing into the VCC supply terminal of the circuit with specified input conditions and open outputs. Input conditions are chosen to guarantee worst case operation unless specified.

I_I Input leakage current The current flowing into an input when the maximum allowed voltage is applied to the input. This parameter guaranteed the minimum breakdown voltage for the input.

I_{IH} Input high current The current flowing into an input when a specified HIGH-level voltage is applied to that input.

I_{IL} Input low current The current flowing out of an input when a specified LOW-level voltage is applied to that input.

I_{OH} Output high current The leakage current flowing into a turned OFF open collector output with a specified HIGH-output voltage applied. For device with a pull-up circuit, the I_{OH} is the current flowing out of an output which is in the HIGH state.

I_{OL} Output LOW current The current flowing into an output which is in the LOW state.

I_{OS} Output short-circuit current The current flowing out of an output which is in the HIGH state when the output is short circuit to ground.

I_{OZH} Output OFF current HIGH The current flowing into a disabled 3-state output with a specified HIGH output voltage applied.

I_{OZL} Output OFF current LOW The current flowing out of a disabled 3-state output with a specified LOW output voltage applied.

Ac Switching Parameters and Definitions

f_{MAX} The maximum clock frequency The maximum input frequency at a clock input for predictable performance. Above this frequency the device may cease to function.

t_{PLH} Propagation delay time The time between the specified reference points on the input and output waveforms with the output changing from the defined LOW level to the defined HIGH level.

t_{PHL} Propagation delay time The time between the specified reference points on the input and output waveforms with the output changing from the defined HIGH level to the defined LOW level.

t_{PHZ} Output disable time from HIGH level of a 3-state output The delay time between the specified reference points on the input- and output-voltage waveforms with the 3-state output changing from the HIGH level to a HIGH impedance OFF state.

t_{PLZ} Output disable time from LOW level of a 3-state output The delay time between the specified reference points on the input and output voltage waveforms with the 3-state output changing from the LOW level to a HIGH impedance OFF state.

t_{PZH} Output enable time to a HIGH level of a 3-state output The delay time between the specified reference points on the input and output voltage waveforms with the 3-state output changing from a HIGH-impedance OFF state to the HIGH level.

t_{PZL} Output enable time to a LOW level of a 3-state output The delay time between the specified reference points on the input and output voltage waveforms with the 3-state output changing from a HIGH-impedance OFF state to the LOW level.

t_h Hold time The interval immediately following the active transition of the timing pulse (usually the clock pulse) or following the transition of the control input to its latching level, during which interval the data to be recognized must be maintained at the input to ensure its continued recognition. A negative hold time indicates that the correct logic level may be released prior to the active transition of the timing pulse and still be recognized.

t_s Setup time The interval immediately preceding the active transition of the timing pulse (usually the clock pulse) or preceding the transition of the control input to its latching level, during which interval the data to be recognized must be maintained at the input to ensure its recognition. A negative setup time indicates that the correct logic level may be initiated sometime after the active transition of the timing pulse and still be recognized.

t_w Pulse width The time between the specified reference points on the leading and trailing edges of a pulse.

t_{rec} Recovery time The time between the reference point on the trailing edge of an asynchronous input control pulse and the reference point on the activating edge of a synchronous (clock) pulse input such that the device will respond to the synchronous input.

B
APPENDIX

List of device numbers

Section	Device #	Function
4	LM11	Operational amplifier
4	TL062	Dual low-power JFET input operational amplifier
8	LM117	3-Terminal adjustable regulator
8	LM120	3-Terminal negative regulator
6	LM122	Precision timer
8	LM123	3-Amp 5-volt positive voltage regulator
4	LM124	Low-power quad operational amplifier
9	LM139	Low-power low-offset quad voltage comparator
9	LM139A	Low-power low-offset quad voltage comparator
8	LM217	3-Terminal adjustable regulator
8	LM220	3-Terminal negative regulator
6	BA222	Monolithic timer
6	LM222	Precision timer
8	LM223	3-Amp 5-Volt positive voltage regulator
4	LM224	Low-power quad operational amplifier
9	LM239	Low-power low-offset quad voltage comparator
9	LM239A	Low-power low-offset quad voltage comparator
8	LM285	Micropower voltage reference diode
4	LM307	Internally compensated monolithic operational amplifier
4	LM308A	Super-gain precision operational amplifier
8	LM317	3-Terminal adjustable regulator
8	LM320L Series	3-Terminal negative regulator

6	LM322	Precision timer
8	LM323	3-Amp, 5-volt positive voltage regulator
4	LM324	Low-power quad operational amplifier
8	LM337	Three-terminal adjustable negative voltage regulator
9	LM339	Low-power low-offset quad voltage comparator
9	LM339A	Low-power low-offset quad voltage comparator
8	LM340 Series	Voltage regulator
8	LM350	Three-terminal adjustable positive voltage regulator
4	353	Wide-bandwidth dual JFET operational amplifier
8	LM377	Three-terminal adjustable negative voltage regulator
5	LM380	Audio power amplifier
5	LM384	5-Watt audio power amplifier
8	LM385	Micropower voltage reference diode
5	LM386	Low-voltage audio power amplifier
5	LM387	Low-noise dual preamplifier
5	LM389	Low-voltage audio power amplifier with NPN transistor array
5	LM390	1-Watt battery-operated audio power amplifier
5	LM391	Audio power driver
8	TL431	Programmable precision reference
6	555	Timer
6	556	Dual timer
6	558	Quad timer
6	LM565	Phase-locked loop
6	LM566	Voltage-controlled oscillator
6	LM567	Tone decoder
6	LM568	Low-power phase-locked loop
8	BA612	Large current driver
4	LM709	Operational amplifier
4	BA718	Dual operational amplifier
8	LM723	Voltage regulator
8	LM723C	Voltage regulator
4	BA728	Dual operational amplifier
4	LM741	Operational amplifier
4	LM747	Dual operational amplifier
4	LM748	Operational amplifier
8	TL780 Series	Three-terminal positive fixed voltage regulator
5	831	Low-voltage audio power amplifier
4	LM833	Dual audio operational amplifier
5	LM833	Dual audio operational amplifier
3	DS1013	Digital delay line

7	TDA1190P	TV sound system
7	LM1201	Video amplifier system
7	LM1202	230-MHz video amplifier SYSTEM
7	LM1205	130-MHz RGB video amplifier system with blanking
7	LM1310	Phase-locked loop FM stereo demodulator
7	MC1350	IF amplifier
7	MC1374	TV modulator circuit
7	MC1377	Color television RGB-to-PAL/NTSC encoder
7	MC1378	Color television composite-video-overlay synchronizer
3	DS1386	Watchdog time-keeper
7	LM1391	Phase-locked loop
7	MC1391P	TV horizontal processor
8	MC1403	Precision low-voltage reference
8	MC1404	Precision low-drift voltage reference
3	MC1408	Eight-bit multiplying D/A converter
10	MC1408	Eight-bit multiplying D/A converter
6	MC1455	Timer
4	LM1458	Dual operational amplifier
3	DS1486	Watchdog time-keeper
8	LM1494L	Linear four-quadrant multiplier
7	LM1496	Balanced modulator/demodulator
3	MC1488	Quad MDTL line driver
3	MC1489	Quad MDTL line receiver
3	MC1508	Eight-bit multiplying D/A converter
10	MC1508	Eight-bit multiplying D/A converter
4	LM1558	Dual operational amplifier
8	LM1594L	Linear four-quadrant multiplier
7	LM1596	Balanced modulator/demodulator
3	DS1602	System timer
3	DS1620	Digital thermometer/thermostat
4	MC1741C	Internally compensated, high-performance operational amplifier
7	LM1823	Video IF amplifier/PLL detector system
5	1875	20-watt audio power amplifier
5	1877	Dual audio power amplifier
7	LM1881	Video sync separator
4	LM1900	Quad amplifier
5	LM1971	μPot™ digitally controlled 62-dB audio attenuator with mute
7	LM2416	Triple 50-MHz CRT driver
7	MC2833	Low-power FM transmitter system

7	LM2889	TV video modulator
4	LM2900	Quad amplifier
8	LM2901	Low-power low-offset quad voltage comparator
8	LM2901A	Low-power low-offset quad voltage comparator
4	LM2902	Low-power quad operational amplifier
8	LM2931 Series	Low-dropout voltage regulator
8	LM2935	Low-dropout dual voltage regulator
8	LP2950	Low-dropout micropower voltage regulator
8	LP2951	Low-dropout micropower voltage regulator
9	CA3000	Dc amplifier
7	CA3011	FM IF amplifier
7	CA3013	IF amplifier/discriminator/AF amplifier
9	CA3054	General-purpose transistor array
7	TDA3190P	TV sound system
4	LM3301	Quad amplifier
8	LM3302	Low-power low-offset quad voltage comparator
8	LM3302A	Low-power low-offset quad voltage comparator
4	MC3303	Quad low-power operational amplifier
7	MC3356	Wideband FSK receiver
7	MC3357	Low-power FM IF
7	MC3359	High-gain low-power narrowband FM IF
7	MC3362	Low-power dual-conversion FM receiver
7	MC3363	Low-power dual-conversion FM receiver
7	MC3371	Low-power narrowband FM IF
7	MC3372	Low-power narrowband FM IF
4	MC3403	Quad low-power operational amplifier
3	MC3448A	Quad tri-state bus transceiver with termination networks
3	MC3453	Quad line driver with common inhibit input
6	MC3456	Dual timer
4	MC3476	Low-cost programmable operational amplifier
3	MC3487	Quad line driver with tri-state outputs
6	MC3556	Dual timer
4	LM3900	Quad amplifier
6	LM3905	Precision timer
6	3909	LED flasher/oscillator
8	MC3999T	Automotive high-side driver switch
2	4000	Dual three-input NOR gate plus inverter
2	4001	Quad two-input NOR gate
2	4002	Dual four-input NOR gate
2	4006	18-Stage static shift register
2	4007	Dual complementary pair plus inverter
2	4008	4-bit full adder
2	4009	Hex inverter/buffer

2	4010	Hex buffer (noninverting)
2	4011	Quad two-input NAND gate
2	4012	Dual four-input NAND gate
2	4013	BM dual-D flip-flop
2	4014	Eight-stage static shift register
2	4015	Dual 4-bit static register
2	4016	Quad bilateral switch
10	4016	Quad bilateral switch
2	4017	Decade counter/divider with 10 decoded outputs
2	4018	Presettable divide-by-n counter
2	4019	Quad and/or select gate
2	4020	14-stage ripple carry binary counter
2	4021	8-stage static shift register
2	4022	Divide-by-8 counter/divider with 8 decoded outputs
2	4023	Triple 3-input NAND gate
2	4024	Seven-stage ripple-carry binary counter
2	4025	Triple 3-input NOR gate
2	4027	Dual JK master/slave flip-flop
2	4028	BCD-to-decimal decoder
2	4029	Presettable binary/decade up/down counter
2	4030	Quad Exclusive-OR gate
2	4031	64-stage static shift register
2	4032	Triple serial adder
2	4034	8-stage 3-state bidirectional parallel/serial input/output bus register
2	4038	Triple serial adder
2	4040	14-stage ripple-carry binary counter
2	4041	Quad true/complement buffer
2	4042	Quad clocked D latch
2	4043	Quad 3-state NOR R/S latches
2	4044	Quad 3-state NAND R/S latches
2	4046	Micropower phase-locked loop
10	4046	Micropower phase-locked loop
2	4047	Low-power monostable/astable multivibrator
2	4048	Tri-state expandable eight-function eight-input gate
2	4049	Hex inverting buffer
2	4050	Hex inverting buffer
2	4051	Single 8-channel analog multiplexer/demultiplexer
2	4052	Dual 4-channel multiplexer/demultiplexer
2	4053	Triple two-channel analog multiplexer/demultiplexer

10	4053	Triple two-channel analog multiplexer/demultiplexer
2	4060	14-stage ripple carry binary counter
2	4066	Quad bilateral switch
10	4066	Quad bilateral switch
2	4067	16-channel analog multiplexer/demultiplexer
10	4067	16-channel analog multiplexer/demultiplexer
2	4069	Hex inverter
2	4070	Quad 2-input Exclusive-OR gate
2	4071	Quad 2-input OR buffered B-series gate
2	4072	Dual 4-input gate
2	4073	Double-buffered 3-input NAND gate
2	4075	Double-buffered triple 3-input NOR gate
2	4076	Four-bit D-type register with tri-state outputs0
2	4077	Quad 2-input Exclusive-NOR gate
2	4081	Quad 2-input AND buffered B-series gate
2	4082	Dual 4-input AND gate
2	4093	Quad 2-input NAND Schmitt trigger
2	4097	8-channel analog multiplexer/demultiplexer
10	4097	8-channel analog multiplexer/demultiplexer
2	4106b	Hex Schmitt trigger
2	4174b	Hex D-type flip-flop
2	4175b	Quad D-type flip-flop
2	4194b	4-Bit bidirectional universal shift register
2	4415	Quad precision timer/driver
10	4415	Quad precision timer/driver
2	4504b	Hex level shifter for TTL-to-CMOS or CMOS-to-CMOS
2	4510	BCD up/down counter
2	4516	Binary up/down counter
2	4527	BCD rate multiplier
4	MC4558	Dual wide-bandwidth operational amplifier
4	BA4560	Dual high-slew-rate operational amplifier
4	BA4561	Dual high-slew-rate operational amplifier
4	MC4741	Differential-input operational amplifier
3	MM5368	CMOS oscillator/divider
10	MM5368	CMOS oscillator/divider
6	MC5369	17-Stage oscillator/driver
3	MM5450	LED display driver
3	MM5484	16-Segment LED display driver
4	BA6110	Voltage-controlled operational amplifier
4	LM6161	High-speed operational amplifier
7	LM6171	High-speed low-power low-distortion voltage-feedback amplifier

9	BA6209	Reversible motor driver
4	LM6261	High-speed operational amplifier
4	LM6361	High-speed operational amplifier
1	7400	Quad 2-input NAND gate
1	7401	Quad 2-input NAND gate
1	7402	Quad 2-input NOR gate
1	7403	Quad 2-input NAND gate with open collector outputs
1	7404	Hex inverter
1	7405	Hex inverter with open collector outputs
1	7406	Hex inverter buffer/driver with open collector outputs
1	7407	Hex buffer/driver with open collector outputs
1	7408	Quad 2-input AND gate
1	7410	Triple 3-input NAND gate
1	7411	Triple 3-input AND gate
1	7412	Triple 3-input NAND gate
1	7413	Dual 4-input NAND Schmitt trigger
1	7414	Hex Schmitt trigger
1	7415	Triple 3-input AND gate with open collector outputs
1	7416	Hex inverter buffer/driver with open collector outputs
1	7417	Hex buffer/driver with open collector outputs
1	7420	Dual 4-input NAND gate
1	7421	Dual 4-input AND gate
1	7422	Dual 4-input NAND gate with open collector outputs
1	7425	Dual 4-input NOR gate with strobe
1	7426	Quad 2-input NAND gate with open collector outputs
1	7427	Triple 3-input NOR gate
1	7428	Quad 2-input NOR buffer
1	7430	8-input NAND gate
1	7432	Quad 2-input OR gate
1	7437	Quad 2-input NAND buffer
1	7438	Quad 2-input NAND buffer, open collector
1	7440	Quad 2 input NAND buffer, open collector
1	7442	BCD-to-decimal decoder
1	7445	BCD-to-decimal decoder/driver with open collector outputs
1	7446	BCD-to-7 segment decoder/driver
1	7447	BCD-to-7 segment decoder/driver
1	7451	Dual 2-wide 2-input AND/OR/INVERT gate

1	7453	Expandable 4-wide, 2-input AND/OR/INVERT gate
1	7454	4-wide, 2- and 3-input AND/OR/INVERT gate
1	7455	Expandable 2-wide, 4-input AND/OR/INVERT gate
1	7664	4-2-32 input AND/OR/INVERT gate
1	7465	4-2-32 input AND/OR/ INVERT gate with open collector output
1	7473	Dual JK flip-flop
1	7474	Dual D-type flip-flop
1	7475	Dual 2-bit transparent latch
1	7476	Dual JK flip-flop
1	7478	Dual JK edge-triggered flip-flop
1	7483	4-bit full adder
1	7485	4-bit magnitude comparator
1	7486	Quad 2-input Exclusive-OR gate
1	7489	64-bit random-access memory with open collector outputs
1	7490	Decade counter
1	7492	Divide-by-12 counter
10	7555	CMOS timer
8	7805	5-volt voltage regulator
8	7812	12-volt voltage regulator
8	7815	15-volt voltage regulator
8	MC7900 Series	Negative fixed-voltage regulator
10	BA9201	8-bit D/A converter with latch
10	BA9201F	8-bit D/A converter with latch
10	MC10319	High-speed 8-bit A/D flash converter
7	MC13001XP	Monomax® black-and-white TV subsytem
7	MC13030	Dual-conversion AM receiver
5	MC13060	Mini-watt audio output amplifier
7	MC13141	Low-power dc- to 1.8-GHz LNA and mixer
7	MC13142	Low-power dc- to 1.8-GHz LNA, mixer, and VCO
7	MC13175	UHF FM/AM transmitter
7	MC13176	UHF FM/AM transmitter
3	14410	Tone encoder
3	14415	Timer/driver
3	14490	Contact bounce eliminator
8	MC33128	Power management controller
8	MC33023	High-speed single-ended PWM controller
8	MC33063A	Dc-to-dc converter control circuit
4	MC33102	Dual sleep-mode operational amplifier
8	MC33161	Universal voltage monitor
8	MC33166	Power switching regulator

8	MC33269	Low-dropout positive-voltage regulator
8	MC34023	High-speed single-ended PWM controller
8	MC34063A	Dc-to-dc converter control circuit
5	MC34119	Low-power audio amplifier
8	MC34161	Universal voltage monitor
8	MC34166	Power switching regulator
7	MC44002	Chroma 4 video processor
1	74107	Dual JK flip-flop
1	74109	Dual JK positive edge-triggered flip-flop
1	74112	Dual JK edge-triggered flip-flop
1	74113	Dual JK edge-triggered flip-flop
1	74114	Dual JK edge-triggered flip-flop
1	74121	Monostable multivibrator
1	74122	Retriggerable monostable multivibrator
1	74123	Dual retriggerable monostable multivibrator
1	74125	Quad 3-state buffer
1	74126	Quad tri-state buffer
1	74132	Quad 2-input nanf schmitt trigger
1	74133	13-input NAND gate
1	74134	12-input NAND gate with tri-state output
1	74135	Quad Exclusive OR/NOR gate
1	74136	Quad 2-input Exclusive-OR gate with open collector outputs
1	74138	1-of-8 Decoder/demultiplexer
1	74139	Dual 1-of-4 decoder/demultiplexer
1	74145	BCD-to-decimal decoder/driver with open collector outputs
1	74147	10-line to 4-line priority encoder
1	74148	Eight-input priority encoder
1	74150	16-input multiplexer
1	74151	8-input multiplexer
1	74153	Dual 4-line to 1-line multiplexer
1	74154	1-of-16 decoder/demultiplexer
1	74155	Dual 2-line to 4-line decoder/demultiplexer
1	74156	Dual 2-line to 4-line decoder/demultiplexer (o.c.)
1	74157	Quad 2-input data selector/multiplexer (noninverted)
1	74158	Quad 2-input data selector/multiplexer (inverted)
1	74160	DEC decade counter
1	74161	4-Bit binary counter
1	74162	BCD decade counter
1	74163	4-Bit binary counter
1	74164	8-Bit serial-in, parallel-out shift register

1	74165	8-Bit serial/parallel-in serial-out shift register
1	74166	8-Bit serial/ parallel-in serial-out shift register
1	74168	4-Bit up/down synchronous counter
1	74169	4-Bit up/down synchronous counter
1	74173	Quad D-type flip-flop with tri-state outputs
1	74174	Hex D flip-flop
1	74175	Quad D flip-flop
1	74180	9-Bit odd/even parity generator/checker
1	74181	4-Bit arithmetic logic unit
1	74182	Carry: look-ahead generator
1	74190	Presettable BCD/decade up/down counter
1	74191	Presettable 4-bit binary up/down counter
1	74192	Presettable 4-bit binary up/down counter
1	74193	Presettable BCD/decade up/down counter
1	74194	4-Bit directional universal shift register
1	74195	4-Bit parallel-access shift register
1	74196	Presettable decade ripple counter
1	74197	Presettable 4-bit binary ripple counter
1	74240	Octal inverter/line driver with tri-state outputs
1	74241	Octal inverter/line driver with tri-state outputs
1	74242	Quad bus transceiver with tri-state outputs
1	74243	Quad bus transceiver with tri-state outputs
1	74244	Octal inverter/line drive with tri-state outputs
1	74245	Octal bidirectional transceiver with tri-state inputs/outputs
1	74251	8-input multiplexer (3-state)
1	74253	Dual 4-input multiplexer with tri-state outputs
1	74256	Dual four-bit addressable latch
1	74257	Quad 2-line to 1-line data selector/multiplexer (3-state)
1	74258	Quad 2-line to 1-line data selector/multiplexer (3-state)
1	74259	8-bit addressable latch
1	74266	Quad 2-input Exclusive NOR gate with open collector outputs
1	74269	8-Bit bidirectional binary counter
1	74273	Octal D-type flip-flop
1	74279A	Quad SR latch
1	74280	9-Bit odd/even parity generator/checker
1	74283	4-Bit full adder with fast carry
1	74289	65-Bit random-access memory (o.c.)
1	74290	Decade counter
1	74293	Four-bit binary ripple counter
1	74299	8-Bit universal shift/storage register

1	74323	8-Bit universal shift/storage register
1	74350	4-Bit shifter with 3-state outputs
1	74352	Dual 4-input multiplexer
1	74353	Dual 4-input multiplexer
1	74365	Hex buffer/driver (3-state)
1	74366	Hex inverter buffer with tri-state outputs
1	74367	Hex buffer/driver with tri-state outputs
1	74368	Hex inverter/buffer (3-state)
1	74373	Octal transparent latch with tri-state outputs
1	74374	Octal D-type flip-flop with tri-state outputs
1	74375	Dual two-bit transparent latch
1	74375	Octal D flip-flop with clock enable
1	74378	Hex D flip-flop with clock enable
1	74379	Quad D-type flip-flop with clock enable
1	74381	Four-bit arithmetic logic unit
1	74382	Four-bit arithmetic logic unit
1	74390	Dual-decade ripple counter
1	74398	Quad two-port register
1	74521	Eight-bit identity comparator
1	74533	Octal transparent latch with tri-state outputs
1	74534	Octal D-type flip-flop with tri-state outputs
1	74568	BCD/decade up/down synchronous counter (tri-state)
1	74569	4-bit binary up/down synchronous counter (tri-state)
3	DS75494	Hex digit driver
10	MC145040	8-bit A/D converter with serial interface
10	MC145041	8-bit A/D converter with serial interface

Index

Index of devices

549